T0138229

# Tides of History

# Tides of History

*Ocean Science and Her Majesty's Navy*

MICHAEL S. REIDY

*The University of Chicago Press    Chicago and London*

**MICHAEL S. REIDY** is assistant professor of history and philosophy
at Montana State University and coauthor, with Alan G. Gross and
Joseph E. Harmon, of *Communicating Science: The Scientific Article from
the 17th Century to the Present.*

The University of Chicago Press, Chicago 60637
The University of Chicago Press, Ltd., London
© 2008 by The University of Chicago
All rights reserved. Published 2008
Printed in the United States of America

17  16  15  14  13  12  11  10  09  08      1  2  3  4  5
ISBN-13: 978-0-226-70932-1 (cloth)
ISBN-10: 0-226-70932-9 (cloth)

Library of Congress Cataloging-in-Publication Data

Reidy, Michael S.
    Tides of history : ocean science and Her Majesty's Navy / Michael
S. Reidy.
        p.   cm.
    Includes bibliographical references and index.
    ISBN-13: 978-0-226-70932-1 (cloth : alk. paper)
    ISBN-10: 0-226-70932-9 (cloth : alk. paper)   1. Great Britain.
Royal Navy—History.   2. Oceanography—Great Britain—Research—
History.   I. Title.
    VA454.R25   2007
    359.00941—dc22

                                            2007028060

*For my parents,*
*Dorothy and Jim*

And indeed nothing is easier for a man who has, as the phrase goes, "followed the sea" with reverence and affection, than to evoke the great spirit of the past upon the lower reaches of the Thames. The tidal current runs to and fro in its unceasing service, crowded with memories of men and ships it had borne to the rest of home or to the battles of the sea. It had known and served all the men of whom the nation is proud, from Sir Francis Drake to Sir John Franklin, knights all, titled and untitled—the great knights-errant of the sea. It had borne all the ships whose names are like jewels flashing in the night of time.... Hunters for gold or pursuers of fame, they all had gone out on that stream, bearing the sword, and often the torch, messengers of the might within the land, bearers of a spark from the sacred fire. What greatness had not floated on the ebb of that river into the mystery of an unknown earth? . . . The dreams of men, the seed of common-wealths, the germs of empires.  JOSEPH CONRAD, *Heart of Darkness*

# Contents

# Acknowledgments

When I accepted my first academic position, a one-year appointment at the University of Oklahoma, I had to scramble to complete my dissertation. Once the semester began, I had everything finished except the acknowledgments, but I was too burned out to write another word. When I told my adviser, who was probably as anxious as I was to see the dissertation finished, she gave me a quizzical look. The acknowledgments should be one of the most rewarding aspects of the project, she said. It gives you the chance to thank everyone that helped you through the entire process. At the time, strung out on caffeine, I did not quite grasp the importance of those words. After years of work turning that dissertation into a book, a process that clearly showed me the collaborative nature of historical scholarship, I can now fully appreciate the joy in thanking the people and institutions that made this research possible. And there is no better place to start than with my adviser, Sally Gregory Kohlstedt. My debt and gratitude to her is too enormous to put into words; perhaps it is enough that her influence can be seen on every page of this book, and that when I attend History of Science Society meetings, she is the first person I look for. I also owe a great deal to my coadviser, Roger Stuewer, whose courses in the history of physics, discussions at his office and at the local tavern, and close reading and editing of my work helped mold me into a historian of science. And to the rest of my dissertation committee and the faculty at the University of Minnesota, including John Beatty, Jennifer Gunn, Alan Shapiro, and Arthur Norberg. I reserve special thanks to Alan Gross, who has helped me become an active researcher, often simply by treating me as an equal, which I am not. I also benefited greatly from my graduate school colleagues, all of them, but especially Erik Conway, John Jackson, Don Opitz, David Sepkoski, and Chris Young. Kai-Henrik Barth and Mark Largent not only read and commented on every word of my dissertation, but also made my graduate life exceedingly fun. If you know Mark, you know what I mean.

Monetary support in the form of research and travel grants from the University of Minnesota Graduate School and the National Science Foundation allowed me to spend an exciting and productive fifteen months abroad. The Department of History and Philosophy of Science, Cambridge University, and the Department of Science and Technology Studies, University College London, granted me visiting studentships and research access to the university and department libraries. Special thanks to Simon Schaffer and Patricia Fara for eye-opening discussions at the Eagle, and to Liba Taub, who always reminded me to wear a bike helmet through the cobbled streets of Cambridge. My eight months there were the most stimulating of my career, and I will never forget the hospitality that I received. I also owe thanks to Joe Cain for helping me get settled in London, and to Paul Hughes, the only person I know who is more interested in the history of tides than I am.

My visits to England, with stops in London, Cambridge, Oxford, Bristol, Liverpool, and Taunton, make the list of people and institutions that helped me far too long. However, I cannot fail to mention the archivists and staff at the Public Records Office in London, the Royal Society of London, the Oxford University Museum of Natural History, the Hydrographic Office in Taunton, the Bristol Records Office, the Society of Merchant Venturers in Bristol, the Sydney Jones Library at the University of Liverpool, the Liverpool Records Office, and Andy King at the Bristol Industrial Museum. I owe special thanks to Jonathan Smith at Trinity College, Cambridge, for all his time and energy in sifting through the Whewell Papers. And finally, to Deanna and Hugh Gallagher, who boarded me, fed me, and kept me up all night playing pinochle.

Michael J. Crowe and Philip Sloan at the University of Notre Dame first introduced me to the history of science and helped me view it as a subject worthy of a career. That my area of research is closely aligned with theirs is no coincidence. My research on American tidology was made possible through a Mellon Fellowship from the American Philosophical Society, and I thank Martin Levitt for his time and helpful advice. The libraries at the University of Minnesota have been a consistent resource throughout my career. I have always found the Owen H. Wangensteen

Historical Library of Biology and Medicine a special place to work and Elaine Challacombe a special person to work with, and I am indebted to the staff there for allowing me to take so many digital photographs (with their equipment, free of charge). An Andrew W. Mellon Travel Fellowship from the University of Oklahoma allowed me to work for a month in their History of Science Collection. Steven Livesey, Peter Barker, and Marilyn Ogilvie have always been extremely helpful every time I have made it down to Norman. A National Science Foundation Small Grant in Training and Research enabled me to explore the field of historical geography and, more important, introduced me to some outstanding historical geographers, including Bob Wilson and Arn Keeling. I can't remember all of our conversations at the 320 Ranch, but I am sure they were interesting.

At Montana State University, I have found a congenial atmosphere and a challenging intellectual environment. I received helpful comments and advice from Prasanta Bandyopadhyay, Kirk Branch, Rob Campbell, Susan Cohen, Kristen Intemann, Tim LeCain, Georgina Montgomery, Michelle Maskiell, Mary Murphy, Sara Pritchard, Yanna Yannakakis, and Jan Zauha. Billy Smith has become my confidant, and although he is not that good at chess, he certainly has sharpened my poker skills. Robert Rydell read every page of the manuscript in its early stages and kept me from making some major blunders, but above all, he has been a mentor and guide as I (he?) maneuvered my way through tenure. And special thanks to Brett Walker, who not only commented extensively on several drafts of my manuscript, but has also become a best friend in the process. He is a prolific researcher and phenomenal cyclist, and although I am not sure how he does it all, I do know that I owe him a great deal.

Since the beginning of my time here at Montana State, the university has been more than generous with financial support for my travel and research. This includes a Scholarship and Creativity Grant from the Vice President for Research, Technology Transfer, and Creativity; a National Science Foundation EPSCoR Research Grant; a Research and Creativity Grant from the College of Letters and Science; and, finally, the support of talented staff within the Department of History and Philosophy, particularly Diane Cattrell and Deidre Manry.

History is a deeply collaborative enterprise; whatever I did not gather directly from the archives necessarily has come from reading other scholars' work. Although footnotes are meant to convey my debt to these researchers, the small print in the back of the book does not always do the trick. In particular, I cannot fail to mention the work of Helen Rozwadowski, who most closely shares my own interests in the oceans and has been a faithful and rigorous sounding board; Margaret Deacon, whose seminal work is the starting point for all research on the history of ocean science; Richard Yeo, who has written the best book out there on William Whewell; Anne Secord, whose work has challenged me to look for science outside the usual venues; David Cartwright, whose history of the tidal science was constantly at my side; and Bernie Lightman and Keith Benson, who have consistently supported me and my work from the beginning of my career. Some of these people I know well, others I have never met, but all have contributed significantly to my research.

As the dedication of this book suggests, I owe much to my parents, whose love, support, and encouragement every step along the way has made this work possible. My debt to them reaches far beyond the monetary (although the Toyota Camry certainly helped). I am truly lucky to have such wonderful parents. Throughout the process, my sister Michelle has shared my successes and failures, and as I attempted to transform a rather abstruse dissertation into a readable book, I constantly had her in my mind as an audience. She is the most amazing woman I know. And to my brother Jim, whose love of science and teaching fostered my own interests and ambitions, and his son Jimmy, one cool dude who will never read this book. I also owe a great debt to my extended family, including Dave Brown and Kate Hudak, both of whom I love and respect beyond compare.

Among other things, my brother Jim also introduced me to the ocean's tides. Commercial fishing with him off Kasilof, Alaska, in the Cook Inlet for eight years certainly taught me a thing or two about the power of the ocean. Special thanks to the fishing crew in Alaska, especially Jerry and Marcia (and Shawn and Chatelle) Schooley, Tim Mycroft, and all the boys of Camp Nematode, especially Matt Desmond and Damien Gaul. I also would like to thank the staff at the University of Chicago Press, including Christie Henry, Catherine Rice, Pete Beatty, and Erik Carlson, who have been exceedingly supportive and most of all patient. The anonymous readers for the University of Chicago Press offered excellent criticism, and Alice Bennett did a marvelous job meticulously copyediting the entire manuscript. I acquired illustrations from numerous sources, including the Royal Society of London, Trinity College, Cambridge, the Bristol Industrial Museum, the Henry Madden Library at California State University–Fresno, the United Kingdom Hydrographic Office, the Royal Geographical Society, the National Archives Image Library, Kew, and the libraries at the University of Minnesota. Although I am responsible for any faults in the text, it was these people and institutions that made the research and writing possible.

# Introduction: The Littoral in Science and History

For the boundary between sea and land is the most fleeting and transitory feature of the earth, and the sea is forever repeating its encroachments upon the continents. It rises and falls like a great tide, sometimes engulfing half a continent in its flood, reluctant in its ebb, moving in a rhythm mysterious and infinitely deliberate. RACHEL CARSON, *THE SEA AROUND US*

The boundary between land and sea is constantly on the move, rising and falling with the tide each day, obeying the powers of the moon each month, varying slightly on average each year, and since the last ice age, rising higher and higher each century. The littoral, from the Latin for shoreline, is hardly a boundary at all. As Rachel Carson described it, it is "fleeting," "mysterious," and most of all "transitory."[1] Yet it is also where life dawned and civilization first became entrenched, where fisherman pick their nets, tourists board ferries, and imperialists disembark to conquer nations. It brings fear and fortune, and both life and death are ever-present. Indeed, for those living and working near the British coast, the littoral can prove exceedingly dangerous. In October 1953 an extraordinary high tide in the river Thames overshot its assigned boundaries and drowned over three hundred people in and around east London. Although the tide had been predicted—through advanced mathematical techniques and powerful mechanical computers—nothing could be done. The tides were racing up the estuary more forcefully each year, a product of land subsidence, rising sea levels, and human quayside

1

development. The astounding growth of London in the previous two centuries had made the situation all the more precarious. What once had been a sprawling floodplain now housed the largest city in Europe, with many residents within the environs of the tidal Thames. The solution was Great Britain's most extensive and expensive technological fix, the Thames Barrier, costing upward of £500 million when completed in 1982. Great Britain had finally tamed its greatest natural feature, making the largest and most important tidal river in the Kingdom devoid of any tide.[2]

Before civil engineers succeeded in ridding the Thames of its most dangerous tides, natural philosophers had attempted their own solution— controlling the Thames through accurate tidal observation, reduction, analysis, and prediction. In doing so they uncovered what Pierre-Simon Laplace had called "the thorniest problem in physical astronomy." Their methods entailed an intricate study of the ocean's tidal rhythms, a project that began in the Thames estuary but soon spread to encompass all the earth's oceans. To chart this movement in the study of the tides, from the shoreline to the largely unexplored marine frontier, is to uncover an unwritten chapter in British history. The tides are attractive to the scientist because of their sheer complexity. They interest the historian because scientists attempted to understand this complexity at a time when coastal shipping could make or break a nation, when "ruling the waves" meant ruling a vast empire, and when nations considered the rim of the ocean part of the spoils of war.

The rivers, estuaries, and coastlines of Great Britain tied the nation together, helping define its relationship to other nations and to its natural environment. Ports were the launching place for England's naval fleet and the area most in need of protection during wartime. The littoral also was the source of Britain's energy—it provided food, protection, and cheap transportation—but before the mid-eighteenth century it occupied the minds only of those who lived and worked near its incessant ebb and flow. Fashion followed the aristocracy, and the aristocracy traveled inland. This began to change as improved road transport to the capital, steamship transport from London, and changing notions of the effects of seaside air brought first the aristocracy, then the middle and lower classes to the seaside.[3] Sea bathing grew popular as therapy. In 1804 the London physician A. P. Buchan promoted the medicinal effects of the ocean in his *Practical Observations concerning Sea-Bathing*, contrasting the turbulence of the city with the salubrious effects of the seashore. By then scores of accounts, from medical case studies to popular travelers' tales, had turned the attention of the elite toward Britain's ever-changing coastline.[4] In the first half of the nineteenth century, the littoral began to occupy the minds of a growing number of painters, writers, natural philosophers, collectors, and antiquarians, who flocked to the coast to study the shoreline exposed at low tide.[5]

As the coast became fashionable, the intertidal zone became a preferred geological laboratory for studying the earth's past, leading to a complex revision of both space and time. The first impression of the seashore was its horizontal dimension. The ocean stretched to the horizon, an unbroken expanse that both separated Britain from the rest of the world and connected it through trade routes to its ever-expanding possessions. The horizontal dimension, however, easily gave way to the vertical. Someone walking on the beach six hours after high tide could find the topography completely transformed. Where there had been a dark, quivering ocean, one now found dry land underfoot; Britain grew with each receding tide, only to shrink again as the ocean advanced.

As geographical space was contemplated anew, so too was geological time. In the long span of the earth's history, change is difficult to perceive. Mountains form and erode, glaciers advance and retreat, the earth's rotation slows.[6] Yet on the shores surrounding Britain, one could observe firsthand the slowly changing division between land and sea and the steady processes of accumulation and erosion. While the changing coastline testified to the immensity of geological time, the tides on the seashore connected one with the daily rhythms of the sun and moon. Indeed, "time" and "tide" have always been close metaphorically as well as alliteratively. The paradoxical juxtaposition of geological and human time seemed exceptionally clear where the land and sea met, and it was often in the littoral that geologists formulated their theories of terrestrial change. Charles Lyell, one of the founders of modern geology, relied heavily on the sea for his theories, especially on the variation in its levels and the destructive energy of its tidal currents.[7] The unchanging rhythms of the sea combined with the slowly changing nature of the coast offered Lyell evidence of the immensity of geological time and the slow but steady processes needed for his uniformitarian framework. During his visits to the intertidal zone, he also found physical proof for his ruminations—fossilized shells embedded in the exposed cliffs near the sea. Lyell was so taken with the force of the ocean that he considered its erosive power one of the main mechanisms for geological change on land.

In the second quarter of the nineteenth century, natural philosophers in England turned to the seaside as a strategic space for establishing the credibility of their emerging disciplines. It is not surprising that they

Figure I.1. Frontispiece to Charles Lyell's *Principles of Geology*, published in 1830. The columns of the Temple of Serapis clearly illustrate the changing levels of the land and sea, and thus the changing nature of the littoral in human time. University of Minnesota Libraries.

found it so ripe for theorizing. Fascination with the littoral followed from Britain's island geography, where no one lives more than seventy miles from the coast. Surrounded by water with convenient seaports around its coasts, Britain brimmed with commerce and creativity. By the early nineteenth century, its system of river and coastal navigation surpassed those of all other European nations.[8] As vessels vied to drop off their wares and set sail for distant ports, the littoral helped define the nation's cultural, political, and economic identity and also shaped the science pursued on its shores. While Lyell and other geologists looked to the intermingling of the land and sea to support their theories, other natural philosophers focused on the physics of the intertidal zone. Dangerous shoals, shifting sandbanks, submerged rocks, fierce eddies, wild currents, and awesome tides severely hampered the transfer of people and goods between sea and shore, and naval and commercial vessels often foundered on the coast rather than anchoring safely in port.

Most of the major commercial and military ports in Britain are estuary ports, where the tide is the most significant factor for shipping and naval operations. London, Bristol, Liverpool, Edinburgh, Glasgow, and Hull, along with the naval dockyards at Woolwich and Deptford, are all typical British ports, where access depends entirely on the rise and fall of the sea.[9] In the late eighteenth and early nineteenth centuries, moreover, innovations in iron and steam made vessels heavier so they drew more water, further limiting river and coastal access. A few rivers, such as the Clyde, which reached the bustling trading city of Glasgow, were cut off for iron-hulled ships. Navigators could reach other ports, such as Newcastle, only during extreme high water, which lasted less than an hour. Indeed, all the harbors on the east coast of Britain depended entirely on the rise of the tide to admit large vessels. This included the river Thames leading to London, the world's largest port. A report by Robert Stevenson and Son, Civil Engineers, to the Admiralty in May 1834 cogently summed up the situation: "The want of a Harbour, which can be taken at all times of the tide, is severely felt along the whole range of coast from the Orkney Islands to the Straits of Dover; and fatal accidents have frequently occurred from the danger of uncertainty which attend the navigation of vessels."[10]

Robert Stevenson was mainly concerned with maritime shipping, but Dover was also a strategic military space for the British fleet. Captain James Anderson of the Royal Navy, therefore, wanted the tides in the Straits of Dover studied for military purposes, "circumstances which, I have reason to fear, have not been hitherto sufficiently attended to; but which would prove of the utmost importance, especially on expeditions where much boat service must be had recourse to; and in disembarking troops at a particular point, or in making an attack upon vessels at anchor during the darkness of the night; when a want of the necessary knowledge of the tides, or as it has often been called 'a mistake in reckoning them,' might be productive of the most fatal consequences."[11] Anderson attached a footnote referring to a mistake Lord Nelson had made in calculating the tides at the Straits of Dover "as a reason why the boats sent in by him to attack the French flotilla in Bologna Bay, in 1801, did not get up with the enemy till long after the appointed time." Knowledge of the tides proved a commercial and military necessity, and natural philosophers tried to portray themselves as experts who could advance Britain's maritime success.

As citizens of an empire built almost exclusively on trade and naval might, British mariners were well aware that vessels, cargo, and crew were being lost owing to insufficient tidal data. Interest in the science of the tides did not spring ab ovo from the minds of natural philosophers. What the British navy mastered after the mid-nineteenth century—the ability to sail unencumbered throughout the world's oceans—began with the official hydrographic surveys of the mid-eighteenth century, occasioned by the need to sail safely around the treacherous coasts of the English Channel and the North Sea. The same holds for the study of the tides, research that began in the private sector on the coasts of Britain and only slowly reached the Royal Navy and the natural philosophers. As the official hydrographic surveys expanded from the seas around Britain to the vast extent of the Atlantic, Indian, Arctic, and Pacific oceans, so too did the study of the tides. The machinery of empire required as its lubricant a science of the sea, essential for any overseas expansion of trade or successful military campaign.

## From the Littoral to the Ocean

In the first half of the nineteenth century, the British learned to manage the oceans; in the second half, they ruled over large portions of the oceans' rims. Extending their reconnaissance from the littoral outward, the British Admiralty, maritime community, and scientific elite collaborated to bring order to the world's seas, estuaries, and rivers. These experts transformed the vast emptiness of the oceans into an ordered and bounded grid, inscribed with isolines of all kinds—tidal, magnetic, thermal, and barometric—in areas uncharted and on coasts unseen. In the process, science expanded from a local undertaking receiving parsimonious state support and embracing only sporadic communication among philosophers of different nations to worldwide and relatively well-financed research by a hierarchy of practitioners working with sophisticated instruments. At the center of this transformation, the modern scientist emerged. In the first half of the century, natural philosophers engaged in natural philosophy; in the second half, scientists practiced science.

This rise of the geophysical sciences in the early Victorian era, and the concomitant rise of the modern scientist, represents a significant chapter in the centuries-old attempt by European nations to master the oceans by organizing space and time. As voyages of European discovery and trade grew in political and economic importance, the need for advanced navigational techniques directed much of the practice of science. The search to find longitude at sea, in particular, had eluded mariners and astronomers for centuries, and it become the most pressing scientific and technological quest of the eighteenth century. The science of astronomy was almost entirely linked to that cause, as were the pocketbooks of kings and the lives of whalers, merchants, and naval men. The perfection of the lunar-distance method for determining longitude, one of the great feats of eighteenth-century celestial mechanics, spawned an array of scientific and technological advances. Astronomers mapped the starry vault and accurately accounted for the motion of the moon.

As England industrialized, the scientific and artisan communities become more closely linked, leading to advances in both scientific instruments and precision measurement. Significantly, the search for longitude was aided by liberal government support, which included building palatial observatories and awarding generous grants and lavish prizes. After astronomers and artisans perfected the lunar-distance method and the marine timepiece, the emphasis shifted to the ocean itself—its tides, currents, and changing magnetic field. These too were made part of the grid, as isomaps of all types overlapped with lines of latitude and longitude to circumscribe the ocean and make it safe for trade and travel.

Many European countries contributed to these advances, but Britain emerged from the Napoleonic Wars as the dominant maritime power, controlling the waves and many of the world's strategic islands and coastlines. Its security and prosperity depended on being able to navigate the oceans and dominate distant coasts and channels. Efficient trade required that the British Admiralty provide correct soundings, tide tables, and sailing directions to its mariners, both military and commercial. And where it conferred military or economic advantage, science profited. As the Royal Navy became the protector of maritime commerce, the Admiralty increasingly turned to science to aid in overseas expansion. The result was a coordinated effort to advance the science of the sea undertaken from a powerful, global perspective. By midcentury science moved from being practiced locally to being practiced globally, placing it squarely within European expansionism and situating scientists as dependable purveyors of knowledge for the good of the state.

Alexander von Humboldt, a Prussian explorer and cosmopolitan man of letters, contributed significantly to this change in the nature and scope of scientific inquiry.[12] He provided the initial model and methods, calling for observations from around the world to determine the interrelations of diverse phenomena in nature. His ambitious program required organization that went far beyond individual practitioners and came to fruition in Britain, where the Admiralty and the scientific elite struck up a mutually beneficial relationship. Profits to the empire were extensive. Safe navigation required correct theories of the earth's magnetism, tides, and ocean currents, while environmental studies such as biogeography and economic botany were valuable for commerce. The new sciences, in turn, profited from Britain's voyages of exploration and its willingness to commit people, equipment, and money to studying the world's lands and oceans. With British sailors, surveyors, and scientists distributed throughout the world, a network of observers was finally able to answer Humboldt's call for coordinated observations.

Through his twenty-year research project on the tides, William Whewell, Cambridge professor and eminent man of science, envisioned the world as a global laboratory. In June 1835 thousands of seamen, surveyors, dockyard officials, amateur observers, and local savants measured the tides every fifteen minutes, day and night, for two weeks as part of Whewell's "great tide experiment." Nine countries participated, with close to seven hundred tidal stations contributing data. For the first time in history, savants measured, tabulated, graphed, and charted the tides around the world at exactly the same time and under the same astronomical influences. Whewell created a map to show how the tides progressed through the entire Atlantic and onto the shores (and into strategic ports, inlets, estuaries, and rivers) of the major maritime nations and their colonial possessions. His revolutionary approach to tidology became a prominent way to practice science in the next several decades.

The significance of Whewell's research project, particularly his use of global observations and his cooperation with the British Admiralty, was not lost on his contemporaries. John Herschel, George Biddell Airy, James David Forbes, James Clark Ross—virtually all of the British scientific elite—participated in similar long-term multinational studies ranging from terrestrial magnetism to global barometric fluctuations. They tackled any subject where global measurements could help chart geophysical laws and expand British influence. Scientists graphed rule and rationality onto a seemingly chaotic region of the globe, defining a bounded space that could now be crossed and recrossed by Her Majesty's vessels, just as their own isolines so elegantly traced and retraced the globe.

Whewell boasted that his global tide experiment included the most "multiplied and extensive observations yet encountered in science." Sitting quietly in his landlocked rooms in Trinity College, Cambridge, he undertook to systematize and stabilize this jumbled confusion, transforming it into a meaningful assemblage for the British Admiralty. This process, from observation to isomap, represents an important shift in reconceptualizing the world's oceans. It is no coincidence that Whewell both organized global geophysical initiatives and helped popularize graphical representations of scientific data. Yet graphs and isomaps were only the most transparent of Whewell's instruments. Sifting through the piles of observational data on his desk, Whewell was himself at the center of an interweaving set of political, economic, and cultural practices that flowed together to define the ocean as a scientific environment. He carried these practices with him as he established the role of the ocean in his science and constructed his science for organizing the ocean. He relied on them to create a geographic space that the maritime powers of Europe could then control.

While this book gives priority to the analysis of geographic *space*, it also emphasizes *process:* the ebb and flow of ideas, interests, and intents that gave rise to a new conception of the ocean and to a role for the new scientist. William Whewell coined the term "scientist" in 1833, at the dawn of his tidal studies. He then spent much of his intellectual career formulating what it meant to do science and who could be considered a scientist. Whewell's tidal work at Trinity College, however, only put the final touches on a systematizing effort that involved scores of practitioners connected by the world's oceans and coastlines. From the beginning, taming the tides was a collaborative affair that required broad support. To produce a map that eventually would be dispersed to navies across Europe, Whewell relied on British scientific institutions, the British Admiralty, and a group I have termed associate laborers, including dockyard officials, harbormasters, expert calculators, tide table makers, and professional military men associated with the trigonometric and coastal surveys of the British Navy. Some worked for the Admiralty Hydrographic Office, but not as captains on great sea voyages and surveys, like James Clark Ross or Robert Fitzroy. Rather, they worked alongside these men, reducing the data, finishing the charts, and calculating the astronomical tables. It was within this collaboration of interests and ideas, by people who differed in class, profession, and gender, that Whewell delineated the social and intellectual role of the modern scientist.

By midcentury, scientists could accurately predict the tides in the major British ports and had produced tide tables for Britain's colonial possessions. The sheer mass of the undertaking reinforced the notion that scientific experts could translate the unknown into knowledge of use to the British Admiralty. The ability to control the sea-lanes, ports, and estuaries of the world conferred remarkable power on the burgeoning imperial nation and heightened the status of the newly defined scientist. As scientists regulated and managed the ocean on paper, the British

Admiralty used the physical ocean to transport troops, wealth, and ultimately British culture to the ends of the world. The perceived ability to order the world's oceans therefore helped shape our modern attitudes toward science and the scientist. It was this perception as articulated by Whewell and others in the first half of the nineteenth century that, in the second half of the century, helped make imperial domination possible.

## The Tides as History

Rather than a history of tidal theory, this book is a history of British scientific culture during the transition from industrialization to empire, when understanding the sea became important politically, economically, and strategically.[13] The tides serve as a potent case study for analyzing the organization, support, and rise of modern science and the modern scientist. We can examine the collaborative nature of scientific practice and uncover the contributions of the silent and often invisible majority.[14] As we follow the process of tidal analysis, the labor of these associates becomes eminently visible: in the technologies they built, the observations they made, the calculations and reductions they performed, and the tables and graphs they constructed. The history of tidal research in Britain also highlights the role of science in projecting power to distant lands, the essence of nineteenth-century imperialism.[15] Studies of imperialism and science have continued to be terracentric,[16] but tidal research suggests that a critical aspect of British imperialism occurred not in the large landmasses of Europe, India, South Africa, or Australia but in the nautical spaces between them.

The narrative of this story is simultaneously temporal and geographic. It reaches from the ports and estuaries of Britain's coastlines to the wide expanse of the world's oceans, from Britain's coastal trade in the eighteenth century to the foreign trade of the nineteenth. Beginning in the late seventeenth century and advancing through the age of enlightenment, the opening chapter, "Philosophers, Mariners, Tides," traces England's initial commitment to the study of the littoral environment. After an initial surge of interest in tidal theory following the founding of the Royal Society of London, Isaac Newton formulated a general theory of the tides based on his law of universal gravitation. Edmond Halley then searched out the tides, beginning in the Thames estuary and extending his study to the Atlantic Ocean as master and commander of the *Paramore*. Natural philosophers in England, however, failed to build on these initial successes. The study of the sea shifted to the Continent, especially France, where Pierre-Simon Laplace and others made significant advances on Newton's achievements, formulating a hydrodynamic theory that still serves as the basis of tidal analysis. Yet owing to the complexity of the calculations involved, a workable hydrodynamic approach had to await advances in mechanical computing.

Throughout the eighteenth century, British naval captains and Admiralty surveyors, military and colonial administrators, dockyard engineers and tide table makers sustained the study of the tides. Newton's general theory marshaled several sets of tidal observations for support, but as explorers spread across the globe, his theory rarely held, or was held up, as an explanation. Rather than securing his achievements, the maritime community, working largely beyond Newton's gaze, made Newtonian science a matter of contention. Naval captains such as Henry More and James Cook gathered observations of tides that contradicted Newton's theory, while others such as Charles Vallancey, a colonial engineer, and John Abram, a teacher of navigation, offered theories that were explicitly anti-Newtonian. After the initial success of Newton and Halley in describing the general features of the tides, there was no systematic study of tidal theory in Britain until the second quarter of the nineteenth century. First the tides themselves had to change.

Set against the backdrop of Britain's expanding coastal trade, chapter 2, "The Bounded Thames," describes the environmental changes that led to a renewed interest in tidal theory. While geological changes can raise the level of the sea and increase its tides a few centimeters each century, man-made changes can be much more dramatic, contributing several feet in a matter of decades.[17] Beginning in the late eighteenth century and continuing unabated through the first third of the nineteenth, the Thames was radically transformed. Civil engineers straightened, dredged, and banked the river and otherwise modified its course to control the depth of the water, the rise and fall of the tide, and the silting of the channel. Chapter 2 begins with the physical bounding of the Thames to demonstrate the dramatic changes to the river itself. It then traces the initial study of the tides as a local and practical problem, the province of those interested in the physical aspects of the coastline for their commercial significance. The chapter thus takes into account both the changes to the natural environment of the river and the myriad of almanac publishers, surveyors, and tide table makers who attempted to calculate those changes for the safety of mariners. The choice of the tides as a legitimate research topic was anthropogenic, caused by human changes to the physical topography of Britain's rivers and coastlines.

Figure I.2. A view of the crowded and banked river Thames in front of the Customhouse. Reproduced from Joseph G. Broodbank, *History of the Port of London,* 2 vols. (London: Daniel O'Connor, 1921). University of Minnesota Libraries.

The environmental changes to the river Thames induced early Victorian natural philosophers to study the theoretical aspects of tidal science so as to produce accurate tidal predictions. Chapter 3, "Dessiou's Claim," analyzes this sustained attempt to predict the tides on Britain's coasts. Rather than focusing solely on theory and observational data—both exceedingly important—this chapter follows the study of the tides from the less glamorous position of calculation. The overarching narrative of events remains the same, a transition from its beginnings as a local problem to its appropriation by the scientific elite as a vital topic of scientific research. But it follows recent historiography in drawing attention to the calculators themselves, similar to John Brewer's analysis of the relation between war and the English state. By focusing on "bookkeeping not battles, with ink stained fingers rather than bloody arms," Brewer inverted the traditional narrative of England's unprecedented achievements in the late eighteenth century.[18] The heroes of Brewer's texts were not commanders on the high seas or diplomats in the halls of government but office clerks laboring away in accounting firms, organizing and administrating the English fiscal state. Recent scholarship has only begun to extend this approach to the equally remarkable rise of the physical sciences in the late eighteenth and early nineteenth centuries.[19] A focus on highbrow ruminations rather than ink-stained fingers still distorts the history of scientific research.

Chapter 3 focuses on the legal claim made by Joseph Foss Dessiou, the first state-funded tide calculator in Britain, against the proprietors of the *Nautical Almanac*, who used his tidal reductions without proper compensation. But its larger aim is to underscore the significance of associate laborers to Victorian physical astronomy more generally. In the study of the tides, the field's early years parallel the founding of the British Association for the Advancement of Science as well as the reform of the Royal Society. Yet modern tidal analysis did not start with these elite scientific institutions; it had much more humble beginnings. British industrialization rapidly made London the largest port in the world, and publishing London tide tables became a competitive and lucrative business. Charles Knight, a prolific publisher and writer, suggested to Henry Brougham, founder of the Society for the Diffusion of Useful Knowledge, that the Society include an almanac among its inexpensive publications (first published in 1828 as the *British Almanac*). After mariners questioned the accuracy of these initial tide tables, John William Lubbock, a Cambridge graduate and successful London banker, stepped in to calculate the major tidal constants for the port. Yet the ownership of the tables published in the *British Almanac* for 1829 was highly contested. The tables were calculated from observations gathered for twenty years by dockyard officials and foremen; they were reduced by Dessiou, an expert calculator working at the Hydrographic Office; and they were published at the suggestion, expense, and reputation of Knight, all under the coordination of Brougham and the aegis of the Society for the Diffusion of Useful Knowledge. From the beginning, the study of the tides was a collaborative and hierarchical affair begun by associate laborers themselves.

While these early practitioners saw the study of the tides as local and practical, the theorists approached the subject as a state would, making it a major research agenda for the Admiralty, the British Association, and the Royal Society of London.[20] Chapter 4, "'Tidology,'" follows the founding of the research field from the point of view of these institutions and the scientific elite. In the early 1830s Lubbock, while contributing to the tide tables published for the London docks, also introduced William Whewell, his former tutor at Trinity College, Cambridge, to the study of the tides. Whewell was searching for a scientific subject for active research, and this chapter follows his entry into tidal studies, from his coining of the term "tidology" in 1830 to his overt shaping of the field based on his own studies in the history and philosophy of science.

William Whewell has been a central figure of study for historians, philosophers, and sociologists for over a century.[21] Master of Trinity College for twenty-five years, he published on a wide range of sciences, on architecture, poetry, and religion, and on a large number of other topics in more popular reviews and sermons. He is best known for his multivolume *History of the Inductive Sciences* (1837) and his equally impressive *Philosophy of the Inductive Sciences* (1840), significant texts in the history and philosophy of science. His life spanned the course of specialization in the sciences, and he wrote extensively on the place of science and the scientist in the broader culture. But no researcher has yet comprehensively examined Whewell's tidal investigations in relation to his other scientific pursuits. Whewell's most extensive research on the tides occurred while he was writing his *History* and formulating his *Philosophy*, and his extensive research spanned nearly two decades and included over thirty-five publications (fourteen in the prestigious *Philosophical Transactions of the Royal Society of London*). An in-depth analysis of the centrality of Whewell's tidology in relation to his other metascientific investigations revamps Whewell scholarship and resituates him as a researcher in the vanguard of the dominant scientific practice of his time.[22]

From his writing of textbooks and his early excursions into crystallography and meteorology, Whewell formulated a methodological approach to science. He would later publish this method in his *History* and *Philosophy*, but in the early 1830s he used the approach heuristically, as a way of quickly advancing a field of research. Whewell's powerful position within the political and scientific communities, combined with his emerging methodology geared toward a burgeoning field of research, allowed him to promote and organize tidology as a vital project for both theorists and the state. His early work in tidology also taught him valuable lessons concerning the discovery process, including the difficulty of connecting facts with theory, the disparate ways of testing those theories, and the proper methods of data analysis and representation. All these topics reappeared in nuanced form at the end of the decade in his *History* and *Philosophy*. The practice of tidology helped him formulate what it meant to do science and placed him at the forefront of the discussion on the proper social and intellectual role of the scientist.

After the founding of tidology as a significant field of research has been described from the perspective of both the associate laborers and the elite theorists, the next two chapters analyze the practice of tidology through the 1840s. They are linked through Whewell's career as a tidologist and through the analysis of tidology as practiced by socially diverse teams of researchers. Chapter 5, "The Tide Crusade," examines the multinational tidal experiments of 1834 and 1835 to demonstrate the collaborative nature of research in physical astronomy. In particular, it highlights the relationships between the Admiralty and science, between departments within the Royal Navy and institutions in British science, and between the elite theorists in Britain and the numerous scientific servicemen and commissioned and noncommissioned officers of the Royal Navy.

Lubbock and earlier researchers viewed the tides temporally; they worked with long-term observations that corresponded with the ancient and everlasting motions of the heavens. Lubbock's research was also local, confined to London and other major ports in Britain. Whewell, by contrast, viewed the tides spatially, as a problem to be solved through expansive geographical space rather than extensive astronomical time. Whewell advanced tidology as a prototypical spatial science by co-opting the methods of Alexander von Humboldt and applying them through the most pervasive military institution in the world, the British Admiralty. His systematizing effort involved networks of observers stationed around the globe and set the stage for further involvement of the Admiralty in large-scale geophysical initiatives.

Chapter 6, "Calculated Collaborations," analyzes Whewell's close collaboration with those practitioners he termed "subordinate labourers," a diverse group who worked alongside the scientific servicemen and scientific elite. In particular, it focuses on the contributions of expert calculators to large-scale geophysical research. Not only were these diligent men of numbers the first to bring the data to the natural philosophers, but they first applied the results to the construction of tables, and thus to the testing of theory. Their ingenuity with difficult calculations made them indispensable for Whewell's tidology. Though they began as calculators Whewell desired "to keep at work," they actively contributed to his tidology: they tested his theories, advanced his methods, and suggested new avenues of research.

The final two chapters trace the validation and ratification of the spatial turn in science, including the consolidation of its methods, organization, and collaboration. When global research became prominent in Britain, the specialization and popularization of science caused a transformation in its social organization. Who could actively participate was open to debate. Chapter 7, "Creating Space for the "'Scientist,'" extends Whewell's spatial and disciplinary shaping of the field of tidology to his equally ambitious shaping of the "scientist," or full-time devotee. Science for Whewell was a hierarchical affair; the elite theorists deserved recognition above the calculators and tide table makers who were paid for their work. He purposefully delineated the study of the tides as "the last great bastion of physical astronomy," the peculiar province of theorists, accessible only to those with advanced mathematical training. Thus he contributed to a definition of the scientist that effectively excluded associate laborers. Yet Whewell himself was the son of a skilled carpenter, and his relatively humble beginnings, combined with the transparent input of associates to the theory of the tides, informed his negotiation of the intellectual and social role of the scientist. Whewell's practice of tidology and his larger historical vision of how scientists made discoveries contributed to his definition. Rather than making a single inductive inference, the scientist was involved in a historical process that nurtured a specialized field of research through its many stages. Large-scale and historically informed research was for Whewell the ultimate sign of a scientist.[23]

Definitions of the modern scientist were formed squarely within the military's growing appetite for intelligence and control. Through his twenty-year research project on the tides, Whewell helped transform science into an international and hierarchical undertaking that relied heavily on government participation and support. Whewell's approach to the study of the tides as a spatial science began on the coasts of Britain but quickly extended outward to include nine countries and upward of seven hundred tide stations worldwide. He then campaigned for the next two decades to have the Admiralty fund an expedition to search out the tides in the deep ocean. In this endeavor, Whewell *made* the tides global, and though a revolutionary approach to tidology, by midcentury his methods were common to science. The conclusion examines the incentives that pushed governments to set up networks of observers in terrestrial magnetism, meteorology, and the study of the ocean itself. This move from science practiced in the laboratory to science practiced over the whole globe was a defining feature of nineteenth-century ocean science. "The Tides of Empire" therefore offers a new perspective from which to view George Biddell Airy's work on the tides, John Herschel's and Robert Fitzroy's work in meteorology, the intense study of the earth's magnetism, the beginning of oceanography, and ultimately the rise and institutionalization of multinational collaborations, leading to the International Polar and Geophysical Years.

Historians have recently become fascinated with the ocean.[24] This book extends current work by examining the complex ways humans have used science and technology to reconceptualize the oceanic environment. It explores how scientists and sailors perceived, interacted with, and valued the ocean at the dawn of European imperialism, expanding on the approaches environmental historians have used to examine the wilderness. Indeed, the book looks at how people value space more generally, raising questions usually confined to historical and cultural geography, environmental history, and the history of cartography.[25] This analysis, then, describes both the practice of science in a new global environment and the way Victorian scientists reconceptualized that environment through the practice of science.

Whewell and his contemporaries inscribed the ocean with order and meaning. They graphed and charted it onto maps of all types, disseminated to navies around the world. The product of Whewell's researches, the isotidal map of the world, passed easily between the man of science and the men of the sea. But the most potent force to emerge was the central position of science and the new scientist in translating the unknown for the good of the state. The ocean was a space no longer reserved for mariners, a harrowing abyss to be crossed with trepidation. It became a place that scientists ordered and helped control. The world's oceans, ports, and estuaries were the nodes of imperial advance, and science, the means by which power and systematic knowledge were extended around the globe. By situating tidal studies within its geographic, social, political, and economic contexts, this book examines how Victorians practiced global geophysical research, an important incubator for the professional configuration of the modern scientist.

ONE

# Philosophers, Mariners, Tides

At every rising and every setting of the moon the sea violently covers the coast far and wide, sending forth its surge . . . ; and once this same surge has been drawn back it lays the beaches bare, and simultaneously mixes the pure outpourings of the rivers with an abundance of brine, and swells them with its waves. As the moon passes by without delay, the sea recedes and leaves these outpourings in their original state of purity.  VENERABLE BEDE, *TIME AND ITS RECKONING* (725)

The ocean came to the inhabitants of the island of Britain with a history, one deeply rooted in the biblical stories of the Creation and Flood. To early medieval scholars, the ocean was boundless, infinite, literally and figuratively unfathomable. As Alain Corbin has noted, it "represented the realm of the unfinished, a vibrating, vague extension of chaos [that] symbolized the disorder that preceded civilization."[1] The biblical Flood gave its waters a sinister motive, as the tides rose unbounded and chaos returned to the land. As early as the eighth century, however, the Venerable Bede linked the rising of the tide with the phases of the moon. For Bede, the littoral was an edge of the ocean that could be studied, the one region that seemed to follow rational laws. The moon held sway over the violent waters near land and helped return the island's estuaries and rivers to their original purity. By juxtaposing the rational laws of the coastline with the seemingly chaotic nature of the ocean, Bede grafted a Christian recovery narrative onto the strand, defining a geographical region that could now bring light and clarity to a dark and dangerous realm.[2]

Familiar only with the imperceptible tides of the Mediterranean, the scholars of antiquity showed relatively little interest in the tides until the Greeks expanded eastward and the Romans westward, where they encountered the ebb and flow of the world's oceans. The inhabitants of Britain, however, flanked on all sides by massive bodies of water, were geographically ordained to grasp the tides' significance. The land Shakespeare referred to as a "precious stone set in the silver sea" possessed unusually violent and often deadly tides, in some areas reaching upward of thirty-five feet.[3] Their destructive action materially affected the river and coastal areas where the early inhabitants worked, lived, and traveled. Originating far out in the deep Atlantic, the oceanic tide separates into two wave trains as it washes onto the shores of Britain's southwest coast. These wave trains then interfere as they travel around the island to produce a bewildering variation. On the east coast of Scotland, for instance, single tides can be found (one high tide and one low tide each day), while only a short distance north or south, double or even triple tides appear. The tide that travels up the Thames estuary, moreover, demonstrates constructive interference, where the two wave trains coincide almost perfectly. This not only produces unusually high tides in Britain's largest port but also masks the laws that produce them.

The extraordinary and irregular tides on England's east coast are further complicated by the island's topography. The southeast of England is sinking, a portion of the precious stone inexorably returning to the silver sea. As the ocean slowly encroaches, the tides rise higher on Britain's shores and ebb more dramatically through its estuaries and rivers. Everything from the extent and timing of Britain's coastal trade to the strategic defense of London itself therefore has relied on a thorough knowledge of the state of the tide. Although the moon's effect on the tides was still being debated a thousand years after Bede, in the seventeenth and eighteenth centuries and even into the nineteenth, the significance of the tides for Britain's commercial and military success certainly was not.

Studying the littoral environment, with the aim of imposing the rational patterns of England's coastlines on the deep and chaotic arena of the world's oceans, became a central commitment of physical astronomy. From the scientific revolution to the Enlightenment, natural philosophers and mariners attempted to extend Bede's recovery narrative from the land to the sea. Astronomy and practical navigation came together as sailors began taking measurements of terrestrial magnetism, the depth and salinity of the oceans, and the flora and fauna of newly explored lands. Owing to its practical significance for trading nations, the motion of tides and tidal streams soon became an important realm of investigation, culminating in the work of Isaac Newton and Edmond Halley, both highly respected philosophers of their day. In many respects, however, these early advances in tidal theory represent a false beginning.

While these early natural philosophers brought the tides under the rule of rationality and even expanded that rationality to parts of the world's oceans, they failed to translate their knowledge into practice that would benefit mariners. No systematic study of the tides followed, and the methods used to predict the tides remained a mystery. Throughout the eighteenth century the study of the tides in Britain was largely sustained not by eminent natural philosophers, but by naval captains, military and colonial administrators, dockyard engineers, and tide table makers. Thus an analysis of tidal knowledge before the Victorian era must discuss local calculators, navigation manuals, manuals of seamanship, and the myriad of individuals who published observations and theories that contradicted the theoretical developments within the emerging scientific community. Precisely because Newton's formulation appeared to be incorrect, a renewed interest in tidal theory brought in fresh observations from around the globe. Though the problems with Newton's general explanation became more apparent as the eighteenth century progressed, the institutional mechanisms necessary for a solution had yet to be adopted. No formal link existed between the Admiralty and the scientific community. George III founded the Hydrographic Office, which served as the research and development wing within the Admiralty, only in the last years of the eighteenth century, and the first Admiralty charts appeared only in 1800.[4] Throughout the period, the Admiralty failed to become vested in the theoretical study of the tides, and since the number of wrecked vessels remained relatively small, so too did the knowledge of the world's oceans.

## The Tides and the Early Royal Society

Although knowledge of the tides was of immediate practical concern for commerce and travel and even as a source of power, it was desperately inadequate up to the sixteenth century. The first wave of research on tides appeared on the Continent, but the mechanisms used to explain their cause varied considerably. Galileo Galilei proposed an argument based on the earth's movement; Johannes Kepler held fast to an attractive hypotheses similar to that of magnetic attraction; and René Descartes offered a compression theory that related the tides to the moon's passage over the oceans.[5] The moon was known to control the tides in some fashion, but none of these early researchers had a theory that successfully explained tidal observations. The first advances in the theoretical understanding of the tides came in England in the last quarter of the seventeenth century, after the founding of the Royal Society of London and the beginning of the *Philosophical Transactions of the Royal Society of London*, a private publication headed by its permanent secretary, Henry Oldenburg.

Robert Moray, first president of the Royal Society, brought the study of the tides to the attention of the nascent scientific community by publishing a notice in the fourth number of the *Philosophical Transactions* in June 1666. Moray described two peculiarities of the tides around the coast of Britain but stopped short of offering an explanation; his entry was simply a narrative of events.[6] The first extended account appeared two months later from John Wallis, one of England's leading mathematicians. In contrast to Moray's practical account, Wallis's treatment was overtly theoretical, based on Galileo's theory of the earth's movement. Two decades before Newton's *Principia* appeared, Wallis accounted for the tides by invoking what later became enshrined as Newton's first law of motion: "That a Body in motion is apt to continue its motion, and that in the same degree of celerity, unless hindred by some contrary Impediment; (like a Body at rest, to continue so, unless by some sufficient mover, put into motion)."[7] To connect the motion of the tides to the moon's periodicity, something Galileo had failed to do, Wallis suggested that the earth and moon be viewed as one body with "one common center of Gravity." Wallis announced that this "surmise," combined with the earth's annual and diurnal movements, was the true cause of the daily tides.

Wallis's theory evoked quite a stir within the newly formed Royal Society, and he was forced to respond to criticisms with a call for the accumulation of observations, setting out vague directions that he believed any "Understanding Person" living near the sea could easily follow.[8] The president of the Society, Moray, was much more cautious—and ambitious. Moray suggested the construction of tidal observatories equipped with the proper tide gauges, and to ensure that no errors crept into the investigation, he added a set of queries and a printed table that observers could use to register their observations.[9]

The first organized but far from systematic collection of tidal data began. "An Account of Some Observations Made by Mr. Samuel Colepresse at and nigh Plimoth" contained a direct answer to Moray's queries, along

( 263 )   Num. 16.

# PHILOSOPHICAL
## TRANSACTIONS.

Munday, August 6. 1666.

## The Contents.

An Essay of Dr. Wallis, exhibiting his Hypothesis about the Flux and Reflux of the Sea, taken from the consideration of the Common Center of Gravity of the Earth and Moon ; together with an Appendix of the same, containing an Answer to some Objections, made by severall Persons against that Hypothesis. Some Animadversions of the same Author upon Master Hobs's late Book, De Principiis & Ratiocinatione Geometrarum.

### An Essay
## Of Dr. John Wallis, exhibiting his Hypothesis about the Flux and Reflux of the Sea.

Ow abstruse a subject in Philosophy, the *Flux and Reflux of the Sea* hath proved hitherto, and how much the same hath in all Ages perplexed the Minds even of the best of *Naturalists*, when they have attempted to render an Account of the Cause thereof, is needless here to represent. It may perhaps be to more purpose, to take notice, that all the deficiencies, found in the *Theories* or *Hypotheses*, formerly invented for that End , have not been able to deterre the Ingenious of *this* Age from making farther search into that Matter: Among whom that Eminent Mathematician Dr. *John Wallis*, following his happy *Genius* for advancing reall Philosophy, hath made i a part of his later Inquiries and Studies , to contrive and de duce a certain Hypothesis concerning that *Phænomenon*, take

N n                                     from

Figure 1.1. Title page of John Wallis's "Hypotheses about the Flux and Reflux of the Sea, Taken from the Consideration of the Common Center of Gravity of the Earth and Moon," the first publication on the theory of the tides to appear in a scientific journal. *Philosophical Transactions* 1 (1666): 263.

with a table registered to Moray's specifications. Colepresse's findings, however, did not support Wallis's theory. Nor did the registry of Henry Powle, who informed Oldenburg that the year's highest tides in the Severn occurred near the equinoxes, in March and September, not in February and November as Wallis had suggested. Likewise, Joseph Childrey sent his "Animadversions on Dr. Wallis's Hypothesis of Tides" to the Royal Society as a direct refutation of Wallis's theory. Childrey argued that the distance of the moon from the earth also affected the daily tides, adding yet another cause to their variability. Within two years of the founding of the *Philosophical Transactions*, Oldenburg had received enough letters to warrant his own entry, "An Account of Several Engagements for Observing Tydes." He stated approvingly that the virtuosos of England had taken great care to "direct and recommend" observations, having lodged "very particular directions" with pilots and mariners in remote stations of the globe. Those reports so far received, however, differed so considerably that Oldenburg could make little of them.[10]

The most prevalent entry in the early *Philosophical Transactions* was this type of empirical observation, in keeping with the Royal Society's Baconian aim of compiling a history of nature. The initial entry by Moray, and later entries by Powle, Colepresse, and Childrey, as well as the multitude of unpublished letters sent to Oldenburg, offered little more than historical, firsthand narratives of specific tides as they were observed on the coasts of England. These fit well within the experimental tradition advocated by Robert Boyle and others—Moray even called them "experiments"—and demonstrated no generality. They stayed clear of causal explanations, aiming only to collect data accurately through experiment and observation—to catalog "matters of fact." These entries testify to the Royal Society's faith that once enough observations were collected, truth would follow. In lamenting the incongruity of his own theory with direct observations, Wallis noted, "Were the matter of fact agreed upon, it is not likely, that several hypotheses should so differ."

These disparate observations of the tides did allow researchers to advance "schemes" for constructing tide tables, professing to make the coastline safer for England's growing merchant fleet. Rather than narrating discrete historical events, these researchers relied on anonymous tidal observations, often expressly incomplete. Through "tryals"—meaning mathematical manipulations—they hit upon a "common rule" that could then be generalized from one port to the whole of Europe. Henry Philips, for instance, a respected writer on navigation, published a predictive table based on a rule he devised that related the tides at London Bridge to the position of the moon.[11] He refused to divulge how he devised his rule, but the process was overtly mathematical, and its strength lay in its generality. After noting the rule's agreement with his own tidal observations, Philips maintained that "the like you may do for any other Port or place."

Astronomer Edmond Halley's first publication on the tides was also a rule that mariners could use to predict the tides. Charles Davenport had published a table of the times of high water for the Bay of Tonkin in the South China Sea, offering practical advice on when it was safe to enter the estuary. In an appendix to Davenport's account, Halley used the data "to bring the hitherto unaccountable irregularity of these tides to a certain rule."[12] He constructed a figure similar to the one published earlier by Philips that he then used to compute the time and height of high water based on the position of the moon. That his rule accounted for Davenport's observations was enough for Halley; like those before him, he stopped well short of offering any reason why his scheme worked. Halley regretted that "this is a task too hard for my undertaking, especially when I consider how little we have been able to establish a Genuine and satisfactory Theory of the Tides, found upon our own Coast, of which wee have had so long Experience." Halley's analysis fit squarely between observational and theoretical accounts of the tides. Like other published tables, his was practical in scope, claiming generality and predictive accuracy.

The Bay of Tonkin in modern Vietnam was of scientific interest because of its unusual single tides, but most early tide table makers sought to predict the tides closer to home. The focal point was often London Bridge. In 1675, three years before Halley's first publication on the tides, King Charles founded the Royal Observatory at Greenwich and appointed John Flamsteed his first astronomer royal at an annual stipend of £100. In England it was the first paid position for a natural philosopher outside the universities, and Flamsteed, a careful observer, was a propitious choice. He spent his entire career observing the positions of stars and updating existing star catalogs, culminating in his *Historia Coelestis Britannica*, published posthumously in 1725. As the king's astronomer with the express charge of aiding navigation, Flamsteed also turned his observational acumen to the flux and reflux of the sea. Because the production of tide tables was based heavily on observation and calculation— no workable theory yet existed to guide the process—Flamsteed first attempted to unravel the secret methods used by other tide table makers. He began with the previous tables published for London Bridge, both those in local almanacs and the one published by Henry Philips in the *Philosophical Transactions*. Though he failed to ascertain their methods,

A Table of the time of High Tide for this present year *Anno* 1668.
*London*-Bridge : where *M.* denotes *Morning*, and *P. Afternoon.*

| | Jan. | Febr. | Mar. | April | May | June | July | Aug. | Sept. | Octo. | Nov. | Dec. |
|---|---|---|---|---|---|---|---|---|---|---|---|---|
| | H. M | I. M. | H. M | H. M. | H. M. | H. M. | H. M. | H. M. | H. M. | H. M. | H. M. | H. M. |
| 1 | 12 | 2 P 13 | 11 49 | P 6 3 | P 2 34 | 18 4 | 4 30 | 5 P | 5 5 | 54 6 | 36 8 | 18 8 36 |
| 2 | 1 P 2 | 3 11 2 | 45 3 | 57 4 | 44 | 51 4 | 59 5 | 32 6 | 37 7 | 27 9 | 20 9 | 33 |
| 3 | 2 3 | 3 56 3 | 29 4 | 22 4 | 39 5 | 20 5 | 25 6 | 2 7 | 23 8 | 32 10 | 24 10 | 32 |
| 4 | 3 37 | 4 30 4 | 8 4 | 35 5 | 11 5 | 48 5 | 52 6 | 39 8 | 28 9 | 43 11 | 30 11 | 36 |
| 5 | 4 2 | 5 14 | 40 5 | 24 5 | 40 6 | 18 6 | 23 7 | 25 9 | 41 10 | 54 12 | 31 12 | 4 |
| 6 | 4 57 | 5 29 5 | 9 5 | 55 6 | 12 6 | 54 7 | 18 28 | 11 2 | 12 3 | M 31 | M 40 | |
| 7 | 5 26 | 5 58 5 | 38 6 | 30 6 | 49 7 | 37 7 | 47 9 | 43 12 | 24 M 3 | 1 32 1 | 43 | |
| 8 | 5 55 | 6 30 6 | 9 7 | 13 7 | 28 8 | 27 8 | 46 11 | 7 M 24 | 1 10 2 | 27 2 | 37 | |
| 9 | 6 2 | 7 11 6 | 49 8 | 5 8 | 21 9 | 27 9 | 57 12 | 30 1 | 35 2 | 9 3 | 18 3 | 28 |
| 10 | 7 5 | 7 59 7 | 32 8 | 58 9 | 18 10 | 39 11 | 19 M | 30 2 | 34 3 | 1 4 | 2 4 | 10 |
| 11 | 7 4 | 53 8 | 28 9 | 58 10 | 19 11 | 53 12 | 46 1 | 48 3 | 24 | 45 4 | 37 4 | 41 |
| 12 | 8 3 | 58 9 | 28 11 | 2 11 | 26 1 | 16 M | 46 2 | 53 4 | 6 4 | 23 5 | 8 5 | 10 |
| 13 | 9 33 | 1 7 10 | 33 12 | 6 12 | 34 1 M 16 | 2 6 3 | 45 4 | 39 4 | 57 5 | 37 5 | 37 | |
| 14 | 10 15 | 12 16 11 | 40 M | 6 M 34 | 2 27 3 | 15 4 | 24 5 | 11 5 | 27 6 | 6 6 | 4 | |
| 15 | 11 4 | M 16 12 | 44 1 | 10 1 | 44 3 | 33 4 | 4 4 | 56 5 | 41 5 | 59 6 | 42 6 | 35 |
| 16 | 12 54 | 1 18 M | 44 2 | 13 2 | 5 4 | 22 4 | 41 5 | 25 6 | 14 3 | 34 7 | 21 7 | 12 |
| 17 | M 54 | 2 18 1 | 44 3 | 9 3 | 49 5 | 1 5 | 55 5 | 54 6 | 54 7 | 16 8 | 57 7 | 58 |
| 18 | 1 56 | 3 8 2 | 37 4 | 3 4 | 36 5 | 24 5 | 43 6 | 29 7 | 4 8 | 68 8 | 56 8 | 49 |
| 19 | 2 50 | 3 49 3 | 27 4 | 45 5 | 1 6 | 36 6 | 15 7 | 11 8 | 34 8 | 57 9 | 54 9 | 55 |
| 20 | 3 36 | 4 24 4 | 13 5 | 23 5 | 52 6 | 44 6 | 53 8 | 4 9 | 35 10 | 54 10 | 58 11 | 5 |
| 21 | 4 13 | 4 57 4 | 50 6 | 3 6 | 33 7 | 25 7 | 35 9 | 0 10 | 40 10 | 55 12 | 9 12 | 30 |
| 22 | 4 44 | 5 30 5 | 26 6 | 48 7 | 17 8 | 16 8 | 28 10 | 4 11 | 43 11 | 58 1 P 17 | 1 P 52 | |
| 23 | 5 13 | 6 5 6 | 4 7 | 42 8 | 9 7 | 9 27 11 | 1 12 44 | 12 55 2 | 26 3 | 2 | | |
| 24 | 5 41 | 6 48 6 | 51 8 | 38 9 | 0 10 | 10 10 | 33 12 16 | 1 P 4 2 P | 1 3 | 27 3 | 58 | |
| 25 | 6 13 | 7 46 7 | 48 0 | 35 9 | 58 11 | 12 11 | 32 1 P 16 | 2 34 2 | 58 4 | 17 4 | 41 | |
| 26 | 6 58 | 8 53 8 | 53 10 | 44 10 | 58 12 | 18 12 48 | 1 23 | 32 3 | 48 4 | 59 5 | 15 | |
| 27 | 7 49 | 10 10 10 | 2 11 | 47 12 | 1 1 P 21 | 1 P 49 3 | 14 4 | 4 4 | 32 5 | 36 5 | 47 | |
| 28 | 8 55 | 11 28 11 | 12 12 | 49 1 P | 2 19 2 | 43 3 | 43 4 | 41 5 | 11 6 | 14 6 | 19 | |
| 29 | 10 16 | 12 43 12 | 20 1 | 47 2 | 3 11 3 | 27 4 | 18 5 | 18 5 | 42 6 | 55 7 | 0 | |
| 30 | 11 44 | 1 P 21 2 | 28 2 | 55 3 | 54 4 | 16 4 | 30 5 | 52 6 | 31 7 | 43 7 | 45 | |
| 31 | 1 P 2 | 2 16 | 3 40 | | 4 | 37 5 | 21 | 7 21 | | 8 36 | | |

These things I have found to fall out right at *London* for many years, and so I suppose they may in other places. If the difference be not so much between the Neap-tides and Spring-tides in other places, the Diameter must be divided into fewer parts.

As for the higest Tides to happen two or three dayes after the full Moon, I have not made much observation of it, and see little reason for it, but the time thereof agrees herewith. And high Spring-tides are not alwayes alike; this year I have not observed any. I should be glad to hear, how these rules hold in other places, that so this *true time* of the Tides may be more punctually known.

*April* 6. 1668.                                                   *An*

Figure 1.2. Tide table for London Bridge published in the *Philosophical Transactions* by Henry Philips. Philips did not divulge the rules by which he devised this table. "A Letter Written to Dr. John Wallis by Mr. Henry Philips, Containing His Observations about the True Time of the Tides," *Philosophical Transactions* 3 (1668): 659.

Figure 1.3. Portrait of John Flamsteed (1646—1719). Charles II appointed Flamsteed the first royal astronomer in 1675. A relentless observer, Flamsteed ran the Royal Observatory at Greenwich, completed in 1676, which he equipped with his own instruments and staffed with his own computers. He began observing the tides in the summer of 1682, during his travels to and from the observatory, and began publishing tide tables for London the following year. Reproduced by permission of the Royal Society of London.

his own meticulous observations proved that these earlier tables were erroneous to the point of being dangerous.

Flamsteed worked in Greenwich but lived in London, and he spent much of his time traveling by boat between the two, taking special note of the tides in the river Thames. With the help of an "ingenious Friend," he spent the summer of 1682 making tide observations, forming predictive tables, and testing his results. "Considering how much the River of Thames is frequented by Shipping, and how long it has been the Chief Place of Commerce in these parts of the World," Flamsteed warned, "one would think our Seamens Accounts of the Tides should be very exact, and their Opinions concerning them Rational; whereas if they be enquired into, nothing will be found more Erroneous and Idle."[13] He especially criticized the "tide Tables of our Almanacks" for contributing to false ideas. These almanacs merely added three hours to the time of the moon's southing, a mathematical manipulation that agreed rather well for tides at the full moon and new moon but not throughout the month. Flamsteed noted that Mr. Booker, whose tide tables for London Bridge were reputedly the most accurate, simply subtracted an hour for the quarter tides, a rough approximation at best. Flamsteed therefore produced his own tables, published annually between 1683 and 1688. Although he asserted that his predictions could be used "for any other Port in His Majesties Dominions or Neighbouring Countrys, by only subtracting or adding" a constant amount, his limited travels did not warrant this claim. A few friends, no matter how ingenious, did not provide the observations to test this generalization, similar to those of the almanac makers he was attempting to refute.

Edmond Halley made this exact point in print, and the disagreement between the two strong-willed astronomers foreshadowed debates that resurfaced (and were in some respects resolved) in the Victorian era. In fact, the later debates were in part precipitated by a biography of Flamsteed that the astronomer Francis Baily published in the mid-1830s. Baily used the character of Flamsteed to reassert the fundamental role of the observer and calculator in physical and observational astronomy, a point rebutted by natural philosophers such as William Whewell who defended the more theoretical aspects of science. Both in the late seventeenth century and in the mid-nineteenth, the debate centered on the process of scientific discovery and the highly contentious issue of intellectual property.

Flamsteed was a respected natural philosopher, the king's astronomer, but his reputation rested almost exclusively on his meticulous astronomical observations. His insistence on the accurate reduction of data brought him into conflict with the leading theorists of his day, including Edmond Halley, in many respects his archrival. At times vituperative, the disagreements between them reached a peak only after 1704, when both had ended their forays into tidal analysis. Isaac Newton was by that time the most esteemed natural philosopher in England, perhaps the world, and he had requested observations of the moon to perfect his lunar theory. Flamsteed had made these observations as astronomer royal but had refused to publish them. Newton, with Halley as his bulldog,

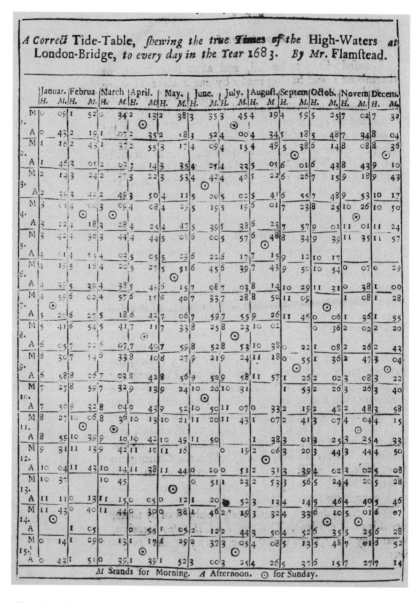

Figure 1.4. John Flamsteed published tide tables for London Bridge annually from 1683 to 1688. The astronomer Edmond Halley criticized these tables in print, leading to a row between the two strong-willed academics that lasted throughout their lives. John Flamsteed, "A Correct Tide Table, Shewing the True Times of the High-Waters at London-Bridge, to Every Day in the Year 1683," *Philosophical Transactions* 13 (1683): 10.

chided Flamsteed for lagging in their publication, insisting that the observations were not Flamsteed's at all. The astronomer royal had acquired them as a public servant paid by the Crown, and both Newton and Halley insisted they belonged in the public domain. Flamsteed, however, had taken up the job with very little backing from the state, despite the title "royal" attached to his position: he provided his own instruments, hired his own computers, and covered all the other expenses of running the observatory. In his mind, the observations were his private property to do with as he wished. His job did not entail providing raw data for speculative natural philosophers; it was his responsibility to ensure that observations were carefully made and efficiently reduced. And the acquisition and reduction of data, he insisted, took time. The row culminated in Newton and Halley's stealing the observations from under Flamsteed's nose and publishing them in 1712 as an unauthorized edition that Flamsteed promptly gathered up and burned.[14]

The acrimonious dispute between Halley and Flamsteed had begun with the subject of the tides as early as 1684. Halley's attitude concerning the role of observation and calculation dramatically altered that year. After he published his report on the tides in the Bay of Tonkin—a paper that had focused specifically on the role of observation and calculation at the expense of theory—he received a nine-page document that not only changed his views on the tides but eventually transformed all of physical astronomy. The manuscript, titled *De Motu Corporum in Gyrum* (On the Motion of Bodies in an Orbit), came from Isaac Newton and proved that an inverse-square law was responsible for the elliptical paths of the planets. Halley, at work on celestial mechanics at just that time, realized the treatise held revolutionary potential.[15] He also recognized that the new law could be used to correct Flamsteed's tide tables, and as editor of the *Philosophical Transactions* from 1685 to 1693, he was in a position to write a scathing critique of Flamsteed's previous work. These published remarks initiated the bitter correspondence between the two and spawned a resentment that would last throughout Flamsteed's life. Though the debate was disguised as a disagreement over tidal knowledge, the underlying tension concerned the role of observation and calculation in relation to theory. For Halley, now equipped with what would become the most powerful theory in physics, observations of the tides were useful only if they were incorporated into Newton's general theory of universal gravitation.

Figure 1.5. Philosopher and self-styled mariner Edmond Halley took to the sea for three voyages on the *Paramore* between 1698 and 1701. From these voyages he constructed the first visual representations of the earth's magnetic and tidal phenomena. Reproduced by permission of the Royal Society of London.

## Edmond Halley and the Beginning of Tidal Analysis

Edmond Halley's name will forever be attached to the theory of tides, not from his own tinkering with the tides at Tonkin or from his early attempts at tidal prediction, but because of the next two phases of his intellectual career. The first phase comprised his role in Newton's transforming *De Motu* into the greatest mathematical text ever written, the *Philosophiae Naturalis Principia Mathematica*; the second was his work as master and commander of the *Paramore*.[16]

Between 1684, the year Halley first viewed *De Motu*, and 1687, the year Newton published the *Principia*, Halley encouraged—or rather, prodded and cajoled—Newton to finish the text. Once the work was completed, he saw the manuscript through the press as editor and financier. Though the text contained much that excited the young astronomer, Halley was especially interested in Newton's achievements concerning the world's oceans. Halley had been to sea at age twenty, spending a year on the small island of St. Helena off the west coast of Africa. During his journey, he had weathered tempests in the deep ocean and battled tidal streams and currents closer to shore. His early interest in astronomy, along with his fascination with questions in nautical science, prepared him to appreciate Newton's groundbreaking theory of universal gravitation, especially as it related to the tides.

Newton's explanation of the tides, deduced from his theory of universal gravitation, correctly described the physical causes of the tides as due to the attraction of the world's oceans by the dual action of the sun and moon. His account, however, like his lunar theory, was incomplete.[17] Rather than using his new theory of universal gravitation to definitively explain the tides, Newton used the tides to exemplify his theory of universal gravitation. Indeed, he did not include much observational proof for universal gravitation in the *Principia* beyond his explanation of the tides and a few observations of the pendulum. And his use of the tides in this manner was suspect. He based his theory on tidal observations made at Bristol, which has some of the largest tide variations in the world, along with other data that were, to put it mildly, well chosen.[18] Moreover, Newton used tidal data to determine the mass of the moon, used the mass of the moon as a variable in his law of universal gravitation, and used his law of universal gravitation to explain the tides, a circular argument ridiculed in later work by both John William Lubbock and William Whewell.[19] After Newton, the tides as they actually appeared on the coasts of Europe remained as inexplicable as ever.

Once the *Principia* was published, Halley presented an ornately bound copy to James II, with a separate letter outlining Newton's theory of the tides. That Halley extracted the section on the tides rather than, say, Newton's lunar theory or his explanation of comets is not surprising. The king was an experienced naval commander, and as Halley correctly surmised, he was exceedingly interested in the tides. Owing to the publication of Wallis's theory some twenty years earlier, the study of the tides had also generated heated debates within the Royal Society, and Halley believed Newton's treatise finally laid all this to rest. As publisher and financier of the *Principia*, moreover, Halley had an additional incentive to espouse Newton's explanation of the tides. Perhaps more than any other natural phenomenon, the tides were the area where Newton's theory could be shown to address practical problems of navigation. By extracting the section on the tides, Halley could demonstrate both the practical application of Newton's theory and the veracity of Newton's claims. He used the tides, in other words, to boost the sales of Newton's otherwise extremely abstruse text.

Newton's *Principia* was a difficult book, to say the least; only a few mathematicians throughout Europe could follow his reasoning. Halley's separate explanation of the tides made Newton's theory of the tides accessible, and Halley reprinted his letter to the king in the *Philosophical Transactions* in March 1697.[20] He offered, for the first time for most readers, the basis of the modern explanation of the tides and the foundation for all subsequent work. The world's oceans, Halley explained, were pulled by the gravitational attraction of the moon and sun, producing an oval ocean over a spherical earth, with one bulge directly under the moon and sun and another on the opposite side of the earth. The bulges were largest when the sun and moon were in conjunction and in opposition, compounding their pull on the world's oceans. These were termed "spring tides," since they seemed to spring upon the coasts and estuaries of Europe. At quadrature, however, the attraction of the sun and moon worked against each other to produce the lowest bulges of the oceans, the "neap tides," the smallest tides of the month. As the earth rotated, the two bulges traveled over its landmasses, causing two floods and two ebbs each day. This explained both why the highest tides occurred during the full moon and new moon and the lowest tides occurred at quadrature, when the sun and moon were separated by ninety degrees. Newton's explanation, noted Halley, "has been observ'd to hold true on the coast of England, at Bristol by Capt. Sturmy, and at Plymouth by Mr. Colepresse." He further explained how great diversities could arise owing to particular circumstances of each place: "But it would be endless to account all the particular Solutions, which are easie Corollaries of this Hypothesis." That Newton's theory explained even the observations at the port of Tonkin in the South China Sea—observations so recalcitrant to generality that even Halley had failed to deduce them—demonstrated, according to Halley, "the excellency of this Doctrine."

Direct observational verification of Newton's law of universal gravitation would have to wait well over half a century, with the return of Halley's comet in 1747 and the expeditions sent to Peru and Lapland between 1735 and 1744 to determine the shape of the earth. The tides, however, represented the one area where the practical relevance of Newton's theory could be demonstrated to a relatively wide audience.

II. *The true Theory of the Tides, extracted from hat admired Treatise of* Mr. Isaac Newton, *Intituled,* Philosophiæ Naturalis Principia Mathematica; *being a Discourse presented with that Book to the late King* James, *by* Mr. Edmund Halley.

*IT may, perhaps, seem strange, that this Paper, being no other than a partile Account of a Book long since published, and whereof a fuller Extract was given in* Numb. 187. *of these Transactions, should again appear here: but the Desires of several honourable Persons, which could not be withstood, have obliged us to insert it here, for the sake of such, who being less knowing in Mathematical Matters; and therefore, not daring to adventure on the Author himself, are notwithstanding, very curious to be informed of the Causes of Things ; particularly of so general and extraordinary* Phænomena, *as are those of the* Tides. *Now this Paper having been drawn up for the late King* James's *Use, (in whose Reign the Book was published) and having given good Satisfaction to those that got Copies of it ; it is hoped the Savans of the higher Form will indulge us this liberty we take to gratifie their Inferiours in point of Science ; and not be offended, that we here insist more largely upon* Mr. Newton's *Theory of the* Tides, *which, how plain and easie soever we find, is very little understood by the common Reader.*

The sole Principle upon which this Author proceeds to explain most of the great and surprising Appearances of Nature, is no other than that of *Gravity*, whereby in the Earth all Bodies have a tendency towards its Centre ;

X x x         as

Figure 1.6. Edmond Halley published a separate explanation of Newton's theory of the tides in the *Philosophical Transactions* of the Royal Society of London nine years after he saw the *Principia* through the press. It simplified Newton's explanation and was based on a previous letter that Halley had written to King James. Edmond Halley, "The True Theory of the Tides, Extracted from That Admired Treatise of Mr. Isaac Newton," *Philosophical Transactions* 19 (1696): 445.

This explains, for instance, Halley's paper presented to the Royal Society entitled "Discourse Tending to Prove at What Time and Place, Julius Cesar Made His First Descent upon Britain." The paper was based on the regularity of the tides and seemingly was of only historical and military significance (a matter of defense against invasion), but Halley used it to demonstrate the predictive power of Newton's theory, albeit in a rather peculiar way. He calculated when a historical event had happened rather than when a future natural phenomenon would occur.[21] By recreating an event that took place over sixteen hundred years earlier, Halley demonstrated that the tides could help in "fixing and ascertaining the Times of memorial Actions, when omitted or not duly delivered by the Historian." Reducing even history to the exact laws of mechanics was one of the most cogent examples of the newly afforded power of physics. Even Newton was impressed.

Newton's theory was powerful enough to awaken the ghost of Caesar, but alas, it did not explain the actual tides observed on the coasts of England. The theory was correct in principle and worked well for tides that were influenced only by the paths of the sun and moon, but few tides exhibited such a simple relationship. As Halley well knew, Newton's theory needed to be refined, and after the publication of the *Principia*, his professional and scientific interests turned increasingly to the sea.

When Halley first set sail to search out the tides in the summer of 1688, London was already the largest city in Europe, well on its way to becoming the center of trade in the Western world.[22] The growth of the city was astounding, especially its east side, whose wharves, quays, and docks attracted a steady stream of mariners, artisans, and skilled craftsmen associated with coastal shipping. The Thames also housed the shipbuilding industry and ports used by the fledgling British navy. The commercial and military significance of the Thames made knowledge of its approaches critical. The river was one of the most used, most dangerous, and surprisingly, least known areas of the coastal seas. Very few navigational aids had been published, and mariners trusted those only with extreme caution. John Seller, a well-known surveyor and hydrographer, had published his *English Pilot* in 1671, compiled largely from old Dutch charts. Halley, along with numerous mariners and cartographers, considered it inaccurate and lamented the debt it owed to the Dutch, Britain's maritime and military rivals. The king assigned Captain Greenville Collins, an experienced naval officer, the sole duty of breaking this British reliance on Dutch navigational charts. His first commission was to survey the coasts of England, Scotland, and Wales, but progress was slow, and after eight years of work, mariners greeted his *Great Britain's Coasting Pilot* of 1693 with little applause.

No reliable charts, therefore, existed for the east coast of England, including the approach to London. Under orders from the Crown, Halley charted the Thames estuary and English Channel in the summer of 1688, the year after publication of the *Principia*. Very little is known about this early survey, though political and economic considerations were its main impetus and a dash of espionage was most certainly involved. On his return, Halley produced a chart of the Thames estuary that he presented to the Royal Society of London.[23] The chart is now lost, but the major soundings and tidal data appeared again in subsequent charts that Halley published in the first few years of the eighteenth century, incorporating additional observations from the voyages of the *Paramore*, the vessel the Crown had furnished to Halley to sail through the Atlantic Ocean.

## The Last Philosopher-Commander

After Newton's *Principia* appeared, Halley proved himself much more than a philosophical landlubber with a passing interest in the sea. He spent the next decade engrossed in problems of physical astronomy in relation to navigation, undertaking three voyages in the *Paramore* at the turn of the century. His attempt to extend the usefulness of Newton's great achievements from England's shores and estuaries to all of the Atlantic Ocean formed perhaps the most exciting chapter of his distinguished scientific career. Impressed with Halley's abilities and his interest in questions of navigation, the king appointed him to command the *Paramore* with orders to sail the Atlantic trade routes and measure the earth's magnetic field. Sailors had long used the magnetic compass to find their way in the deep ocean, based on the principle that a magnetized needle points roughly to the magnetic pole. The principle, however, demonstrated some peculiar inconsistencies. The earth's magnetism seemed to vary considerably, to wander through time and space. Some of the perturbations were local, stemming from iron in mountains or from magnetic storms, while others seemed more general, changing throughout the decades and the seasons, even daily and hourly. How and why these variations occurred remained a mystery. Between November 1698 and June 1700, Halley made two voyages to chart the variation of the magnetic needle in an attempt to bring the earth's magnetic properties to rule.

Halley met with difficulties almost immediately after he set sail on his first voyage in October 1698. The *Paramore* was relatively small, a fifty-two foot pink,[24] with a flat bottom to allow for close surveying of estuaries and coastlines. It was not properly balanced, however, and as the crew attempted to put in at Portsmouth for repairs, the strong tides in the English Channel combined with heavy gales to prevent it from entering the port. This was an ominous beginning to a voyage that had all the ingredients of a swashbuckling adventure, with devastating storms, constant danger from pirates, military operations, and a near mutiny that led to charges of treason. Halley had no training in the navy or as a mariner, and he was assigned his position as master and commander partly to ward off obstructiveness from mariners under his command.[25] The ruse failed. The crew was unrelentingly hostile toward the philosopher-commander. His lieutenant, "because perhaps I have not the whole Sea dictionary so perfect as he," represented Halley to the ship's company as incompetent.[26] After realizing that the crew had been directing the vessel on a course contrary to his orders, Halley was forced to end his voyage early, terminating his scientific expedition at Barbados.

Halley returned to the Atlantic after several months ashore, but only after convincing the Admiralty that he was still the right man for the task. Sixty years before the famous voyages of Captain Cook, Halley sailed farther south than any other Westerner on an explicitly scientific mission. His voyage represents an early link between the English state and the scientific understanding of the ocean. Halley published a map of the magnetic variation over the Atlantic in 1701, after his return, using isogonic lines (lines of equal magnetic variation), a new and remarkably powerful graphic device used in subsequent studies of terrestrial magnetism. Alexander von Humboldt adopted Halley's graphical techniques in the early nineteenth century, expanding their use to most other questions in physical astronomy, and they served as a potent addition to Whewell's tidal studies in the 1830s.

The navigational hazards Halley encountered in getting from the Thames to the Atlantic, including strong gales and powerful tides and tidal streams, prompted his third and final voyage, again with the support of the Crown. In many respects it imitated his first voyage, charting the same area in the summer of 1688. He remained in the English Channel and North Sea, but this time he focused exclusively on the dangerous tides and tidal streams in the approach to the Thames estuary. He scoured the English Channel for tide observations throughout the summer of 1701, anchoring his vessel around the southern and eastern coasts of England and on the western coasts of the Low Countries and France.[27] Both regions were pregnant with military and commercial significance, and his voyages foreshadowed the connection of science to commercial and imperial expansion so evident in the scientific voyages of a century later.[28] The state supplied the vessel, the crew, and the equipment and found the results valuable for trade and national defense.[29] When Whewell furthered the relationship between natural philosophy and the Admiralty 130 years later, he was extending a legacy that reached back at least to Halley.

Halley's advances in extending the laws of the tides outward from the coasts and estuaries of Britain were extraordinary. He undertook the first methodical magnetic survey of the Atlantic and used his observations to construct a map that mariners could use. He also made the first comprehensive study of the tides and tidal currents in the Thames estuary, and following his visual representation of terrestrial magnetism, he created a map to represent their directions in the English Channel. It was the first tidal chart of its kind and included a detailed analysis of the approach to Plymouth Harbor, which would become the most important port in the defense of the country once France became England's major military threat.[30] Perhaps most important, Halley's visual displays were unique for their synoptic integration of the vertical with the horizontal. His work on the trade winds of the world's oceans, for instance, was directly analogous to his research on tides and terrestrial magnetism. He defined winds as "the Stream or Current of the Air" and said they were caused by astronomical effects, most notably the "Suns Beams upon the Air and Water," not by the motion of the earth itself, as earlier philosophers had claimed. He then published a "Scheme, shewing at one view all the various Tracts and Courses of these Winds" that he considered far more useful than "any verbal description whatsoever."[31] Such visual representations seem commonplace today, but they were revolutionary in the late seventeenth century, and they helped initiate a vertical and horizontal orientation to understanding the ocean, reaching from the sun and moon, through the trade winds and magnetic properties of the atmosphere, all the way down to the tidal currents on the ocean's surface.

Halley's work done in the name of the safety of navigators endeared him to both the Admiralty and the British Crown. His intellectual command of the new natural philosophy, combined with his interest in winds, tides, and the magnetism of the world's oceans, led to other official and unofficial duties for the state. Queen Anne, for instance, sent him on several diplomatic missions to the Continent to proffer advice on the fortification of seaports. Halley was well prepared to give such

# THE
## Defcription
### AND
# USES
Of a New and Correct
# SEA-CHART
Of the Weftern and Southern
# OCEAN,
Shewing the Variations of the
*COMPASS.*

THE Projection of this *Chart* is what is commonly called *Mercator's*; but from its particular Ufe in *Navigation*, ought rather to be named the *Nautical*; as being the only true and fufficient *CHART* for the Sea. It is fuppofed, that all fuch as take Charge of Ships in long Voyages, are fo far acquaint-ed with its Ufe, as not to need any Directions here. I fhall only take the Liberty to affure the Reader, that having taken all poffible Care, as well from Aftronomical Obfervations, as Journals, to afcertain the Situa-tion and Form of this *Chart*, as to its principal Parts, and the Dimenfions of the feveral Oceans ; he is not to expect that we fhould defcend to all the Particularities neceffary for the Coafter, our Scale not permitting it. What is here properly New, is the *Curve-Lines* drawn over the feveral Seas, to fhew the Degrees of the *Variation* of the *Magnetical Needle*, or *Sea Compafs* : which are defign'd according to what I my felf found in the *Weftern* and *Southern* Oceans, in a Voyage I purpofely made at the Publick Charge in the Year of our Lord 1700.

That this may be the better under-ftood, the curious Mariner is defired to obferve, that in this *Chart* the Double Line paffing near *Barmudas*, the Cape *Verde Ifles*, and Saint *He-lena* every where divides the *Eaft* and *Weft Variation* in this *Ocean*, and that on the whole Coaft of *Europe* and *Africa* the *Variation* is Wefterly, as on the more Northerly Coafts of *A-merica*, but on the more Southerly Parts of *America* 'tis Eafterly. The Degrees of *Variation*, or how much the Compafs declines from the true North on either Side is reckoned by the Number of the Lines on each fide the double Curve, which I call *Line of No Variation*; on each fifth and tenth is diftinguifhed in its Stroak, and numbered accordingly, fo that in what Place foever your Ship is, you find the *Variation* by Infpection.

That this may be the fuller un-derftood, take thefe Examples. At *Madera* the *Variation* is 3 and ½d. Weft ; at *Barbadoes* 5½d. Eaft ; at *An-tekos* 9d. Weft ; at Cape *Race* in *Newfoundland* 14½ Weft ; at the Mouth of *Rio de Plata* 18d. Eaft, &c. And this may fuffice by way of Defcription.

As to the Ufes of this *Chart*, they will eafily be underftood, efpecially by fuch as are acquainted with the Azimuth Compafs, to be, to correct the Courfe of Ships at Sea : For if the Variation of the Compafs be not allowed, all Reckonings muft be fo far erroneous : And in contin-ed Cloudy Weather, or where the Mariner is not provided to obferve this Variation duly, the *Chart* will readily fhew him what Allowances

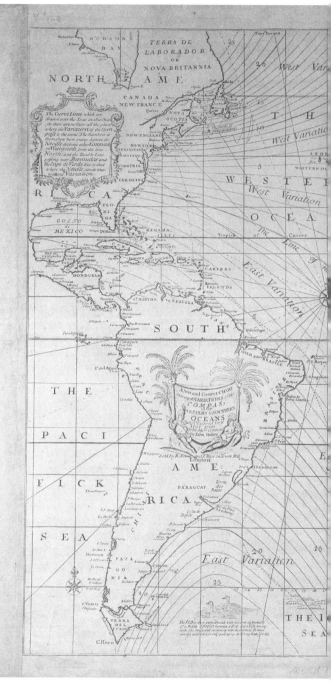

he muſt make for this Default of his Compaſs, and thereby rectify his Journal.

But this Correction of the Courſe is in no caſe ſo neceſſary as in running down on a Parallel *Eaſt* or *Weſt* to hit a *Port*: For if being in your Latitude at the Diſtance of 70 or 80 Leagues, you allow not the Variation, but ſteer Eaſt or Weſt by Compaſs, you ſhall fall to the Northwards or Southwards of your Port on each 19 Leagues of Diſtance, one Mile for each Degree of Variation, which may produce very dangerous Errors, where the Variation is conſiderable; for Inſtance, having a good Obſervation in Latitude 49d. 40m. about 80 Leag. without *Scilly*, and not conſidering that there is 8 Degrees Weſt Variation, I ſteer away *Eaſt* by Compaſs for the Channel; but making my way truly *E. 8d. N.* when I come up with *Scilly*, inſtead of being 3 or 4 Leagues to the South thereof, I ſhall find my ſelf as much to the Northward: And this Evil will be more or leſs according to the Diſtance you ſail in the Parallel. The Rule to apply it is, That to keep your Parallel truly, you go ſo many Degrees to the Southward of the *Eaſt*, and Northward of the *Weſt*, as in the *Weſt* Variation; but contrariwiſe, ſo many Degrees to the Northwards of the *Eaſt*, and Southwards of the *Weſt*, as there is Eaſt *Variation*.

A further Uſe is in many Caſes to eſtimate the Longitude at Sea thereby; for where the *Curves* run nearly *North* and *South*, and are thick together, as about *Cape Bona Eſperance*, it gives a very good Indication of the Diſtance of the Land to Ships come from far; for there the Variation alters a Degree to each two Degrees of Longitude nearly; as may be ſeen in the *Chart*. But in this Weſtern Ocean, between *Europe* and the *North America*, the Curves lying nearly Eaſt and Weſt, cannot be ſerviceable for this Purpoſe.

This Chart, as I ſaid, was made by Obſervation of the Year 1700, but it muſt be noted, that there is a perpetual tho' ſlow Change in the *Variation* almoſt every where, which will make it neceſſary in time to alter the whole Syſtem: at preſent it may ſuffice to advertiſe that about *C. Bona Eſperance*, the Weſt Variation encreaſes at the Rate of about a Degree in 9 Years. In our Channel it encreaſes a Degree in ſeven Years, but ſlower the nearer the Equinoctial Line; as on the *Guinea* Coaſt a Degree in 11 or 12 Years. On the *American ſide* the *Weſt Variation* alters but little; and the *Eaſt Variation* on the *Southern America* decreaſes, the more Southerly the faſter; the *Line of No Variation* moving gradually towards it.

I ſhall need to ſay no more about it, but let it commend it ſelf, and all knowing Mariners are deſired to lend their Aſſiſtance and Informations, towards the perfecting of this uſeful Work. And if by undoubted Obſervations it be found in any Part defective, the Notes of it will be received with all grateful Acknowledgement, and the Chart corrected accordingly.
E. HALLEY.

This CHART is to be ſold by *William Mount*, and *Thomas Page* on *Tower-Hill*.

Figure 1.7. Edmond Halley's magnetic chart of the Atlantic (1701). Reproduced by permission of the Royal Geographical Society, London.

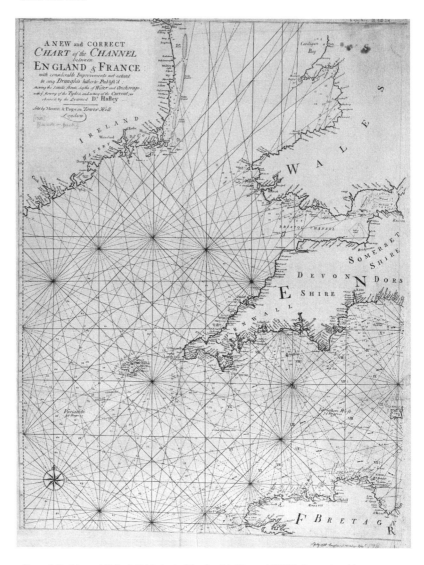

Figure 1.8. Edmond Halley's tidal chart of the English Channel (1702). Reproduced by permission of the Royal Geographical Society, London.

advice, since he had studied French port fortifications, probably un-known to the French, while surveying the English Channel.[32] His work also earned Halley the Savilian chair of geometry at Oxford in 1704, the post formerly held by John Wallis, and eventually a position as second astronomer royal, a Crown appointment he received in 1720 after

Flamsteed's death. He was now officially working for the state, in a paid position with the sole purpose of investigating the science of the sea for the protection of mariners.

Halley's personal command of the *Paramore* was less successful. After the near mutiny, natural philosophers no longer were allowed to captain ships. They became supernumeraries on oceanic voyages of discovery, visitors who rarely had control of when and where they could take measurements. The scientific study of the sea after Halley therefore remained episodic, unconnected, and often useless. As with the initial investigations in the Royal Society, advances on Halley's achievements would have to wait more than 130 years, until William Whewell expanded on Halley's use of isogonic lines and numerous scholars took up the study of terrestrial magnetism and meteorology over the world's oceans. George Biddell Airy would even publish a treatise on the landing of Julius Caesar, also from an analysis of tidal data. But all these approaches lay dormant for over a century. This says much about Halley but even more about the foundation needed to enhance knowledge of the ocean.

The subject of the tides was in many respects unique. Most topics in the history of science progressed from observation to laws of the phenomenon to causal laws. Kepler, for instance, used Tycho Brahe's observations of Mars to determine the mathematical laws governing the motion of planets. These laws of the phenomenon, in turn, were required for (and finally explained by) Newton's theory of universal gravitation. The tides, however, represented one instance where the causal law, Newton's theory, was well known before the laws of the phenomenon had been sufficiently spelled out. In other respects, however, the study of the tides followed much the same pattern as other questions in physical astronomy. Though Halley had established a link between science and the state, the *Paramore* was a lone vessel in an expansive ocean. His three voyages allowed Halley to grasp the significance of a synoptic view for questions in physical astronomy, but this in turn required much more data than a single researcher on a single vessel could gather. Interest in the tides was directly proportional to the number of shipwrecks on British shores and the consequent loss of revenue for the state. It lay undeveloped as long as it remained a question of theory. Only after the topic reached into the heart of the commercial and military sector of society did it obtain the needed encouragement.

## Tidal Prediction After Newton and Halley

Newton correctly explained the general causes of the tides, but nowhere in the *Principia* did he describe how to use his theory to produce tide tables. Halley, likewise, had published a separate, nonmathematical account of Newton's explanation, but like other tide table makers before him, he refused to divulge his own methods of prediction. He focused instead on searching out tidal phenomena throughout the oceans in hopes of refining Newton's general theory. He produced valuable charts that were widely used by mariners, but this did not bring the tides to a useful rule. Despite these early advances in theory, mariners still had no way of predicting the tides in the ports of England, much less the estuaries and coasts around the world. The subject therefore continued to occupy the minds of the most respected mathematicians in Europe. In 1738 the French Academy of Sciences proposed the theory of the tides as a prize competition. Four mathematicians shared the award in 1740: Daniel Bernoulli, then professor of anatomy and botany at Basel; Colin Maclaurin, professor of mathematics in Edinburgh; Leonard Euler, professor of mathematics at St. Petersburg; and Antoine Cavalleri, a Jesuit who based his theory on Cartesian vortices.[33] Of the four prizewinning essays, only Bernoulli's included tables that enabled researchers to make predictions for individual ports, a decisive advance on Newton's theory.

A hundred years later, the early Victorians used not Newton's theory of the tides but the variant that Bernoulli worked out for this competition, which they referred to as the equilibrium theory. The equilibrium theory outlined the essential mathematical formulas needed to predict the tides for each day of the year based on the positions of the sun and moon. The most important variable for each port was the "vulgar establishment," the mean hour of high water after the full moon. From this vulgar establishment, researchers could use Bernoulli's analysis to produce a table that predicted the times of the daily tide according to the position of the moon during the month. Bernoulli's theory, however, was based on an earth completely covered with water; no continents or estuaries existed on either Newton's or Bernoulli's mathematical sphere. Thus Bernoulli's theory could not account for ports such as London, situated not only in the English Channel, but thirty miles up the river Thames. The situation of the port and its communication with the open sea caused variations in the time of high water that refused to submit to theory. These local circumstances did not change, however, and thus they could be supposed to always produce the same effect. A "corrected establishment" therefore was required that was peculiar to each port and could be determined only through direct observation.[34] Finding the corrected establishment for each port became the most important task in the creation of tide tables. Bernoulli had calculated tables for some of the required corrections, but he did not sufficiently explain how they were to be used. For the port of London, the corrected establishment was found only in the second quarter of the nineteenth century, through the work of Joseph Foss Dessiou and John William Lubbock, working under the auspices of the Society for the Diffusion of Useful Knowledge.

Bernoulli's treatise was inaccessible to most natural philosophers, much less the average mariner. The mathematics was difficult to penetrate even for the mightiest intellect, and no translation from the French existed until Lubbock's "Account of the 'Traité sur le flux et reflux de la mer,'" published more than ninety years later. Bernoulli's treatise significantly advanced the methods of tidal prediction, but using those methods still demanded both theoretical refinement and intense and laborious calculations. Moreover, to find the corrected establishment, one also had to have a large number of observations encompassing several years to calculate the important variables specific to each port. Observations of the tides were a valued commodity, however, and once taken they were guarded as private property. The inaccessibility of Bernoulli's methods combined with the lack of long-term observations forced mariners to rely on navigation manuals such as J. W. Norie's *Practical Navigation* or John Hamilton Moore's *Practical Navigator*.[35] These texts were part instruction, part tabulation, and part hints, judgments, and personal anecdotes. They usually began with decimal arithmetic and trigonometry, moved to geography and the art of surveying, and ended with the more difficult aspects of applied astronomy for use in navigation. Intended for lay readers, the section on astronomy was usually simplified to a fault, covering the nature of parallax, refraction, the theory of the winds and tides, and finally the methods open to sailors for ascertaining their latitude and longitude by the sun, stars, and moon.

Taking Norie's *Practical Navigation* as an example, the section on the tides explained, in the simplest terms, the Newton-Bernoulli equilibrium theory. Norie stressed that the theory assumed that water of the same depth covered the entire earth, and he described the corrections required for its practical application, what he termed the "common method." The common method entailed finding several variables that the mariner needed to calculate, including the year of the lunar cycle and the moon's age.[36] From these variables, and several appended tables, one could then determine the time of the tide at any given place on the coast. Norie also included a table that listed the vulgar establishment at over six hundred ports along the coasts of Britain. In theory, one could begin with the vulgar establishment, use Norie's prescription—he included examples— to find the corrected establishment, and from there find the time of high water. In practice, it never quite worked out that way.

Part of the problem arose from the professed aim of navigation. "Navigation," wrote Norie, "is that art which instructs the Mariner in what manner to conduct a ship through the wide and tactless ocean, from one part to another, with the greatest safety, and in the shortest time possible."[37] Except for a short section on soundings and an equally brief section on the tides, Norie's text would help a navigator only in the open ocean far from the sight of land. The instructions were worthless for finding the time of the tide close to shore. Seamanship, however, was quite different. Its goal was the safety of the ship while sailing in and out of port, setting the mariner safely on the open ocean, where the art of navigation commenced. Most mariners, therefore, turned not to Norie's tables but to manuals on seamanship such as William Hutchinson's *Treatise on Practical Seamanship*, which dealt almost exclusively with the problems posed by tides and winds.

William Hutchinson was a merchant navy man, and his experiences as a sailor were gathered in his *Practical Seamanship*, first published in 1777. Experience and observation had taught Hutchinson a great deal about the tides. He was appointed master of the Liverpool docks in 1760, a position he held for more than twenty years. He kept a daily register of the tides from 1764 to 1783, observations that would later be used fruitfully by both Lubbock and Whewell. As dockmaster in a dangerous port, he also was in a good position to witness many shipwrecks that were due to reckoning the tides by the "common method" described by Norie and others. Two-thirds of the way from Newcastle to London, Hutchinson related, one was "amongst dangerous shoals, and intricate channels... and the ships are as large as the shoal channels will admit them to get through with the flow of the tide, which requires to be known to a great exactness." Similarly near Liverpool, Hutchinson warned, "I have observed ships coming in at neap tides about the quarters of the moon, when instead of meeting with high water, as expected by the common way of reckoning, they have found it about a quarter ebb, that for want of water enough they have often struck or come aground and laid upon the bar, when lots of great damage has often been the consequence."[38] To guard against such mishaps, Hutchinson counseled that "every prudent diligent officer, should endeavour to get all the help he can come at from tide tables, books, charts &c, to make himself as well acquainted as possible with the tides." This was all the mariner could do, but it was

Figure 1.9. Wreck of the *Boorhampooter* off of the southeast coast of England, with the cliffs of Dover in the background. From the *Illustrated London News* 3 (28 October 1843): 277.

usually not enough. The "tables, books, charts &c" were based on the "common methods" that Hutchinson had just warned against.[39]

The resources mariners could use to find the time and depth of the tide for the major ports in Britain thus were numerous. First, they could consult local and privately produced tide tables, a growing cottage industry in the late eighteenth and early nineteenth centuries. The tide table makers were careful to keep their methods secret, and though based loosely on Bernoulli's published methods, the tables were only as good as the years of observations they were based on. Second, mariners could consult navigation manuals, such as Norie's, which simplified Bernoulli's methods but did allow one to make a rough approximation of the tides in each port if one could follow the mathematics. Third, mariners consulted manuals on seamanship, which rarely included tide tables but did offer pointers on how to navigate particularly dangerous stretches of the coast. And last, mariners and pilots could rely on the port itself. Because the first three options often led to erroneous calculations of the tides, port watchmen often indicated when it was safe for navigators to enter by observing the tides directly. At Leith, for instance, the watchman signaled the depth of water by hoisting brightly colored balls. At Aberdeen, the watchman hoisted a flag when the water was deep enough to allow "ordinary vessels" safe passage.[40] These visual cues, however, had their own problems, ranging from drunk or otherwise inattentive watchmen to the fickleness of the tides themselves.

More often than not, mariners in the mid-eighteenth century pursued all these avenues at once. In the end, it was the intuitive knowledge of the pilot, his ability to combine these methods with his own knowledge of the peculiar actions of the local tides, that brought a vessel safely to port. And of course not all vessels arrived safely. As ships grew larger and their hulls sank deeper, mariners increasingly called for accurate methods of determining the tides, which entailed finding a theory that explained their action throughout the coasts. Yet no theorists in Britain turned to this study. Except for the work of Bernoulli and other mathematicians on the Continent, tidal theory had hit a dead end.

## Tidal Theory in the Enlightenment

Tidal studies continued to be of interest in the second half of the eighteenth century, but advances in tidal theory, like advances in other branches of astronomy, moved across the Channel.[41] The standard interpretation, quite plausible at first, suggests that lack of interest in the tides followed from Newton's momentous achievements; as one historian put it, there existed a "vague belief that all that needed to be known about the tides was known."[42] This is certainly true if one focuses exclusively on eminent men of science. But the study of the tides in Britain throughout the century was sustained largely by naval captains, military and colonial administrators, teachers of navigation, and local tide table makers. Rather than *lack* of interest owing to Newton's *correct* formulation, these sources suggest a continued interest in tidal theory precisely because Newton's formulation appeared to be incorrect.

Entries in the *Philosophical Transactions* during the eighteenth century focus on the incongruity of Newton's general theory with particular observations, contributions that came largely from the maritime community beyond Newton's gaze. Murdoch Mackenzie, a surveyor for the Admiralty, observed that "mankind seemed to imagine a thorough Knowledge of the Tides might be obtained from an attentive Consideration of the Principles [that Newton] had established, without the Trouble of further Observations."[43] Yet as most seamen realized, the tides were actually "as inexplicable and as little known as ever." Mackenzie's description of the tides on the coast of the Orkney Islands was typical of other entries in this period, which were far more quantitative than qualitative, offering measurements and descriptions, not explanations.

The expertise of those in the maritime community was in reporting measurements collected from around the globe; they offered practical knowledge acquired from their travels for others to apply to theory. The authors invariably called for more systematic tidal observations, but the administrative framework needed to support such a systematic effort did not yet exist.[44]

Astronomer Nevil Maskelyne's first-person account of the tides around the island of St. Helena during his voyage to measure the transit of Venus in 1761 was unique because the island was in the middle of the Atlantic.[45] Maskelyne published the first edition of the *Nautical Almanac* in 1767, was astronomer royal from 1765 to 1811, and spent his life interested in physical astronomy. But he ended with the disappointing disclaimer found in so many published accounts of the tides: "I shall say nothing with respect to any conclusions that may be drawn from the above observations."[46] Captain Henry More's entry the same year, "Observations of the Tides in the Straits of Gibraltar," began in a similar manner.[47] He recited observations made during the time he was employed as an officer in the Royal Navy, sixteen years of which were spent at Gibraltar. He admitted that his aim was practical: to figure out the reasons for the difficulty in navigating the straits. He ended, however, by comparing his results for Gibraltar with other seemingly analogous observations between Portsmouth and the Isle of Wight, analogous because neither followed known laws. "Whether this theory carries with it any valuable degree of reality or not, I hope a proper time will come, when it may be ascertained, by employing fit persons for that purpose."[48] Fit persons needed to be employed because More could not fit his observations into the Newton-Bernoulli equilibrium theory of the tides.

Two accounts of the tides by the famous explorer James Cook read like journal entries filled with misadventure. "An Account of the Flowing of the Tides in the South Sea" explicitly declined to offer any theory; it simply listed the tides for sixteen ports where Cook had made observations at Maskelyne's request.[49] Though his second entry, "Of the Tides in the South Seas," moved beyond the mere tabulation of results, it also fell short of any theoretical conclusions.[50] He related how his intuitive knowledge of the tides, "founded on a notion, very general indeed among seamen, but not confirmed by any thing which had yet fallen under my observations, that the night-tide is higher than the day-tide" helped his crew escape from certain peril after their vessel struck a coral reef. He ended where another theorist could begin: "nor can I assign any other cause for this difference in the rise and fall of the tide, and therefore must leave it to those who are better versed in this subject." As it turned out, and as

Cook probably well knew, no one was better versed. That the night tides were higher than the day tides, a phenomenon Whewell later termed the diurnal inequality, had yet to be incorporated into tidal analysis.

These published accounts were similar to those of the late seventeenth century. Their main purpose was to demonstrate the deficiency of the accepted theory (now Newton's instead of Wallis's), without offering theoretical conclusions of their own. But one difference is apparent: most entries came from naval officers either stationed abroad or recently returned from voyages of trade or discovery. As the century progressed and such voyages increased, faith that the Newtonian theory could predict the action of the tides declined. Though most contributors restricted their entries to tabulating observations, they also hinted at the dangers of accepting a scientific theory when it failed to account for observational data.

Owing to mariners' interest in the tides, the best source of information can be found not in scientific treatises but in the holdings of the Board of Longitude, a clearinghouse for scientific material sent to the Admiralty before the establishment of the Hydrographic Office. The material is as diverse as its senders. In 1789 Charles Vallancey, a colonial engineer, wrote to the Board of Longitude from Dublin, expressing his surprise that so many errors existed in the Irish almanacs for the tides around the entire coast of Ireland, particularly Dublin Bay. London almanac publishers, furthermore, copied and propagated these same errors. "I had been eye witness to vessels being lost in our Southern Ports in consequence of these publications."[51] He therefore requested that the Marquis of Buckingham have observations made around the coast of Ireland. The marquis, in turn, hired Mr. MacMahon, "a good mathematician, a practical astronomer, and an excellent mechanick," to take observations of the tides for twelve months, but only in the port of Dublin.[52] MacMahon also constructed an "aqua-meter," which he placed in Dublin Bay, in addition to a portable meter that Vallancey believed would "become the furniture of most ships." After all this work, even Charles Vallancey was surprised by his own conclusions: that the tides in Ireland seemed to follow entirely different laws than those in England. That the Irish should abide by their own laws seemed to be supported by astronomy.

Other letters sent to the Board of Longitude in the eighteenth and early nineteenth centuries included published tide tables. Those tables published by Mr. Colley, a teacher of navigation, and John Abram, a teacher of astronomy, allowed one to calculate the time of high tide at all ports in "England, Wales, Scotland, Ireland and the coasts of Europe." According to Abram, all one needed was "one observation accurately made in moderate weather. " Like Philips and Halley in the seventeenth century, Colley and Abram would not divulge their methods of calculation—unless, of course, they were offered some remuneration. The Board of Longitude had offered an exorbitant prize (£20,000) for the solution to the longitude problem, and most correspondents would disclose their theories of the tides only if the Board offered a similar inducement. Lazarus Cohen, for instance, only hinted at his own theory, "for the good of the public as well as for my own remuneration," but kept the rest secret until properly paid. Edward Dean, a pastor in Cheshire, gave an interesting but highly implausible theory of the tides based on the existence of a large sea in the center of the earth, for which he received no payment whatever.[53]

The five letters from Eliza Maria O'Shea are appealing not for her theory but for the way she proposed it. She reported to the Board that her brother Thomas had solved the problem of the tides that had "puzzled so many wise men."[54] After they asked for more technical details concerning her brother's theory and invited him to meet with the superintendent of the Board, she said that her brother was very ill, confined to his bed, and therefore "deprived the honour of attending you."[55] In his stead, she would explain his ideas in writing. After the Board again inquired about further details of the theory, Eliza O'Shea protested that her brother was too ill to write another letter, but she again offered to give further details to the best of her power. The theory was unique, based on the premise that the tides are caused not by the moon but by the attraction and repulsion of the earth itself. But uniqueness does not always translate into correctness, and the Board never replied to O'Shea's last letter. The Board may have also realized that Thomas probably never existed; O'Shea believed that only by such a ruse would her views be heard.

Captain Walter Forman of the Royal Navy wrote the Lords of the Admiralty enclosing a lengthy pamphlet entitled *A New Theory of the Tides Shewing What Is the Immediate Cause of the Phenomenon; and Which Has hitherto Been Overlooked by Philosophers.*[56] On the title page, Forman quoted not Newton or Bernoulli, but himself: "Though the Moon's attraction is the primary cause, it is neither the sole nor the immediate cause of the Phenomenon; and it is because they have confined their views to this one principle, that philosophers have never been able to make out the True theory of the Tides." The true principle rested on the elasticity of water, a point Forman spent most of his treatise trying to demonstrate. The rest was a polemic against the Newtonian theory. "Not only is this hypothesis altogether inadequate, but, in many instances, the very effects it must necessarily be supposed to produce are directly the reverse of what actually takes place in nature."[57] According to Forman, Newton's theory failed to explain the same tides in different parts of the world as well as the very different tides in the same part of the world.

The importance is not that Forman misunderstood Newton's theory. He did not. He is quite correct in his negative evaluation of the equilibrium theory. Lazarus Cohen, Edward Dean, and Eliza Maria O'Shea, along with mariners such as Maskelyne, Cook, and Forman, correctly observed that Newton's theory left much to be explained. Newton had outlined a general theory of the tides, yet once naval officers with interests in the tides navigated the globe and local practitioners reported differences that should not occur according to Newton's explanation, they consistently threw out the baby with the ebbing tidal water. Forman, Maskelyne, and others wrote with authority about the Straits of Banca and the port of Minto, the Island of St. Helena and the West Indies. The letters sent to the Board of Longitude demonstrated that many people were intrigued by the extremely complicated tides around the world; they had made observations, postulated theories, and often published their results. Surprisingly, they almost always disagreed with Newton's theory and used both the irregular tides in Britain and the variable tides around the world to demonstrate its inadequacy. At the height of the Enlightenment—the period usually associated with the application of Newtonian thought to both science and society—faith in Newton's theory had bottomed out, and it would not rise again until the work of John William Lubbock in the late 1820s.

## Pierre Simon Laplace and the Hydrodynamic Approach

While the theoretical study of the tides ebbed in England throughout the eighteenth century, it continued to flow across the Channel, matching the Continental achievements in the physical sciences more generally. A host of distinguished French philosophers were attracted to tidal theory, including Jacques and Dominique Cassini, Joseph-Jérôme Le Français de Lalande, and the highly influential Pierre-Simon, Marquis de Laplace. While the English wrestled with the Newtonian equilibrium theory of the tides, in France little was invested in what was considered an English approach. The French began with Descartes and attempted to advance the Cartesian theory to the point where it could predict the seemingly anomalous tides rolling in from the Atlantic Ocean.[58]

Although the starting point might have differed, the organization of tidal science in France looked very similar to that in England. Each time a vessel was grounded on France's coasts, government called for more careful measurements of the tides to safeguard the economic interests of the state. As with the Royal Society of London, the founding of the Royal Academy in Paris in 1666 served to direct the initial investigations. Modeled on an amalgamation of the Accademia del Cimento of Florence and the Royal Society of London, the French Academy was distinct from its Italian predecessor and its English counterpart. Its membership was limited to a set of academicians from various disciplines of science. The initial list of fifteen members, or academicians, included seven geometers and three astronomers, testifying that the Academy focused heavily on mathematics and astronomy in relation to navigation.[59]

Whereas the Royal Society of London received little financial encouragement from the Crown, the French Academy was from the beginning directly connected to the state. The king paid the academicians a professional salary, and they were expected to focus on topics that would help the state. Jean-Baptiste Colbert, France's powerful finance minister under Louis XIV, began awarding prizes for solutions to particularly tricky navigational problems, a tradition that continued throughout the eighteenth century. The Academy offered prizes for the most efficient arrangement of masts for oceangoing vessels in 1727, the proper shape and installation of hourglasses in 1725, the best form of anchor in 1737, problems of terrestrial magnetism in 1743 and 1746, and the theory of the tides in 1740.

Direct access to the state's treasury had its advantages. Louis XIV, the Sun King, attracted some of the top minds in science to Paris, including bringing Dominique Cassini from Italy to run the Royal Observatory. Cassini was an ardent Cartesian, and he spent laborious hours attempting to advance Descartes's vortex theory of the tides. Louis XIV also commissioned massive geodetic surveys of his extensive territory, which included the charge of observing the tides on France's Atlantic coast. As part of the national survey, Jean Picard and Philippe de La Hire initiated a set of tide observations at Brest in 1679, a few years before Flamsteed began similar observations in the Thames estuary.[60] In 1701, moreover, the Academy issued a "Memorandum on the Manner in Which Tides in the Ports Should be Observed," thus instigating the first organized assault to gather observational data on the French coasts. Within two decades, tide observations were reinstituted at Brest and initiated at all the major French ports with direct access to the Atlantic. The Academy served as the center of tidal calculation, where the astronomer Jacques Cassini, the son of Dominique and his successor at the Royal Observatory, took charge of analyzing the data, focusing particularly on the observations at Brest. By the end of the eighteenth century, these constituted the longest set of tidal observations in Europe, attracting the attention of several distinguished members of the French Academy, including the esteemed academician Pierre-Simon Laplace.[61]

Owing to his phenomenal advances in probability theory, game theory, celestial mechanics, and the differential and integral calculus, his contemporaries considered Laplace the greatest natural philosopher in Europe. He is known in the history of science largely for his multivolume *Traité de mécanique céleste*, which generalized Newtonian mechanics. But Laplace's influence also extended to the political realm. He is largely responsible for founding the Bureau of Longitude, and his early and active participation in the Royal Academy made him the premier natural philosopher working for the state.

Laplace's advances in tidal theory were as revolutionary as the work of Newton.[62] Yet unlike the renown that Laplace had achieved among his contemporaries for his advances in Newtonian mechanics, his fame as the founder of modern tidal analysis had to wait until well after his death. Laplace's essentially correct hydrodynamic approach to the tides, much like Newton's earlier hydrostatic approach, failed to alleviate the problems of calculating and predicting the actual tides on the coasts of Europe. Newton and Bernoulli had demonstrated the forces exerted on the tides by the sun and moon, but they had limited their analysis to a hydrostatic response of the ocean to these forces. In working on the tides for over fifty years, Laplace showed that they were a problem of the dynamic motion of fluids, a question of hydrodynamics rather than hydrostatics. Laplace's results first appeared in three massive papers he delivered to the French Academy in 1777 and 1778, and he turned to the tides again in his "Mémoire sur le flux et le reflux de la mer," first read to the Academy in December 1790 but published much later. He reproduced much of this work in the first two volumes of his *Mécanique céleste* in 1799, which also included his now famous tidal equations.

Like other researchers, Laplace began with Bernoulli's advances on the Newtonian theory, but he berated both Newton and Bernoulli for assuming that the oceans were in equilibrium, simply dragged about by the moon. The tides did not follow the moon as a mountain of water, Laplace argued, but oscillated in the form of tidal waves in a basin. In Laplace's formulation, the annual, diurnal, and semidiurnal motions of the earth acted separately to create distinct and periodic oscillations.

He formulated expressions for each oscillation and then added these expressions together to form a composite set of equations to represent the tidal motion of the world's oceans.

Laplace followed these seminal advances with the third and fourth volumes of his *Mécanique céleste* in 1802 and 1805. The last part of the fourth volume and the entire fifth volume appeared much later, between 1823 and 1825, and by then he had acquired a long enough series of tide observations—reinstituted at Brest in 1806 at his suggestion—to compare with his hydrodynamic approach. Writing almost a decade later, William Whewell acknowledged that Laplace, "by treating the tides as a problem of the oscillations rather than of the equilibrium of fluids, undoubtedly introduced the correct view of the real operation of the forces [affecting the tides]."[63] Yet Whewell also chastised Laplace for merely comparing his results with "supposed" critical observations at Brest rather than forming predictive tables of the tides directly from theory. According to Whewell, his theory was never put to an adequate test. In Laplace's defense, the number of human computers at his disposal to digest the massive number of observations and test his theory had severely limited his analysis. He had received assistance from Alexis Bouvard, a calculator at the Bureau of Longitude and later director of the Paris Observatory, but the time-consuming work of creating tide tables directly from theory was beyond Laplace's and Bouvard's interests. Laplace increasingly turned to other investigations.[64] As will be explained more fully in subsequent chapters, adequate computer time proved a major limiting factor in furthering tidal theory throughout the eighteenth and nineteenth centuries, even more of a problem than acquiring observational data or formulating theory.

Other difficulties hampered the influence of Laplace's work as well. He had succeeded in formulating the differential equations from which the explanation of the tides was to be derived, but integrating these equations proved impossible in practice.[65] The science of hydrodynamics was still in its infancy. French mathematicians, most notably Jean-Baptiste-Joseph Fourier, Simeon-Denis Poisson, and Augustin-Louis Cauchy improved the theory of the integration of partial-differential equations. But their solutions used questionable approximations and were valid for only the simplest cases. Furthermore, they disregarded the irregularity of the ocean floor, the effects of wind, and the contours of the coastline, all of which materially affected the tides. The mathematics involved presented so many difficulties that Laplace confined his analysis to a simplified case in which the ocean covered the entire earth at a constant depth of twelve miles. When Nathaniel Bowditch, an American seaman and author of the *American Practical Navigator*, published a highly successful English translation of the first four volumes of the *Mécanique céleste*, he likewise despaired at the mathematics involved. He complained that Laplace often skipped important segments of his proof with the offhand remark, "thus it plainly appears." "I have never come across one of Laplace's 'thus it plainly appears,'" Bowditch grumbled, "without feeling sure that I have hours of hard work before me to fill up the chasm and find out and show how it plainly appears."[66] In the first half of the nineteenth century, the mathematics involved in a hydrodynamic approach to the tides was daunting, too difficult for even the most mathematically adroit to penetrate.

Laplace's work laid the foundation for the hydrodynamic approach, still the modern basis of tidal analysis, but his work had very little immediate effect on tidal prediction. Though it was correct in principle, final solutions to a hydrodynamic approach awaited the advent of modern computers. Thus Laplace's correct theory is eminently important for the history of tidal theory but of little practical importance for the history of tidal prediction in the Victorian era. The British Admiralty, for instance, did not revert to Laplace's theory to make tidal predictions for its ports until well into the twentieth century. The actual techniques for predicting the tides remained with the Newton-Bernoulli equilibrium theory, a research program followed not in France but in Great Britain and the United States. What is apparent is that the foundation of tidal science in France followed a course similar to that in England. Natural philosophers made theoretical advances, called for renewed and continued observations, published announcements in their journals, and relied on professional calculators to advance their arguments. And the results were also the same. Laplace's theoretical conclusions, like Newton's, had little if any direct impact on tidal predictions on the coasts of Europe.

The hydrodynamic approach to studying the tides was not confined solely to France at the turn of the nineteenth century. One natural philosopher in England did take up the study. Thomas Young is best remembered for his contributions to the wave theory of light. But stemming from his less well-known work for the British Admiralty, he also resurrected the theoretical study of the tides in England from over a century of neglect. He served both as a member of the newly reformed Board of Longitude and as superintendent of the *Nautical Almanac*, responsible for the nautical ephemerides (tables of the positions of celestial bodies at regular intervals) published for Britain and its possessions. Young was therefore in the perfect position to advance the study of the oceans, especially tidal theory. Owing to his earlier work in optics, and especially his interest in the analogy between light waves and water waves, Young turned to the tides as a problem of the propagation of waves, just as Laplace had done. In his *Lectures on Natural Philosophy and the Mechanical Arts*, published in 1807, the same work in which he first published his now famous double-slit experiment, he also contributed his first paper on the tides. At a time when the physical sciences were almost entirely dominated by the doctrine of imponderables, especially in England under Newton's influence, Young was the first researcher to demonstrate the wave nature of light.[67] Though Young had initially considered light waves similar to sound waves, he later tentatively suggested that light waves were transverse; that is, the motion of their vibrations was perpendicular to the direction of their propagation, similar to waves in water (such as tide waves in the ocean).

Although Young's initial article did not advance the theory of the tides, it did advance the method of tidal analysis.[68] Before Young, natural philosophers treated the tides as Newton, Bernoulli, and Laplace had done, as operating on an imaginary sphere completely covered with water of equal depth. Young was the first researcher to examine tide waves in canals. In a "Supplement" to the *Encyclopaedia Britannica*, he was also the first researcher to treat the effects of friction on the tides, later used by John William Lubbock and William Whewell to explain the "age" of the tide.[69] Following Halley, he also published a map of "contemporaneous" tides illustrating simultaneous high water in the English Channel and around Great Britain. Though Young used this map simply as an illustration, both Lubbock and Whewell later used similar maps to trace the propagation of the tides throughout the oceans.[70]

Young maintained that his work on the tides was "the best physico-mathematical investigation that [he had] ever attempted," but the tides ultimately proved intractable.[71] Young ended his researches by throwing up his hands and complaining of the inadequacy of direct observations to aid the development of theory: "There is indeed little doubt that if we were provided with a sufficiently correct series of minutely accurate observations on the tides, made, not merely with a view to the times of low and high water, but rather to the *heights at the intermediate times*, we might by degrees, with the assistance of the theory continued in this article, form almost as perfect a set of tables for the motions of the ocean as we have already obtained for those of the celestial bodies which are the more immediate objects of the attention of the practical astronomer."[72] Young advocated measuring not only high and low water—a tricky and costly endeavor owing to the extreme range of the tides on the shores of England—but also the "heights at the intermediate times." He was searching for observations that could be used for a hydrodynamic approach. The enormous problems he encountered in acquiring such observations made his direct successors in tidal theory, Lubbock and Whewell, focus almost exclusively on high water observations. This severely limited the approaches they could take. Equipped only with high water data from a limited number of ports, Lubbock and Whewell opted to use the hydrostatic Newton-Bernoulli equilibrium theory—which was ultimately incorrect.

## Conclusion

The one distinct geographical feature of Britain is its status as an island. Everything on its shores appears a bit smaller in relation to the enormous ocean that surrounds it. "Rivers are little, fields are little, freight cars are little," the historian R. K. Webb acknowledged. "Not even in the mountains of Scotland or Wales will one find true grandeur."[73] But Great Britain does possess at least one natural phenomenon of extraordinary magnitude: its tides. To early modern philosophers, the littoral that separated the land from the sea symbolized the transition from rationality to chaos. Extending rationality from the coastlines to the ocean became one of the premier aims of the new natural philosophy in water-locked Britain. It was part of the spatial extension of scientific knowledge in general that helped define the scientific revolution and usher in the Enlightenment.

The first organized assault on the tides in Britain came with the founding of the Royal Society of London. Robert Moray and Henry Oldenburg sent out queries, tables, and directions for making observations. John Wallis published an overtly theoretical argument based on the motion of the earth, which motivated the first organized collection of tidal observations in Europe, observations Newton then used to support his theory of universal gravitation. As with Wallis's theory, so with Newton's, the validity of the hypothesis rested on empirical verification. The more data that seemed to support Newton's theory, the better it was considered to be, and the nearer it was thought to approximate the truth. That approximation, significantly, had changed. Whereas Wallis had introduced his hypothesis as only a probable conjecture, Halley introduced Newton's hypothesis as "being demonstratively proved, and put past contradiction." Both introduced their treatises as following from the laws of mechanics. Wallis buttressed those laws with a historical narrative of specific events; Newton replaced this narrative with the language of mathematics. In the end, however, even the new powerful mathematics combined with the all-encompassing law of universal gravitation failed to bring the tides to rule.

Tidal analysis in the quarter of a century between Wallis and Newton demonstrates that most of the components that led the early Victorians to their mathematical and theoretical advances were already present at the end of the seventeenth century. Published queries, tide tables, the call for observations, the effect of the moon's parallax and declination, even the Newtonian theory, with its emphasis on mathematics and its justification through direct observation—the Victorians used these same components 150 years later. Indeed, few single researchers would surpass Edmond Halley's accomplishments. From the Thames estuary to the Atlantic Ocean, Halley attempted to extend Newton's analysis with a powerful new synoptic view, publishing his own charts that integrated the vertical and horizontal into questions of physical astronomy. As the commander of the *Paramore*, moreover, Halley helped initiate a collaboration between natural philosophers and the Admiralty that would prove exceedingly fruitful in the following centuries.

Initially, however, very little came from these efforts; no systematic study of the tides followed. The *Philosophical Transactions* was too expensive and aimed at too select an audience to be effective, and thus the number of people exposed to the science of the tides remained small. As the eighteenth century progressed, fewer and fewer intellectuals questioned Newton's theory of universal gravitation, yet more and more began to question its efficacy in explaining the tides. While captains in the Royal Navy, such as James Cook, offered observations of the tides from voyages of trade or discovery that seemed to question Newton's explanation, men such as Charles Vallancey offered theories that directly challenged it.

As Britain's trade routes expanded and its maritime interests intensified, so too did its need for correct methods to predict the tides. The existing tide tables for the major ports of England not only were incorrect, they were dangerous. They were based on secret methods, private observations, and time-consuming calculations. To start from scratch, as John Lubbock, Joseph Foss Dessiou, and the Society for the Diffusion of Useful Knowledge would soon find out, was exceedingly difficult. Though the problems with Newton's general explanation became more apparent, the infrastructure necessary for a direct assault on the solution still lay in the future. It would require cooperation between natural philosophers and the British government, particularly the British Admiralty, and an extension of Halley's analytical tools to the entire globe. It also demanded massive observational data and direct financial support for their reduction. But first, a renewed interest in the science of the tides had to await changes to the actual tides themselves. This occurred not from any natural change to the ocean, but from the industry of humans and their increasing power to encroach on the sea.

TWO

# The Bounded Thames

The time shall come, when, free as sea or wind,
Unbounded Thames shall flow for all mankind;
Whole nations enter at each swelling tide,
And seas but join the regions they divide.
ALEXANDER POPE, *"WINDSOR FORREST"* (1713)

As the American packet the *Leeds* sailed swiftly up the Thames with the surging tide on the afternoon of 26 December 1828, a burst of adrenaline seized the experienced pilot. As the *London Times* reported, the vessel "suddenly" and "most unexpectedly" struck ground in what he had thought was the river's navigable channel. Witnesses heard the frightening sound of cracking seams and buckling copper sheathing and watched helplessly as water rushed into the packet's fractured hull. Within ten minutes the ship was so full of water that no hope remained of its quick removal. The *Leeds,* a mighty vessel of five hundred tons and a potent example of the new American shipbuilding techniques, became yet another obstacle in the river Thames, a "perfect wreck" to be avoided by pilots as much as it was pursued by the fascinated public through the daily press.[1]

The wreck of the *Leeds* on that December afternoon was due less to the pilot's ineptitude than to the increasing difficulties of navigating the Thames estuary. Coastal shipping had gone through dramatic transformations in the preceding decades. Technological advances had allowed iron to be used in ships' hulls as early as 1780, mainly for passenger travel through rivers and estuaries, but in the late 1820s and 1830s this use expanded to oceangoing vessels.[2] The significantly deeper hulls on this larger and heavier class of ship heightened the danger of river and coastal navigation. Moreover, the number of ships plying Britain's coasts and estuaries had grown at a revolutionary pace. In the words of the age's most illustrious poet, the "unbounded Thames" seemed to flow for all mankind, as vessels hoisting the flags of almost every maritime nation entered and exited its turbulent waters. No better example is needed than the *Leeds*, a packet from New York whose Stars and Stripes now pointed toward the Thames as the vessel's fractured hull rested in the middle of the river. Yet Alexander Pope, writing before the mid-eighteenth century, spoke of a river that was unbounded in the physical sense as well. Since the Roman era, only minor shoreline changes had constrained its waters; the Thames flowed freely according to the well-scripted motions of its tides.

Pope would not have recognized the Thames of the early nineteenth century. Human innovation, while producing larger ships with deeper hulls, had also fashioned a significantly different river. Beginning in the second half of the eighteenth century, the river was incessantly bounded as engineers claimed larger and larger sections of its swift waters for their bridges and docks, embankments, jetties, seabreaks, and waterfront warehouses and industries. Man-made structures now bridled the destructive tidal waters of the Thames estuary. Civil engineers became the most influential actors shaping the world's most important river, their construction projects representing powerful statements of human resource and ingenuity. For the enlightened Londoner, this was a sure sign of progress. To build an inland road or canal, hills could be toppled, valleys filled in, the earth flattened, and paths made straight. The earth's waters, however, had seemed immune to change by humans. Though the world's rivers, seas, and oceans were used for transport, all human traces disappeared at once after ships had passed through. Harbors, ports, and dock systems were particularly impressive not because of their magnificent architecture, but because humans had intruded on the sea, shifting the boundary in favor of the land. Guiding the river according to a plan set out by innovative engineers epitomized the power of the most industrious nation in the world. Improving the river meant improving the nation.[3]

To keep the Thames unbounded in Pope's metaphorical sense, therefore, meant ceaselessly bounding the river in a physical sense, a process that began in the second half of the eighteenth century and continued unabated throughout the nineteenth. The object was to normalize and thus control the water's flow, but the unintended effect was to dramatically change the topology and currents of the river itself. When the *Leeds* entered the Thames estuary in 1828, massive engineering projects were under way, including the erection of the new London Bridge, the removal of the old bridge, the building of the Thames Tunnel, and the construction of the Eastern London Docks and the St. Katherine Docks. The changes to the river's shores caused a proportionate change in the velocity of the river's flow and the magnitude of its tides. Signs of Britain's industrial expansion and technological sophistication, these projects precipitated an entirely new tidal regime in the world's most important estuary, bringing changes that even highly experienced pilots and navigators had no way to predict.

The penetration of man-made structures in the littoral also heightened the river's destructive powers, making quayside settlements particularly vulnerable to flooding. Humans became arbitrators in the age-old battle between land and sea, and the sea did not take kindly to reclamation. As encroachment increased, the sea revolted, pouring through the embankments and inundating the countryside, including London itself. Human skill had eclipsed the geological and physical laws that used to determine the river's course, transforming the natural environment into a social and commercial space that required vigilant conservancy.[4] Physical control of the river's contours, banks, and shoals required intellectual control of the science of the intertidal zone.

The natural topography of the river and the indefatigable efforts of humans came together to produce the first sustained interest in the science of the tides in England. Though this interest paralleled the reform movement in the Royal Society, the founding of the British Association for the Advancement of Science, and the Admiralty's increasing involvement in scientific and technological questions, the study of the tides did not start with these scientific societies or military institutions. When studying the rise of worldwide geophysical initiatives in the 1830s, historians have automatically turned to these elite institutions to emphasize the connection between British imperial interests and the growth of modern science. But the study of the tides in the early Victorian era had much humbler beginnings.

Shipowners, bankers, and insurers intent on mitigating the risk to the expanding coastal trade were the first to call for a scientific understanding of the changing rhythms of the tides. Fierce battles were fought in the public press among competing almanac publishers and commercially inclined tide table calculators. The combatants often descended to accusations of thievery, plagiarism, and outright sabotage, and the lessons from these angry clashes were clear enough. British industrialization had catapulted the Thames to the status of the most significant estuary for commercial shipping in Europe, and the port of London became the most crowded dock complex in Britain. This brought urgency to the study of the tides, and the publication of tide tables for the port became a competitive and lucrative business. When the Society for the Diffusion of Useful Knowledge was established in 1828, its founders decided to enter the fray by producing an almanac, complete with a tide table for the port of London for every day of the year. An error by one of their hired calculators ignited a controversy concerning their supposedly scientific approach. The mistakes in calculation forced the Society to actually incorporate natural philosophers into the production of their tables. Only then did the scientific elite in Britain appropriate the study of the tides as a valuable research project.

The first sustained interest in the study of the tides in England began as a local and practical problem, the preserve of tide table makers and almanac writers, those interested in the sea for its importance to Britain's trade. The study of the tides matured into a research project within the scientific community only in late December 1828, when the Society for the Diffusion of Useful Knowledge anxiously enlisted John William Lubbock, a Cambridge graduate and successful London banker, to infuse its tables with scientific legitimacy. A history of the pursuit of science in the late eighteenth and early nineteenth centuries must take into account both the changes to the natural environment of the river itself and the myriad of almanac publishers, surveyors, and tide table makers who attempted to calculate those changes for the safety of mariners. The choice of the tides as a justifiable research topic was as much environmental and commercial as scientific.[5]

## Bridging the Thames

In the Roman era the Thames was much wider, slower, and shallower than it was in the early nineteenth century. The range of the tide was then a mere six feet, compared with the twenty-six feet of the early Victorian era.[6] The Thames followed a winding route through the southeast of England, emptying into a floodplain several miles wide as it meandered toward the North Sea. At each incoming tide, 70 million tons of salt water covered the adjacent lands for over thirty square miles.[7] No one knows who first embanked the river east of London, but doing so dramatically changed the estuary from a sprawling floodplain to a relatively swift, deep, and contained river. The massive embankment restricted the river below London to one-tenth of its original boundaries, transforming the low-lying marshes into fertile fields that could then be farmed to feed the growing population of the capital and its hinterlands.[8]

Modifications to the river continued throughout the medieval period. Large sections of London literally rose out of the waters of the Thames, claimed as dry land by early hydrographic engineers. Each narrowing of the river channel deepened the water and increased its speed. The tide began to flow farther up the river from its source in the North Sea and to purge itself more quickly during its ebb. In the incessant battle between land and sea, the sea now fought a powerful, technologically sophisticated adversary; yet it responded with its own weapons of flooding, changing bottom topography, and unpredictable tidal rhythms. One major factor for the emergence of Victorian tidal studies was a conscious effort to understand the river's response to human containment.

Along with these early embankment projects, the most dramatic changes to the Thames came with the construction of London Bridge, built between 1176 and 1209. Replacing a wooden structure of the late first century, its nineteen arches rested on oval foundations called starlings, between which several waterwheels pumped drinking water for the expanding city. The massive stone structure acted as a dam, splitting the river into two distinct tidal regimes; it obstructed the flow of the tides up the river below the bridge, and it penned the land waters flowing toward the North Sea above the bridge. As the water pooled above the artificial dam, high water rose a foot higher than below the bridge, increasing the propensity for flooding in central London.[9] The water above the bridge froze during extremely cold winters, allowing for grand ice fairs, with skating, dancing, and plenty of drinking.[10]

The bridge also promoted the use of the deep water that pooled below it for loading and unloading goods and passengers. The Pool, as it was called, acted as the center of commerce, the terminal point for larger vessels unable to pass through the bridge's arches. In the eighteenth century these large vessels transferred their wares onto small barges called lighters, which then conveyed the cargo either upriver or to shore. Indeed, London Bridge held a unique vertical and horizontal significance. From Flamsteed and Halley in the seventeenth century to the tide table calculators of the eighteenth, it was the obvious place where the rising and setting of the sea was observed. But it was also where mariners determined zero longitude until 1738, when Greenwich took over as the prime meridian.[11] Thus London Bridge was the zero point for both dimensions, a position that the third dimension, time, amplified in the decades and centuries that followed.

Figure 2.1. Old London Bridge in 1756 before its first major overhaul, from a painting by Samuel Scott. The bridge's nineteen oval starlings impeded the flow of the Thames, producing two tidal regimes, one above the bridge and one below. Reproduced from Sir Joseph G. Broodbank, *History of the Port of London,* 2 vols. (London: Daniel O'Connor, 1921). University of Minnesota Libraries.

Figure 2.2. The frozen Thames in 1814. Old London Bridge, with its nineteen arches in the background, acted as a dam, allowing the river to freeze during extremely cold weather. From the *Illustrated London News* 26 (24 February 1855): 188.

Figure 2.3. The congested Pool as viewed from London Bridge, from a lithograph by W. Parrott (1841). Reproduced from Joseph G. Broodbank, *History of the Port of London*, 2 vols. (London: Daniel O'Connor, 1921). University of Minnesota Libraries.

For well over half a millennium, only London Bridge spanned the Thames near London, and the river had figured out its course. Its bottom topography, currents, and the extent and timing of its tides were all relatively settled by the middle of the eighteenth century. The early reclamation projects took enormous amounts of time, and those living and working in the littoral adapted to the river's new course. Nautical men, especially pilots, kept track of the river's slow response, its changing depth and speed, and the increasing tidal action on its shores. But then something altogether different occurred. Human impact on the Thames quickened so radically that not even that most adaptable of species, with all its technological skill and scientific might, could adjust quickly enough.

As the population of London grew, having only one bridge spanning the river became a problem. London Bridge was crowded with houses and shops, adding to the number of people, horses, and carts congesting its cobbled streets. Though the bridge was under continuous repair, the first major overhaul began in 1758. Engineers removed the waterwheels and built a broad central arch to let larger vessels pass underneath. Unexpectedly, these alterations also lowered the high and low water levels by more than a foot. The wider opening of the central arch occasioned a swifter current, and while the unobstructed tide naturally scoured and deepened the main channel, the swifter stream also lowered the water level and quickened the erosion of the bridge's starlings. These early changes are the origin of the nursery rhyme: London Bridge was indeed falling down.[12]

The effects on the river from the initial construction of London Bridge and subsequent alterations to it were amplified with the construction of each additional bridge. Between 1750 and 1817, six new bridges were built, beginning with Westminster Bridge in 1750 and followed by the stone bridges at Blackfriars in 1769 and Battersea in 1772. The first iron bridge appeared at Vauxhall in 1816, followed in quick succession by two more bridges, Waterloo in 1817 and Southwark in 1819.[13] Massive engineering works, including embankments and seabreaks, accompanied each new bridge, and thousands of tons of mud and clay were displaced. Over a thousand feet of the river's edge was reembanked, for instance, when Robert Mylne constructed Blackfriars Bridge, extending London two hundred feet into what used to be a flowing river.[14] Combined with the damming effect of the bridges themselves, these engineering works dramatically modified the topology, currents, and tidal action of London's main thoroughfare.

As early as 1807, George Rennie, a highly respected civil engineer, noted the harm that the construction of bridges was doing to the nation's greatest river. The result, he feared, would be disastrous for British shipping. By the end of the eighteenth century, the river had already become dangerously overcrowded, with ships at anchor throughout the main channel. Debates about the changing topography of the river took place in the open—on the Strand, in the public press and coffeehouses, and eventually in Parliament. While bridges and embankments were necessary for the expansion of London, they made it hard for the growing merchant fleet to operate freely in an increasingly bounded Thames.

## Navigating the Thames

Surrounded on all sides by water, with an equitable distribution of rivers, England and its commerce seemed to slide easily between land and sea. As both a cause and a consequence of industrialization, England had the world's most advanced system of river and coastal transportation. The coastal trade exploded after the mid-eighteenth century, with coal shipments alone expanding from 1.5 million tons in 1779 to over 5.5 million tons by 1829.[15] Coastal shipping was especially profitable for transporting bulk goods long distances. Coal was by far the most important commodity, brought down from the northeast ports to the entrepôt in London and then transported up the Thames and dispersed throughout the southeast of England.[16]

Britain's industrial strength relied heavily on the coastal transport of raw materials including wool, timber, iron, and coal, and the government was especially keen to collect revenue on all manufacturing and agricultural goods traveling around its coasts. But in the late eighteenth century it also hoped to expand the pool from which the navy recruited its seamen. Because of the notoriously dangerous wind, weather, tides, and tidal streams on the east coast of England and Scotland, the Newcastle collier fleet in particular became renowned for producing the best mariners in the world. It was the trade that had taught Captain James Cook how to sail and was where the *Endeavour*, a converted collier, first went to sea.[17] The Royal Navy relied on the coal traders to guard the British shores from invasion; the coasting trade served as an auxiliary of the navy, training future servicemen.[18] Reciprocally, defending the merchant marine was also the main cause of war; the primary concern of British foreign policy from the mid-eighteenth century onward was defending British shipping interests.[19]

During the Revolutionary and Napoleonic Wars, French naval vessels and privateers wreaked havoc on the British merchant fleet, confiscating some eleven thousand English vessels.[20] What was already a dangerous occupation owing to the natural forces of wind and weather was made doubly dangerous by the seemingly natural paths of European powers toward war. The solution, nothing new to a warring maritime nation like England, was to travel in convoys, seeking safety in numbers. Upward of eight hundred vessels could be seen passing through the English Channel guarded by as many as thirty navy ships.[21] The danger raised the cost of maritime insurance, and Lloyd's insisted that any vessels it insured travel in convoys.

The convoy system overwhelmed the harbors, rivers, and ports along the English coast, especially in the Thames estuary from Gravesend to the congested Pool below London Bridge. Traveling together meant that hundreds of vessels left and arrived at ports at the same time. At Gravesend, the entry into the Thames estuary, hundreds of vessels awaited anxiously for the high tide and favorable winds to carry them up to the London docks. With large vessels at anchor, smaller sailing barges tacking across the stream, and dredgers and other moored vessels presenting constant hazards, the reach could resemble today's bumper cars. Accidents were impossible to avoid, especially in fog or other bad weather. Upriver near the London Docks, the situation grew even more treacherous. Before the turn of the nineteenth century, the port of London contained only dry docks consisting of quays situated alongside the banks to allow ships to berth. Smaller coasters could make the Upper Pool above London Bridge, but larger vessels were relegated to the Lower Pool or confined to midriver, where their cargoes were unloaded onto lighters. Hundreds of lighters then ran from below London to the Upper Pool to unload their cargo at the legal quays. The queue on the river could take weeks, since the required wharf space, lighters, and laborers were quickly overwhelmed.[22] There was often a shortage of coal in London while the river was filled to the brim with it.

Thieves found the overcrowded river a boon.[23] Plundering was rife, as vessels loaded with goods and waiting at anchor were an easy target. Scuffle hunters roamed the docks at night looking for unattended barges and lighters. Armed gangs of river pirates cut smaller vessels loose from their anchors in the middle of the night and then followed them downstream until they drifted ashore. Barge workers often threw heavy articles into the river at high tide, working with "mudlarks" who returned at low tide to loot the exposed littoral.[24] So many vessels were anchored around London Bridge that no amount of policing could stop the plunder. The establishment of the marine police in 1798 was largely ineffective. In 1800 a contemporary estimated that thieves robbed the West India merchants of over £250,000 each year.[25] As one historian of London recently noted, "by the 1790s, overcrowding, delays, and alarming theft levels threatened the Port itself."[26]

The solution in this age of progress was to entirely remake the port of London, and in thirty years civil engineers had completely overhauled the entire system of docks. Beginning with the West India Docks at the turn of the century and ending with the St. Katherine Docks in 1829, by 1830 the port of London built almost half of the four hundred acres of dock space in the entire country.[27] The construction of these wet docks produced what R. J. B. Knight has referred to as the "first industrial complexes of a twentieth century scale."[28] Such a dramatic expansion of the dock system, like the concomitant bridging of the Thames, had unexpected consequences for the river itself.

## Expanding the Port of London

Docks serve as the coupling between sea and land, a stationary boundary within the ever-changing littoral. In a maritime nation like Britain, everything passed through the docks, the middle of an hourglass squeezed between the nautical world of transportation and trade and the terrestrial world of production and consumption. The port of London, like many of the major ports in Britain, is an estuary port, where the constant flux and reflux of the sea determines both the extent and the timing of its trade. Its docks therefore were constructed to deal with the predominant influence of the tides. In dry docks, smaller ships berthed at quays or wharves along the banks while larger vessels anchored in midriver and transferred their cargo to lighters. It was difficult and costly, however, to load and unload large vessels in the middle of the river, where tides with a range of over twenty feet furiously rushed in and out twice daily. Add to that the intense traffic of lighters and barges surrounding each large vessel, and congestion quickly became chaos. The prospect of loading and unloading an entire fleet in this manner led to the idea of the wet dock.

Wet docks were designed to remove the influence of the tide altogether, allowing vessels to stay afloat at all its stages. They consisted of an inland basin with a lock that trapped the high tide, keeping ships afloat while the water in the river receded. The vessels then escaped at a subsequent high tide, when the lock was reopened to communicate with the tidal river. William Vaughan, son of a London merchant and a student of Joseph Priestley's at Warrington Academy, was largely responsible for the set of plans that led to the first wet dock in the port of London. He published a series of tracts on the state of the Thames, including "On Wet Docks, Quays, and Warehouses for the Port of London" in 1793, the same year the Revolutionary Wars with France began.[29] Vaughan estimated that at any given time there were more than 775 ships waiting to anchor at the Upper Pool. Since the Pool was designed for about 500 vessels, they often waited in line for weeks.

Vaughan's scathing assessment of the overcrowding in the river led to an act of Parliament in 1799, the first of its kind, for the construction of a new dock system, the West India Docks at the Isle of Dogs, where the Thames created a U-shaped peninsula east of London. The West India Docks Act of 1799 cited the dangers to river navigation from both the congestion in the main docks and the need for greater depth for the progressively larger class of ships plying the Thames. The West India Docks began receiving vessels in 1802, though the entire system was not completed until 1806. The dock system included massive warehouses to accommodate cargo and high walls to protect the docks from both the high tides of the river and the low dealings of thieves. Even before its completion in 1806, other docks had been designated in quick succession.

Figure 2.4. A view of the river facing the Tower of London, by William Daniell, 1804. Reproduced from Joseph G. Broodbank, *History of the Port of London,* 2 vols. (London: Daniel O'Connor, 1921). University of Minnesota Libraries.

The London Docks Bill in 1801 followed similar legislation. It was so named because its investors wanted to emphasize that their dockyard was closer to the center of London than the West India Docks, a couple of miles downriver at the Isle of Dogs. It did not receive vessels until 1805, but soon after construction began in 1801, it was obvious that the massive engineering works—including adjacent blocks of warehouses— were changing the topography of the river below the Pool. As one of the directors of the docks, William Vaughan also promoted their use, which included ensuring the safety of large vessels. Since vessels entered and left only during high water, Vaughan instituted measurements of the high tides beginning in August 1801. These observations proved essential some twenty years later: both John William Lubbock and William Whewell began with them when they started their theoretical study of the tides in England. William Vaughan therefore is at least partly responsible for setting the investigation of the tides in motion, his aim being to ensure the safe and continued use of a dock system in which he was personally and financially vested.

After the successful construction of the West India and London Docks, further acts of Parliament were passed in quick succession, all based on the overcrowding of the river and the need to protect cargo from both tides and thieves. After three years of construction, the East India Docks

Figure 2.5. The New London Docks at Wapping opened in January 1805. Note the embankment and warehouses on the far side of the river. Even with these new wet docks, the congestion in midriver remained. William Vaughan was one of the directors of the docks, and he initiated measurements of the high tides when construction began in August 1801. From a painting by William Daniel (1803). Reproduced from Joseph G. Broodbank, *History of the Port of London*, 2 vols. (London: Daniel O'Connor, 1921). University of Minnesota Libraries.

opened at the Isle of Dogs in 1806, followed by the Surrey Docks at Rotherhithe in 1807 and the Regent's Canal Docks at Limehouse Basin in 1812. Even these additional wet docks, however, failed to alleviate the overcrowding of the river near London.[30] In many respects they increased the congestion by speeding up loading and unloading to allow a quicker turnaround. Whereas 13,949 vessels had used the port in 1794, by 1824 that number had increased to 23,618.[31] Two additional dock complexes therefore were constructed even closer to the Upper Pool. A bill in Parliament in 1824 called for the construction of a wet dock between the London Docks and the Tower of London, opened in 1828 as the Eastern Docks. The St. Katherine Docks opened a year later, replacing the hospital of the same name, opened in the twelfth century to house female lepers.[32] One of the names listed as a subscriber (financier) was John William Lubbock, who initiated the scientific study of the tides in the river Thames. Like William Vaughan, Lubbock was financially invested in the docks and worked to promote their safety.

The construction boom in the first three decades of the nineteenth century required acts of Parliament, but the docks themselves were not owned or operated by the government. The subscribers were either corporations like the East India Company and the West India Company or wealthy London merchants, mostly bankers and insurers. This same

Figure 2.6. The opening of the St. Katherine Docks on 25 October 1828, from an engraving by
E. Duncan after a painting by W. J. Huggins. John William Lubbock, who resurrected the
theoretical study of the tides in England, was a trustee of the St. Katherine Docks. Reproduced
from Joseph G. Broodbank, *History of the Port of London*, 2 vols. (London: Daniel O'Connor,
1921). University of Minnesota Libraries.

group of corporations and private investors also owned the ships that
used the docks.[33] It made sense to the early Victorians, then, that these
subscribers take primary responsibility for the safety of the river. The
ability to increase the speed of shipping and decrease the risk of shipwreck rested on people like William Vaughan and John William Lubbock. Those who invested their money—the bankers, shipowners, dock
trustees, insurers, and merchants—were concerned above all with limiting risk by operating in a safe and stable environment. After more than
half a century of being bridged, banked, docked, and otherwise bounded,
however, the Thames had become notoriously unsafe and unstable.

## Silting, Flooding, Dredging, and Conservancy

The massive engineering projects on the river's edge required additional
embankments to secure their foundations, ease the transition between
land and river, and accommodate warehouses, custom and trade offices,
and passenger and dock facilities. The human encroachments to the river
concentrated its flow in a narrower course, affecting the width, depth,

Figure 2.7. The construction of the Chelsea Embankment in 1857, showing how land was stolen from the river. *Illustrated London News* 30 (24 January 1857): 67.

and velocity of the navigable channel. With a reduced channel, the tides rose ever higher on the river's banks, threatening low-lying neighborhoods. Flooding, which had always been a problem in the tidal Thames, seemed to increase significantly, especially in central London between Blackfriars and Vauxhall. Examples abound. A day after Christmas in 1806, the tides in the Thames rose to unprecedented levels, overtaking the embankment at Blackfriars Bridge and inundating wharfs, cellars, and warehouses. It then swept through Westminster Hall and adjoining neighborhoods, flooding a large part of the city. Blackfriars again overflowed on 30 December 1814, flooding Windsor Park and again inundating neighboring warehouses and businesses.[34]

To guard against the unprecedented high tides, engineers increased the height and extent of the embankment during the construction of bridges and dock complexes. As soon as the embankments were under way, however, the changing currents of the river fought back, slowly eroding protective walls and ineluctably carrying construction material out to sea. Three years into the construction of the West India Docks and shortly after they began receiving vessels, the newly completed south wall was breached, "not having been carried to a sufficient height to guard against the overwhelming power of the spring tide."[35] As the water tore through its artificial barriers, it carried with it about 150 yards of the embankment and several shiploads of large foundation stones. Similar battles raged during the construction of the St. Katherine Docks, when a larger than expected high tide inundated the entire complex "with such fury and impetuosity that before any efforts could be made to check its progress the whole of the new docks was completely filled with water, exceeding in some places 27 feet in depth."[36] Buildings, warehouses, engine rooms, all were left completely underwater, resurfacing only with the ebbing tide.

The construction of bridges, docks, and embankments added considerable obstructions to the river's flow. Oceanic tidal waters are filled with sediment, including shifting sand, silt, and salt. The high tide brings this sediment upriver from the mouth of the estuary and deposits it in the river during slack tide. If it finds a secure place to rest, around a forming sandbank or a moored vessel, for instance, the ebb tide will not carry it back downriver. As the sediment accumulates, the scouring effect of the ebb is further reduced. Though this is a natural process of tidal rivers, it is made much worse by artificial obstructions. The most obvious are bridges that act as dams, but piers, jetties, abutments, and quays also obstruct the ebbing tide, as do sunken ships and permanently moored vessels such as those used as chapels, hospitals, lighthouses, and storage facilities.

Insufficient depth was a constant problem as the vessels using the port increased in size and tonnage, and continual dredging seemed the only solution. Dredging cleared and widened the navigable channel and strengthened its current, allowing the river to scour itself naturally in its ebb to the sea. An unobstructed river, however, came at a price. After dredging, the tide both rose higher as it rushed unimpeded from the North Sea and fell lower as it ebbed back to the ocean. The result was a more dramatic range of the tide, a lower mean height of the river, and a loss of clearance for large vessels. Thus deeper and more frequent dredging was required as the depth at low water decreased. The constant battle to control the river's regime was a losing proposition, since dredging increased the need to dredge.

Engineers had used dredging as a consistent policy without applying any knowledge of its ultimate effects on the tidal regime. The conservancy of the river before the early Victorian era was dealt with ad hoc, fixing problems rather than preparing for them. The dilapidated charac-

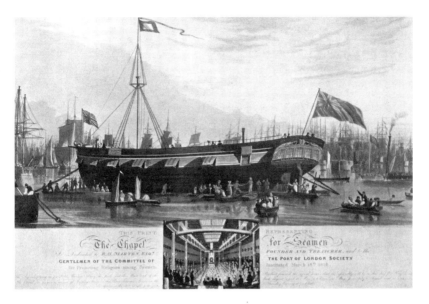

Figure 2.8. The Chapel for Seamen (1818). Moored vessels and other artificial obstructions impeded the river's tidal flow, reducing its ability to scour itself of sediment during its ebb to the sea. Reproduced from Joseph G. Broodbank, *History of the Port of London,* 2 vols. (London: Daniel O'Connor, 1921). University of Minnesota Libraries.

ter of so many of Britain's rivers, however, began to catch the attention of civil engineers in the 1820s, when civil engineering was gaining some recognition as a profession. The Institution of Civil Engineers, founded in 1818, served as both a dinner club and a research center where engineers discussed practical problems associated with the massive engineering works in the city, including those on the river's edge. George Rennie, one of Britain's most celebrated engineers, undertook a comprehensive study of how engineering projects affected the river's regime, specifically focusing on the construction of bridges, docks, and embankments. Practical works in hydraulic engineering "of great magnitude and extent" had been carried out in England, Rennie warned, but "the application of this science to rivers" had made no progress whatever. The laws that governed the waters of the earth's littoral, according to Rennie, were still "involved in mystery."[37]

Rennie's conclusion that humans had overpowered the forces of nature in changing the river's topology and tidal regime had a decidedly modern tone. "The principles upon which the earliest Acts of Parliament were framed for the conservancy of our rivers," Rennie noted, "consisted in deepening, straightening, and embanking them where necessary." No regard was given to how the river itself would react to such alterations. Experience had shown, however, that human encroachments "were liable to perpetual degradation, from the alterations produced in the regimen of the rivers by such artificial works, which frequently augmented instead of remedying the evil, whilst they obstructed the general drainage of the country." Too much dredging and scouring of the main channel could cause a river's complete demise.

Rennie reported his investigations to the meeting of the British Association for the Advancement of Science in 1833, the same year William Whewell first reported on the tides and called for a better understanding of the *science* of hydraulics. But Rennie's report was based on research he had done years earlier, in preparation for the removal of the old London Bridge. The heightening tides themselves had helped erode the starlings of the seven-hundred-year-old bridge, and its removal, as engineers were beginning to comprehend, would radically reshape the tides and currents of the Thames, especially around the main Pool, the heart of British shipping.[38] George Rennie outlined plans for rebuilding London Bridge as early as 1820, but the actual construction was conducted by his son, John Rennie, beginning only in 1824. Like other construction projects on the river's edge, the new bridge altered the topography of the river so much that it severely impeded navigation. Only additional and fortified embankments solved the problem, producing the Thames Embankments of today, the straight and gallant route that allows for such a beautiful view of the river.

The removal of the old bridge caused the tides to flow upriver farther and more swiftly. As Rennie related, "barges, which used formerly to be towed up from Putney to Richmond by horses, are now carried by the current from London Bridge to Richmond in one tide."[39] However, the quickening pace of the river also lowered its average depth both above and below the bridge. It raised the high water level more than a foot, but unfortunately it lowered the low water more than four feet, and many vessels that had formerly gained access to the Pool now found themselves grounded at low water.[40] Thomas Winter, a barge master for thirty years, testified to the Select Committee of Parliament that the currents of the river from Blackfriars Bridge all the way to Vauxhall Bridge had changed drastically, becoming very unpredictable. "Sometimes it is lower, and sometimes there are hills and pools, and near the bridges it becomes very inconvenient to navigate, owing to going into a pool of water, and then into very shallow water."[41] A vessel was likely to hit ground from want of water at one end of the barge while there was twenty feet of clearance on the other end. As Joseph Robinson, a customhouse and commercial agent, related, vessels negotiating the tide above London Bridge were "impeded by those shoals which were not seen or felt before, by their being brought to light in consequence of the receding of the tide." Whereas barges used to be able to sail from Lambeth to London Bridge at all stages of the tide, now they had to wait for high water, an expense of time that increased what barge masters charged for carrying goods.[42] James Walker, the civil engineer employed in 1816 to build Vauxhall Bridge, concluded that it was "quite impossible to look at the river Thames, at low water now, without seeing, as regards to the trade up the river, and the navigation opposite to London itself, that the river is in a state that wants improvement very much."[43] That the *Leeds* from New York could strike ground in the middle of the navigable channel came as little surprise.

The improvements again came in the form of dredging, but there was some question about who was responsible for ridding the Thames of silt. The Corporation of London, the body in charge of the conservancy of the river, often dredged where shoals obstructed navigation, but it refused to do so regularly. It contracted out the conservancy to private companies, which dredged only for gravel that the Corporation could sell as ballast for ships or for other engineering projects.[44] In 1828 one shipowner lambasted the Corporation in the public press for the "very improper state of the river Thames and also the liability of the great loss of property that is likely to take place from the very great accumulation of dirt and filth that now exists, so much so, that a heavy laden ship runs the most imminent risk in entering the pool of London."[45] Signed "A Sufferer," the shipowner's letter was part of a larger chorus of complaints from those dissatisfied with the deteriorating condition of the river.

Shipowning was a risky investment.[46] The merchants, mariners, bankers, and dock financiers needed insurance against risks associated with war and privateering. They turned to the government for protection in the form of surveys and convoys.[47] But they also had to mitigate the risks associated with the river's changing environment, especially its tides. For this, they turned to local tide table makers and almanac publishers. Those with an interest in the coastal shipping trade, insisting on conserving a stable and secure river and coastline, were the ones to make the study of the tides a national objective and tide tables a profitable commodity.

## Tide Tables and Almanacs

Changes on the river's edge quickened in the last half of the 1820s, a flurry of activity that focused largely on the Pool, including the construction of massive dock complexes, additional bridges, and new embankments. The river responded in kind. Its currents, bottom contours, and tidal regime reacted in ways that mariners, engineers, and natural philosophers had trouble figuring out. Privately published tide tables had not introduced these changes into their predictions, nor could they have, for they relied on years of observations and on the now debunked premise that the tidal regime remained constant. Human artifice, ingenuity, and conservancy had permanently changed the river. Yet, as George Rennie's systematic study of river hydraulics demonstrated, Britain's estuaries were also coming under increased scrutiny. The changes to the river itself made the tides more visible.

By the late eighteenth century, mariners could use locally calculated tide tables specific to their estuary or coast to determine the day's tides. George Innes, an amateur astronomer and expert calculator, produced predictions for Aberdeen; Olinthus Gilbert Gregory published tables for Edinburgh and Leith; and George Holden did so for Liverpool. The largest and most lucrative market for tide tables, of course, was London. By the late 1820s, pilots had several tables they could consult, including those by G. W. Butler, Dr. Gregory, Mr. White, and Mr. Bulpit. White published predictions for the port in his *Ephemeris*, an almanac published by the Stationers' Company. His fiercest competitor was Bulpit, whose tables were based on several years of observations made by William Pierce, the dock foreman William Vaughan put in charge of observations at the London Docks. These observations had been made exclusively for the commercial interests of the port, with little knowledge that they would someday be useful for tidal analysis in general. The stimulus was commercial, though their use was eventually scientific. When Lubbock published on the tides in the port of London several years later, he mistakenly assumed that observations of the tides had been taken at the London Docks only beginning in 1805, since those were the observations then extant. But Bulpit had acquired the first four years of observations from Pierce to produce his own tide tables, and he would neither share the observations nor divulge his methods of calculation.

Perhaps the most salient aspect of tidal predictions in the early nineteenth century is the lack of government involvement. The Admiralty did not produce tide tables for British ports, not even for the port of London or the naval dockyards situated downriver on the Thames. That left the publication of tide tables to privately owned almanacs, a power-house of British publishing. As one contemporary noted, almanacs were the "most popular publication in England next to the Bible and the Prayer Book."[48] Many almanacs existed, some published in London and others, of only local interest, in provincial towns. Around the turn of the century, however, the Stationers' Company established a monopoly on almanacs by buying up individual publications and either suppressing them or continuing them under its direction. The company published nearly thirty almanacs, including White's *Ephemeris*, which included tide tables for the port of London. It issued upward of half a million almanacs a year, reaching a large segment of the population, including the working classes.[49]

The upper echelons of British society, including the rising middle class, found it a bit disheartening that such a large segment of the working classes read almanacs. Britain was in turmoil brought on by its rapid industrial and economic development. More and more English men and women were moving to the city to find steady work, swelling the ranks of the working poor. Social norms were transformed as the less seemly aspects of industrializing economies—class consciousness and class antagonism—set the stage for the constitutional revolution in the early Victorian era and the passing of the first Reform Bill in 1832.[50]

By this time the burden of control rested ultimately not on government but on the people themselves. One way the middle classes attempted to curb the growing unrest of the working classes was through educational initiatives, a kind of benevolent despotism underlying the Mechanics' Institutes and Friendly Societies that sprang up all around England and Scotland. The relatively large audience for popular almanacs must be viewed within this larger debate on the education of the working classes. According to British reformers, the almanacs tended to provide the lowest segment of the population with deviant information that verged on blasphemy. *Poor Robin*, one of the almanacs published by the Stationers' Company, was viewed as the worst, filled with indecent material that supposedly imbued the lower classes with treasonous thoughts. Such almanacs, argued the British elite, inflamed superstitions. The *Athenaeum* referred to them as a "vaporous modification of palpable imposture, impudent mendacity, vulgar ignorance, and low obscenity" unchanged since the time of witch burnings.[51] Even the less bawdy publications, the editors of the *Athenaeum* argued, did harm. Both John Tanner's *Ephemeris* for 1678 and John Hamilton Moore's *Almanac* for 1828, for instance, linked the dreadful English weather of December to Saturn's position in the heavens. For the enlightened Englishman, the planets might have been used to explain the weather before the age of Newton and the perfection of astronomy, but to hold to such ideas in modern times was sacrilegious and opposed to the progress of knowledge.

The new middle classes, overconfident in their own notions of progress and utility, applied these concepts to the problems of the working classes. One element to arise from this educational enterprise was the Mechanics' Institutes, the working-class variant of the Literary and Philosophical Societies. A handful of institutes were formed in the first few years of the 1820s, but a turning point came with a popular and widely distributed pamphlet entitled *Practical Observations upon the Education of the People, Addressed to the Working Classes and Their Employers,* first published in the October 1824 *Edinburgh Review.*[52] After its appearance, Mechanics' Institutes spread throughout the British countryside. By the end of the year, eighty had been formed across Britain, and by 1851 over seven hundred existed.[53]

The author of the short treatise was Henry Brougham, a Whig politician in the House of Commons.[54] During his years in Parliament, he focused heavily on the education of the working classes. In his treatise, Brougham urged the working classes to attend Mechanics' Institute meetings, and he also foreshadowed his next major project: "I am not without hopes of seeing formed a Society for promoting the composition, publication, and distribution of cheap and useful works."[55] After a year of deliberation, correspondence, fund-raising, and politicking, Brougham founded the Society for the Diffusion of Useful Knowledge. The original committee elected Brougham chairman, Lord John Russell vice chairman, and Thomas Coates secretary.[56] For a few shillings, anyone could become a member.

Brougham surrounded himself with a talented committee of forty-seven members, including twenty-three members of Parliament.[57] The Society never claimed to promote original research, only to diffuse practical knowledge to a wide audience.[58] As the low cost of each publication revealed, they appealed to the lower classes with the Victorian aims of betterment and improvement. Largely owing to Brougham's leadership and the social and scientific climate of the time, to the Society "useful" often meant the study of natural philosophy, with the aim of glorifying the benevolent Creator.

This was all the more evident from the Society's first publication, its monumental Library of Useful Knowledge. The first series, devoted to natural philosophy, consisted of seventy-two separate numbers.[59] The first number was written by Brougham himself, "On the Objects, Advantages, and Pleasure of Scientific Pursuits."[60] Those that followed were written by natural philosophers of some renown, including George Biddell Airy,

David Brewster, and Augustus de Morgan, and their initial success enabled the Society to expand its publications. Brougham and the other members of the committee decided to change publishers to Charles Knight, the son of a bookseller who became a prolific writer and a respected publisher. Knight joined the Society in early 1827 and began to supervise its publications in July. In early November he proposed what the historian R. K. Webb has referred to as probably "the most successful, if among the least publicized, of the Society's publications."[61] This was the *British Almanac*.

## The *British Almanac* and the Revival of Tidal Theory

Knight was intent on publishing an almanac for both political and economic reasons. He convinced Brougham that a new almanac, if based on scientific principles, would raise both the income of the Society (and himself as publisher) and the level of almanac publications in general—the perfect strings to pull Brougham in. Brougham was immediately enthusiastic but feared it was too late in the year to bring out an almanac. It was already November, and if they were to compete in the market, theirs would have to be published on the same day as the Stationers' Company's almanacs, which arrived with the new year. Knight assured Brougham that he could have an almanac completed if he had "a little help in the scientific matters."[62] Brougham immediately contacted Francis Beaufort, a founding member of the Society, a captain in the Royal Navy, and an experienced surveyor who had risen based on his scientific acumen. Beaufort agreed to supervise the section on astronomy and reviewed many of the articles dealing with navigation.[63] Knight in turn accepted the "charge and risk" of the almanac, with a payment to the Society to be determined after its initial sales.[64]

The first *British Almanac* appeared on New Year's Day 1828, an amazing feat considering the short time they had to gather the material. It consisted of two parts: the *British Almanac* and the *Companion to the Almanac*. The *Almanac* began with two pages devoted to each month. The first page included lists of the feasts, holidays, and noteworthy anniversaries and events, as well as remarks on the weather. The second page was more technical, including, for instance, the method for comparing the clock with the sun, a table of the changes of the moon, and most important here, a tide table for the port of London.

The *Companion* contained information illustrating the *Almanac*, including two sections discussing the theory of the tides. The first section,

"Causes and General Appearances of the Tides," outlined the equilibrium theory as formulated by Newton and improved by Bernoulli. It
was geared toward the common reader and explained why the tides
sometimes appeared irregular on the coasts of Britain: "At some places
there are no tides, at others they rise to a great height: sometimes there
are double tides; and sometimes only one in twenty-four hours."[65] The
shorter section, titled "Common Rules for Finding the Time of High Water," demonstrated, using examples, how mariners could determine the
tides at every port in Britain. A note of caution ended the description:
"From the variations already mentioned, as well as from local causes,
these rules are not perfectly accurate; but they may serve to explain and
exemplify the principles."

The *British Almanac* was highly successful.[66] Only two days after it
had appeared, Brougham advertised its availability by writing scathing
reviews of other almanacs in the *London Times*. He reiterated earlier
remarks on the vulgarity and obscenity found throughout their pages,
just the kind of "open poison" that so affected the minds of the working
classes. "The Company of Stationers have thus, to the present hour,
reigned lords paramount over an important and intrinsically useful
branch of popular knowledge," Brougham preached, "and this authority
and influence they have prostituted to the most degrading purposes."[67]
The *British Almanac*, Brougham announced, was specially prepared with
the help of "zealous and enlightened" natural philosophers, replacing
vulgar ignorance with cultured rationality.

Brougham's unsigned piece in the *London Times* garnered quick and
hostile responses from those in the almanac trade. "Vindex," a member of the Stationers' Company, complained about the incivility of
Brougham's review.[68] He accused the Society for the Diffusion of Useful
Knowledge of "moral and legal" impropriety in stealing material outright from the Stationers' Company and incorporating it into the *British
Almanac*. The charge of plagiarism, tellingly, focused on the process of
calculation. According to Vindex, the editors of the *British Almanac* had
"unfairly availed themselves of the computations, and other labours"
needed for all of the major tables, including the "Notes of the Year" and
the "Law Terms and Returns." But most especially Vindex hit on the columns of "High Water at London," which he claimed were "copied, without acknowledgement, and without the variation of a single figure, from
White's Ephemeris." Vindex detailed the required "skill and care" needed
to compute tide tables, arguing that it would be as "impossible for the
results for a year, by different computers with different rules, to agree
within a minute, or even within five, day by day, as it would be in toss-

ing up a shilling to make it fall with the head uppermost a thousand times in succession." Vindex ended with a question that only slightly hid his disdain for the supposed scientific nature of the supposed stolen tables: "Have computers employed by your 'zealous and enlightened' friends achieved this impossibility?" The interchange that followed in the daily press degenerated into further name calling and accusations of thievery.[69]

The allegations, it turned out, were all true. With little more than a month to prepare, the Society had been forced to use material from other almanacs, especially for the time-consuming calculations needed to produce tide tables. Though this sort of "borrowing" was common practice at the time, the *British Almanac* had boasted of the scientific accuracy of its almanac while producing very little new in either science or accuracy. Knight, embarrassed by the accusations, hired a calculator to produce the tide tables for the next year. They decided to base their methods of calculation on tables published in Andrew Mackay's *Complete Navigator*, a manual for seamen. In the *British Almanac* for 1829, therefore, there appeared not only the high water for London, but also a tide table that covered one hundred ports, mostly in Britain, but also including Calais, Brest, and Amsterdam.

But something again went terribly wrong, a fault that had less to do with the hired calculator than with the intractable nature of calculating the tides themselves. Knight had instructed his computer to base the tide tables on those published by Mackay, which were in turn based on those of Bernoulli. Neither Bernoulli nor Mackay, however, published a "corrected establishment" for the port of London. The "vulgar establishment" is the time of high tide on the day of the full moon and can be calculated from a small series of observations. It can be used to predict the time of high water when the moon is either full or new but gives only a rough approximation for all other days of the month. To determine the time and height of the tide for other days required a corrected establishment that included the distance of the moon from the sun (the moon's right ascension) as an added correction. Furthermore, the tides in most ports, especially derivative tides like those of the port of London, were affected by the force of the sun and moon not on the day of the tide, but at some previous time. The sun and moon acted on the waters of the ocean, and it took some time for the tides to make their way to the ports in question. This "age of the tide" differed for different ports. The corrected establishment, therefore, consisted of the vulgar establishment corrected for the moon's right ascension not on the day of the full moon, but on the day corresponding to it at some previous epoch.

Finding the corrected establishment was tricky; unlike the vulgar establishment, it could not be determined from only a few observations. One could find it, as Joseph Foss Dessiou would later do under John William Lubbock's guidance, by analyzing nineteen years' observations, which was extremely time consuming and labor intensive. This was one reason there existed as many establishments for the port of London as there were published tide tables. Mackay gave an establishment of three hours twenty-two minutes, for instance, while the *Annuaire du Bureau des Longitudes* gave two hours forty-five minutes. The calculator for the *British Almanac* used the time published in Mackay's *Complete Navigator*, a corrected establishment that was incorrect. In addition, the effect attributed to changes in the moon's distance from the sun in Mackay's tables was directly contrary to that given by Bernoulli's table and also directly contrary to observation.[70] The Society's computer copied this error as well.

This was the type of "help in scientific matters" that Knight had worried about and Brougham had insisted Beaufort could remedy. Instead, angry letters flooded the daily newspapers, again highly critical of the "scientific methods" the Society boasted about. A letter from "Detector" suggested that no science was used at all.[71] He assumed that "the subject of the British Almanack has, by this time, become a sore one with the Society for the Diffusion of useful Knowledge," reminding readers of the debates of the previous year. Few circumstances had changed, according to Detector, and he wondered if the Society had made even one calculation for its almanac, "which for the present year was so greatly indebted for its scientific parts to those published by the Company of Stationers, the more especially for their tide table."[72] Admitting that the Stationers' Company had made a mistake by omitting the "Tables of Terms and Returns" in its own almanac, Detector shrugged the omission off as a minor oversight compared with the glaring errors in the *British Almanac*. "The Diffusion Society have miscalculated the tide table in the British Almanack by more than an hour for each day throughout the ensuing year," Detector gloated, "so that for one blunder committed in the Company's almanacks, 365 have been committed in the Diffusion Society's, leaving a balance of blunders against them of 364." Competition between almanac publishers was fierce, and by whatever count, the Society seemed to be losing.

Though these attacks were on target, so too were the Society's responses.[73] The *British Almanac* not only had produced a nice profit for both the Society and its publisher but had led to a considerable improvement over the almanacs of the Stationers' Company.[74] The most indecent, such as *Poor Robin*, were forced to stop publication altogether. As one correspondent wrote, "that drivelling and smutty old sinner" had expired for good, all the while "blaspheming and calling for brandy."[75] Thus, though the Society failed in establishing a scientific basis for their publication, it did end up discrediting the Stationers' Company and ending its monopoly on the sale of almanacs.

The next step for the Society was to finally infuse its tables with actual scientific analysis, to back up its claim with action. Beaufort informed Knight that the only solution was to move toward a better understanding of the tides themselves; only from a correct theory, he argued, could they find the corrected establishment for the port. Beaufort and Knight decided to proceed by adding more scientifically minded individuals to the almanac committee, people who could move beyond Beaufort, a man of practice, to a theoretical understanding of the tides. They turned to someone who was already interested in the tides because of his investments in the port—John William Lubbock, by 1828 a director of the St. Katherine Docks. More important for the Society, Lubbock was also well versed in the mathematics of physical astronomy, having sat for the mathematical tripos at Cambridge. In November 1828, Edward Maltby proposed Lubbock as "a person I think it would be very desirable to have on your Committee." Maltby continued: "Mr. J. W. Lubbock is the only son of Sir John Lubbock; yet, in spite of indifferent health, has devoted himself very much to scientific pursuits, particularly of Mathematics. He was formerly my pupil, then Whewell's at Cambridge—where he took a good degree and probably would have taken a better, if he had not read so much French Mathematics."[76] Maltby's last phrase was meant as a jab, but it is also significant. It was exactly Lubbock's familiarity with French mathematics, especially the writings of Laplace and others concerned with tidal theory, that made him such a valuable member of the committee.

The letter informing Lubbock of his nomination to the Society left no doubt that he was chosen to help perfect the *Almanac*. The letter contained three sentences: "Dear Mr. Lubbock. I beg to inform you that you were yesterday elected a member of the Committee for the diffusion of Useful Knowledge on the nomination of Dr. Maltby and Mr. [James] Mill. I am directed to submit to your revision the accompanying copy of the Companion to the Almanack. Your humble servant. T. Coates."[77] Lubbock was placed immediately on the almanac subcommittee, which met two days after his election. The order of business of that meeting was the accuracy (or inaccuracy) of the tide tables published in the *British Almanac* for that year.[78]

Before the next almanac subcommittee met two weeks later, further critical letters had appeared in the public press, and Lubbock's first task was to figure out exactly what had gone wrong. Beaufort had taken measures to ascertain the corrected establishment at London Bridge by writing a letter to the directors of Trinity House, the body responsible for buoying and lighting rivers and estuaries, but it seems that even Trinity House was not certain of the corrected establishment.[79] Lubbock's initial study suggested that the existing tide tables for the port varied so considerably that "no answer should be made to the statements that had appeared."[80] After determining that the errors resided in the process of calculation, Lubbock further insisted that Knight hire an additional calculator. On Beaufort's recommendation, Knight chose Joseph Foss Dessiou, an experienced calculator working under Beaufort at the Hydrographic Office of the Admiralty. The letters in the public press were unforgiving, written by tide table and almanac publishers who feared having their methods exposed. But for the theoretical investigation of the tides, these letters forced the Society into action, setting the stage for the complete overhaul of tidal analysis.

## Conclusion

As Beaufort, Lubbock, Dessiou, and the rest of the almanac subcommittee met to discuss the propriety of answering the attacks on the Society in the public press, the *Leeds* still lay stranded in the middle of the Thames with a fractured hull, having misjudged the tides only two days earlier. That the public took such an interest in the tides was partly due to the fervent emotions that shipwrecks engendered in the romantic mindset. The British press recounted harrowing tales of shipwrecks, often following the narrative for days, from the first plank seen floating in the sea foam to the last body washed up on shore. The image seared the British consciousness, a reminder that nature was still dominant, a force to be obeyed before it was commanded.

Londoners have always been in constant dialogue with the Thames, the main thoroughfare of the ever-expanding city. It was responsible for London's astounding growth and helped determine the commercial success of the entire country. Yet just as the river shaped the social and economic space of the city, the people of London also shaped the Thames. After the mid-eighteenth century, humans became the most significant intermediaries in the constantly shifting battle between land and sea. The changes to the river Thames, moreover, are representative of the

Figure 2.9. Wreck of the *Premier,* from the *Illustrated London News* 8 (27 January 1844): 53.

transformations occurring all across the British Isles. Between 1820 and 1836, Liverpool's port also underwent an astonishing transformation, as engineers reconstructed the Coburn, Princess, Canning, Clarence, Brunswick, Waterloo, and Victoria docks.[81] In Bristol, the Floating Harbour was finished in 1809, completely transforming the dock system of the expanding city. Though not everyone appreciated the projects designed to bridle the ocean's force, the growth of sea trade in the previous decades required the surge of activity near the coastlines, the most dangerous arena for commercial and naval ventures.

The bias accorded to foreign and long-distance trade in the history of science, focusing too narrowly on captains serving in the Royal Navy or on harrowing voyages of exploration and discovery, has led to a mistaken view of the cause of shipwrecks. Correcting a ship's magnetic compass and finding longitude at sea were two of the more difficult feats of navigation, but being lost at sea caused only a small percentage of the loss of life. The vast majority of shipwrecks occurred close to the shore, if not in the ports and estuaries themselves.[82] The coasts and estuaries of Britain were particularly dangerous, dramatized at the turn of the century in the paintings of J. M. W. Turner. One of his most famous paintings, *A Shipwreck* (1805), engrained the coast in the public's imagination as dark, forbidding rocks contrasted with the tumultuous waves engulfing the victims.

Figure 2.10. J. M. W. Turner, *A Shipwreck* (1805). The British coastline—especially its fearsome rocks and cliffs—was prominent in Turner's paintings, reminding viewers that most shipwrecks happened in the dangerous area of the littoral. Reproduced by permission of the Clore Collection, Tate Gallery, London.

In the early Victorian era, the British lost well over a thousand vessels to shipwreck each year.[83] Loss of life at sea increased so alarmingly that the British Parliament became involved, particularly in the Royal Commission on Shipwrecks of 1836 and the Select Committee on Shipwrecks in 1842. Sir John Barrow, the secretary of the Admiralty, testified that the problems of the magnetic compass had had "the most mischievous of consequences" for large vessels of the Royal Navy, but "in merchant vessels, where there are no guns, unless with an iron cargo, the deviation is scarcely perceptible."[84] Apart from human error, mostly owing to drunkenness of the pilot or captain, Barrow accorded the main cause of shipwreck not to problems with finding longitude or with the vessel's magnetic compass, but to the natural phenomena of tides, winds, and weather.

As subsequent chapters will demonstrate, the tides represented one of the first large-scale geophysical problems to come under scientific scrutiny and analysis. Rather than beginning in the reformed Royal Society, the Royal Astronomical Society, the British Association, or the

British Admiralty, however, the study of the tides in the early Victorian era began in the profitable arena of the British coastal trading routes. It was initiated by insurers and underwriters, dock trustees and shipowners, bankers and merchants intent on mitigating the risk to the coastal shipping trade. The rivers of England had been drastically transformed in the preceding half century as engineers straightened, embanked, dredged, and otherwise modified the river to control the depth of the water, the rise and fall of the tide, and the silting of the stream. These environmental changes, in turn, transformed the river's tidal regime and created an economic niche that tide table makers and almanac publishers quickly filled. Surprisingly, it was the Society for the Diffusion of Useful Knowledge that brought the tides to the attention of the scientific community. The beginnings of the field underscore that in the study of the tides the impetus behind natural philosophers' choice for scientific research began at the quayside. The selection of research topics also adds a dimension to the maxim that science follows the flag. Just as the Thames seemed to "flow for all mankind," so did the study of the tides begin in the estuaries of England before proceeding—like the myriad of vessels sailing from the mighty Thames—out to the world's oceans.

THREE

# Dessiou's Claim

Whether through David Copperfield's experiences while living on a barge in Yarmouth (where "the town and the tide are . . . much mixed up") or through the young gentleman Pip's ill-fated escape with Magwich during ebb tide in the Thames, Charles Dickens detailed early Victorians' close connection with the ebb and flow of the tide.[1] The tides are prevalent in Dickens's novels because they are ever present on the coasts of Great Britain; for those living near the sea, they could alter one's engagements. Those working in a port were not ruled by the coming of dawn or dusk; they did not follow the rhythms of the sun. Rather, they followed the phases of the moon, which dictated the movement of the tides and thus the times of both work and leisure. At extreme hours of the tides, the littoral bustled. At low water, coast dwellers harvested the mudflats, gathering crustaceans, kelp, and anything else of value that the receding waters revealed. As the tide rushed in, so did the many vessels looking for berths, offering constant work for pilots, bargemen, lightermen, and dockworkers. Nowhere did the dynamic flux and reflux of the sea shape daily lives more dramatically than in the bustling center of British commerce, the London docks, the original East End populated by sailors, merchants, and manual laborers. There lived and labored an expert calculator named Joseph Foss Dessiou.

In the mid-1830s, just as Dickens was publishing his highly successful *Pickwick Papers*, one could find Dessiou working in his office close to the docks, within earshot of the ebbing and flowing Thames. Dessiou's hands were not callused from the manual labor of the port; they were ink stained from computing tidal predictions for the month's high tides. The second son of a respected hydrographer, Dessiou went to sea early in his life, traveling to Scandinavia, the Adriatic, Brazil, the West Indies, and Newfoundland.[2] He composed charts and sailing directions of his voyages, and after quitting the sea in 1805, he worked privately as a mapmaker, including producing charts for the firm of William Faden, which supplied the Admiralty. He was appointed naval assistant to the Hydrographic Office in early 1828, was hired by the Society for the Diffusion of Useful Knowledge to calculate tides in 1829, and by 1833 was working on the first public tide tables published in Britain, the *Admiralty Tide Tables*. Dessiou was the first state-funded tide calculator in Britain, and by his death in 1853 he was the world's premier tidal computer, responsible for predicting the tides for all of Britain's major ports.

The science of the tides in the early nineteenth century depended on a combination of advances in observation, theory, *and* computation. Contemporaries and historians alike have focused on the first two, mainly because the interplay of theory and observation conforms to conventional wisdom concerning the advancement of science. Much of that wisdom, not surprisingly, was being defined during the early Victorian era. William Whewell, for one, aligned both his history and his philosophy with what he termed the "fundamental antithesis" between theory and observation. By focusing on the complicated relationship between the two, Whewell was extending a conversation that reached back centuries, at least to Francis Bacon, and helped set the stage for the modern debates in the history and philosophy of science. Calculations—and the calculators themselves—have received far too little attention.

Calculation was an intricate, labor-intensive, and time-consuming job that required continual funding throughout the research project. In the late eighteenth century, most computers worked out of their homes to supplement their full-time work, often as schoolmasters or clergymen. As Mary Croarken has recently demonstrated, their work resembled other cottage industries of the time.[3] Nevil Maskelyne, for instance, used a network of computers distributed throughout England for the tables in the *Nautical Almanac,* the nation's ephemeris that mariners used to find their way in the open ocean. Maskelyne presided over the publication of forty-nine editions of the *Nautical Almanac* between 1765 and 1813, employing upward of thirty-five computers.[4] Though there were few calculators in the eighteenth century, research in positional and physical astronomy required their labor, and their profession grew along with the number of vessels sailing the world's coastlines and oceans.

Focusing on this rather mundane aspect of scientific practice highlights the organization and public patronage of science in the Victorian period. A major impediment to research in physical astronomy was the lack of competent and well-funded calculators, not a shortage of either theory or observations. Newton and Laplace had figured out the theory, and observations flooded into the Hydrographic Office. Difficulties usually arose once a research project, often sustained if not initiated by the calculators themselves, outgrew its confines in the private sphere. Local tide table calculators and almanac publishers guarded the methods and tables they used to produce accurate tide predictions. Between 1829 and 1833, Lubbock and Dessiou transformed these secret methods into public knowledge. Lubbock published them, along with Dessiou's valuable tables, in the *British Almanac* and the *Philosophical Transactions*, and they were quickly adopted by tide table makers around Britain as the foundation for their own predictions.[5] Dessiou's tables, moreover, formed the basis of the first tide tables published in the *Nautical Almanac*, the pride of the British Admiralty. Calculators and the tables they produced played a significant role in the transition of tidal studies from the coordination of the Society for the Diffusion of Useful Knowledge to their administration by the state.

A calculator's job entailed two steps. Once they were obtained, observations in physical astronomy first needed to be "reduced." In the study of the tides, this meant reducing all the observations to a standard zero, figuring in the position of the tide gauge and the geography of the port, and even accounting for the observer's personality. Only then could the "discussion" of the observations commence. In discussing the data, calculators formed tables (or, later, graphs) comparing the observations to other variables, such as the position of the sun and moon or the barometric pressure, to find regularities and, hopefully, lawlike relationships. Since theory guided this second stage of computation, the task of the calculator could be exceedingly tedious. Dessiou's tables relating the height of the tide to wind direction, for instance, entailed a mind-numbing task requiring no special intellectual ability above sifting tirelessly through all the tidal data and then repeating the process with a corresponding meteorological register. Many of the tables Dessiou formed, however, followed little theoretical guidance; they were founded on observation alone, using data that only Dessiou worked through. Sometimes Lubbock guided his work, and other times he was left to his own designs. One of the most significant steps in physical astronomy was comparing observation with theory, an essential job performed almost exclusively by hired calculators. Since calculators dealt directly with the observational data throughout the procedure, they often knew more about what was happening in the research (and why and how) than the natural philosophers themselves.

The history of tidal research in the early Victorian era can be viewed from the point of view of theorists such as Lubbock or from the less glamorous position of calculators such as Dessiou. The overarching narrative remains the same: a transition from its beginnings as a local and practical problem to the subject's appropriation by the scientific elite as a vital topic. Looking at the role of calculation, however, reveals that this transition did not always go smoothly, and the controversies that are otherwise missed turn out to be far more illuminating that the rather simplistic success story of the rise of the field. Questions arose, for instance, concerning the ownership of the tables used as the foundation for tidal predictions. After mariners and almanac publishers questioned the accuracy of the initial tide tables published by the Society for the Diffusion of Useful Knowledge, the Society elected Lubbock to superintend the calculations of the major tidal constants for the port of London. But the intellectual ownership of the tables published in the *British Almanac* for the following year was contested. They were based on observations reduced and calculated by Dessiou, a hired calculator, from rules devised by Lubbock, an elected member of the Society. Yet neither of these men could at that point claim the results as their own. The tables used to predict the tides were published at the suggestion, expense, risk, and reputation of Charles Knight, the publisher of the *British Almanac,* all under the coordination of Henry Brougham and the aegis of the Society for the Diffusion of Useful Knowledge. From the beginning, the study of the tides was both a collaborative and a hierarchical affair that complicated the question of intellectual ownership.

Hidden underneath the debates concerning the rights to the tables festered a far more acrimonious debate concerning payment for intellectual labor. When Charles Knight queried Henry Brougham about the scientific labor needed for the tide tables in the *British Almanac,* Brougham responded: "You shall have help enough. There's Lubbock and Wrottesley and Daniel and Beaufort—you may have your choice of good men for your astronomy and meteorology, your tides and your eclipses. Go to work, and never fear."[6]

Brougham's reply, filled with all the confidence and enthusiasm that came with the dawning of a scientific age, concerned itself with theory, not calculation, and it contrasts with Dessiou's plight as represented in his correspondence with his patrons. "I wish you could do something for poor Dessiou," Beaufort wrote to Thomas Coates, the secretary to the Society for the Diffusion of Useful Knowledge in 1833, referring to Dessiou's escalating financial troubles. "Forced to attend to his business here, he was obliged to hire a person to compute—besides his own nights and mornings which he devoted to the Tides; and he is at this moment in great embarrassment."[7] When Lubbock was brought in to revamp the tide tables for the *British Almanac*, he insisted that Knight hire Dessiou, for he knew the labor involved in research in physical astronomy. Theory and observations were certainly important, but dependable calculators were essential.

The title of this chapter refers to a legal claim made by Joseph Dessiou as he evolved from a calculator working for the *British Almanac* to a calculator working for the state. Even though the proprietors of the *Nautical Almanac* incorporated Dessiou's results into the nation's ephemeris, they continually refused Dessiou payment, and his claim eventually reached top officials in England, including Thomas Spring-Rice, the chancellor of the Exchequer. As late as 1851, Dessiou had yet to receive proper payment. "I am in arrears for Rent, and my Landlord threatens me with an Execution next week if not settled," Dessiou pleaded with Lubbock. "If you can now feel for my distressed situation and favour me with £10, it will save my furniture, relieve my distressed mind, and confer a favour, not to be forgotten."[8] Dessiou's financial difficulties demonstrate how intellectual ownership was intertwined with funding. Both had to do with the transition of the study of the tides from the private sector and the need for tables for the port of London, to the public sphere and the need to predict the tides for all the ports of Britain and its vast possessions. Stuck helplessly in the middle, attempting to save his furniture and ease his mind, labored the indefatigable Dessiou.

## Diffusing Useful Calculations

The study of the tides in England began with competition between almanac publishers. As changes to the river Thames caused difficulties in predicting its tides, heated debates spread into the public press. The Society for the Diffusion of Useful Knowledge had published tide tables in its *British Almanac* for 1828 and 1829. In the first year, the publishers were caught stealing their tables from other almanacs; in the second, they botched the results. Lubbock was one of the directors of the St. Katherine

Docks and therefore was financially vested in their safe use. He was recruited, however, because he was a good mathematician who had taken a personal interest in problems of physical astronomy. Lubbock graduated from Trinity College, Cambridge, as first senior optime in 1825. His earliest papers published in the *Philosophical Transactions of the Royal Society of London* and in the *Transactions of the Cambridge Philosophical Society* dealt exclusively with the latest research in astronomy, especially the theory of the motion of the moon and comets. Although other European nations had perfected Newtonian mechanics, in Newton's own country similar inquiries had faltered. According to George Biddell Airy, the future astronomer royal, "this reproach was removed from us by Mr. Lubbock; and if a scientific character is valuable to a nation, our gratitude is due to him."[9] Lubbock's research was the first to place England on a level "as to our pretensions to original investigations in the highest branches of mathematical philosophy, with the other nations of Europe." Though Airy's remarks must be seen in the context of the decline of science debate in England, his views demonstrate the type of mathematical acumen Lubbock brought to tidal analysis.

Lubbock used these talents to set to work on the theoretical investigation of the tides, the only researcher in England proposing such an agenda.[10] He attempted to acquire the tables that almanac publishers used as the basis for their predictions but was continually denied access to their methods. He was forced to start from scratch, which meant he needed data Dessiou could work with.[11] He suggested to Thomas Coates, the secretary of the Society, that blank "schedules" be sent out to as many observers as possible; they were to fill in the time and height of the tide and send the results directly to Coates. Lubbock realized that owing to the complex nature of the tides many of the returned forms would be useless, but he had a computer to help sift through them all, and at least at this stage of research, quantity, not quality, mattered. But he was already anxious about the cost of such a prodigious amount of work, and he left it to the Almanac Committee to decide on the feasibility of such an approach.[12] Knight's experience with the tides convinced him to spare the required funds at whatever cost. He had forms printed and sent to all the local committees of the Society throughout England and supplied more to the members of the General Committee to forward to local and provincial observers. An organized assault on the tides in England thus began, the first since Robert Moray had sent out similar forms almost 150 years earlier.

The first break came in early May 1829 when a letter from Isaac Solly, chairman of the London Docks Company, reached Lubbock at his

Figure 3.1. A photograph of John William Lubbock (1803-65) late in his life. Reproduced by permission of the Royal Society of London.

home.[13] Solly informed Lubbock that records of the tides had been kept at the London Docks since their opening in 1805, initiated by William Vaughan as one of the directors of the docks. They contained twenty-five years of continuous observations, night and day—exactly what Lubbock needed to calculate the corrected establishment of the port. Long-term observations, if reliable, also would allow him to extend the discussion to the effects of the declination and parallax of the moon, perfecting even Bernoulli's equations and moving far beyond what previous researchers had accomplished. Lubbock was "realizing their utility," and Solly was overjoyed that the tide observations, kept for so long and with such expense, were "likely to become of publick use."[14]

Pressing obligations at his father's bank occupied Lubbock's attention for the next month. He wrote to Dessiou in mid-June 1829 asking him to stop by the bank to talk about the calculations.[15] Dessiou was ill, however, bedridden with gout, and did not meet with Lubbock until the next month, when they first sat down to discuss the best way to proceed.[16] They decided that Dessiou would reduce all the observations for high water and compare them with the moon's position. Simply put, for each month of the year he formed a column of the times of high water, keeping separate the tides that took place on each day of the moon's age. He then added them all together and took the mean. Dessiou also computed the mean time of the moon's southing from the *Nautical Almanac* corresponding to the same dates, and the difference between the two gave the mean time that high water followed the moon's passage over the meridian. In this manner, he formed the essential table needed to predict the London tides, the "Time which the Moon's Passage through the Meridian precedes High Water," enabling Lubbock to place the establishment of the port of London on a firm empirical foundation. Based on over nine thousand observations and including a complete revolution of the moon's cycle (18.6 years), Dessiou's table averaged out all the inequalities found in the original observations owing to the wind, weather, barometric pressure, and human error.[17]

Dessiou had been appointed to the Hydrographic Office in February 1828, originally as a naval assistant whose sole task was to complete sailing directions.[18] His duties had also taught him much about the tides. He had recently completed the *Directions for Navigating in the North Sea* and *Directions for Navigating Throughout the English Channel*, both including intricate details concerning the tides on the coast of Britain.[19] He brought this experience to his collaboration with Lubbock. Lubbock had worked out the necessary calculations based on a detailed study of Bernoulli's equilibrium theory; Dessiou was left with the arduous task of rooting through the data, reducing the results, and forming the table for finding the corrected establishment of the port. He worked seven hours a day for twenty-eight weeks, for a total of 1,176 hours, work he performed out of his home before and after his work at the Hydrographic Office. He effectively held two jobs, one for the Admiralty, the other for the Society, making it necessary for him to hire an assistant for his tidal calculations.[20]

Dessiou suffered from gout most of his adult life. He was often forced to miss work, and in mid-September 1829 he had been absent from the Hydrographic Office for several weeks. His tidal computations also faltered. Lubbock was continually dissatisfied not with the quality of

Dessiou's work but with his speed. "I thought by now Dessiou would have completed the preparatory work he had to do," Lubbock ranted to John Wrottesley, a member of the almanac subcommittee. "I am quite disgusted—if it would not be better to employ some one else I think Mr. Jones who is an excellent computent would very willingly do it."[21] Jones, however, was employed by the *Nautical Almanac,* and they could not recruit him away from John Pond, the astronomer royal and superintendent of the nation's ephemeris. By the first week of October, the whole of the *British Almanac* was in the printer's hands except the section on the tides.[22] Owing to the extreme competition between almanacs, the Society always tried to publish its *British Almanac* on the same day that the Stationers' Company intended to publish its almanacs. For the year 1830 this was set for 23 November 1829. Though Lubbock had "got Dessiou to work again," it was evident that he needed help, and Lubbock hired two other calculators who worked for his father's bank.[23] Even Lubbock was forced to perform some of the computations during the evening, and through their combined efforts the tide tables were with the printer by mid-November and the *Almanac* appeared on schedule.

That Lubbock had taken over the theoretical analysis of the tides for the *Almanac* and *Companion* was readily apparent. In the previous two years, only a tide table for London was given, along with a table of the establishment of over one hundred ports, essentially copied from Mackay's *Complete Navigator.* For the first time, the *British Almanac* for 1830 provided an explanation for "High Water at London," a highly technical description outlining the work of Newton, Bernoulli, and Laplace.[24] A tide table followed listing the establishment for one hundred ports alphabetically from Aberdeen to Yarmouth Road. The ports were the same as those published in the preceding *British Almanac,* but the times were different. "We shall endeavour to verify these statements in succeeding years, as far as possible," Lubbock wrote, "and in the Companion to the Almanac will be given some plain and practical directions, by which persons residing at the Outports may calculate the Tides for themselves."[25] As promised, the *Companion* included a "disquisition" that delineated the connection between the theory and observations of the tides in the Thames estuary. As Lubbock noted, this was the first time this had ever been done.[26]

The Society used a highly effective advertising strategy for the sale of its *Almanac.* Thomas Coates wrote long announcements to the editors of the *Globe, Athenaeum, Morning Chronicle, Literary Gazette,* and *London Times.*[27] Sales in the first month totaled 41,000 copies; the *Companion* sold an additional 17,000 copies.[28] At Lubbock's request, copies were sent

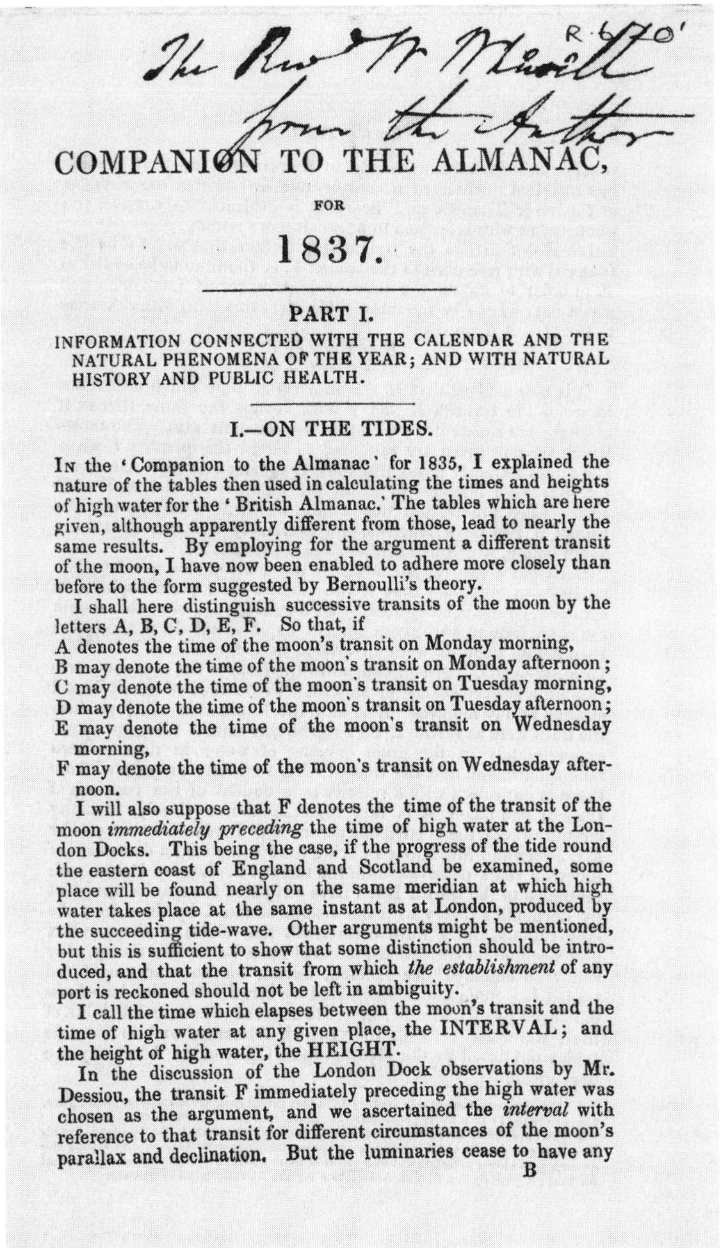

# COMPANION TO THE ALMANAC,

FOR

# 1837.

## PART I.

INFORMATION CONNECTED WITH THE CALENDAR AND THE NATURAL PHENOMENA OF THE YEAR; AND WITH NATURAL HISTORY AND PUBLIC HEALTH.

## I.—ON THE TIDES.

In the 'Companion to the Almanac' for 1835, I explained the nature of the tables then used in calculating the times and heights of high water for the 'British Almanac.' The tables which are here given, although apparently different from those, lead to nearly the same results. By employing for the argument a different transit of the moon, I have now been enabled to adhere more closely than before to the form suggested by Bernoulli's theory.

I shall here distinguish successive transits of the moon by the letters A, B, C, D, E, F. So that, if

A denotes the time of the moon's transit on Monday morning,

B may denote the time of the moon's transit on Monday afternoon;

C may denote the time of the moon's transit on Tuesday morning,

D may denote the time of the moon's transit on Tuesday afternoon;

E may denote the time of the moon's transit on Wednesday morning,

F may denote the time of the moon's transit on Wednesday afternoon.

I will also suppose that F denotes the time of the transit of the moon *immediately preceding* the time of high water at the London Docks. This being the case, if the progress of the tide round the eastern coast of England and Scotland be examined, some place will be found nearly on the same meridian at which high water takes place at the same instant as at London, produced by the succeeding tide-wave. Other arguments might be mentioned, but this is sufficient to show that some distinction should be introduced, and that the transit from which *the establishment* of any port is reckoned should not be left in ambiguity.

I call the time which elapses between the moon's transit and the time of high water at any given place, the INTERVAL; and the height of high water, the HEIGHT.

In the discussion of the London Dock observations by Mr. Dessiou, the transit F immediately preceding the high water was chosen as the argument, and we ascertained the *interval* with reference to that transit for different circumstances of the moon's parallax and declination. But the luminaries cease to have any

B

Figure 3.2. John William Lubbock's discussion of the tides in the *Companion to the Almanac* for 1837, signed by the author and presented to William Whewell. Reproduced from Whewell Papers, R.6.20/1.

to the London Docks, the St. Katherine Docks, and the East India Docks, and a special copy was reserved for Isaac Solly, who had sent Lubbock the London Dock observations.[29] Lubbock hoped that the descriptions of his work would induce others to begin making observations on their own. Lubbock and Beaufort had searched in vain for other long series of tide observations, but none seemed to exist. Once taken, observations were regarded as private property, often handed down from generation to generation for the formation of local tide tables. Lubbock hoped that his and Dessiou's efforts would change this secrecy.

In the meantime, Lubbock set Dessiou to work recalculating the tables already published in the *Almanac* with an additional six thousand observations. Dessiou was both to check the previous results, perhaps refining them, and to find the inequalities owing to the declinations and parallax of the moon. This took 2,100 hours of work spread over the first ten months of 1830. Charles Knight continued to pay Dessiou, assuring Lubbock that "any expense which has been incurred by Mr. Dessiou's calculations, for the purpose of attaining a greater correctness in the Tide Tables of the ensuing years, is most profitably laboured."[30] Dessiou's work, however, was far more costly than Knight had anticipated, and since Dessiou was tending to perfect the tables beyond what Knight thought necessary for a "popular Manual," Knight began to back off on his payments.[31]

Lubbock and Dessiou had worked on the tides for well over a year. They had found the corrected establishment of the port of London and were now perfecting the predictions in relation to the declination and parallax of the moon. Lubbock eventually published the results in the *Philosophical Transactions* in several installments in 1831 and 1832 and was rewarded with the Royal Society's gold medal. But at this point, in 1830, questions arose concerning the intellectual ownership of the tables. At first the debate revolved around the most appropriate venue for publishing Dessiou's work. Knight believed that the calculations, though valuable as a foundation for the results published in the *Almanac*, were far too abstruse for the *Companion*. He thus felt that even though the work was done for the Society and funded largely through him, its publication belonged elsewhere.[32] Lubbock, as the only researcher on the tides in England, hoped he could use the calculations as a foundation for a paper to be presented to the Royal Society. He had made the tides his own special research topic, but the results were not yet his to do with as he pleased.

In May 1830 Lubbock attended an almanac subcommittee meeting to present his case. He rarely attended these meetings, but he realized the importance of a personal request. The committee, in turn, refused to make any rash decisions and referred the question back to Knight, who again said he considered the results too recondite.[33] Lubbock wrote to Coates in June, eager "for the permission to lay before the Royal Society the results obtained by Mr. Dessiou."[34] Up to that point at least, tidal studies were not Lubbock's prerogative but belonged as much to Knight, Beaufort, Brougham, and Dessiou. Only after the committee agreed in writing to allow Lubbock to use the tables could he publish his analysis in the *Philosophical Transactions*.

Lubbock had scored what he believed was a small victory, but in reality he had caused innumerable difficulties for himself and for Dessiou. Only Knight would come out a winner. The problem that arose, one it seems Lubbock had not counted on, centered on the funding of Dessiou's calculations. Now that Knight had offered the intellectual property to Lubbock he no longer shouldered the burden of funding Dessiou's time-consuming and costly computations (and most certainly this was the reason for his generosity). The *Almanac* for 1831 was to be used in all government offices, and Knight was making a handsome profit from its sales. He did not want to see his earnings wash away with the tides. It seemed to Knight that now that Dessiou had computed the necessary tables to be used in the *British Almanac*, Knight's job as patron was complete. Since Lubbock was the one using the calculations for his own benefit, he should pay Dessiou.

Lurking beneath Knight's decision to let Lubbock publish the results of Dessiou's labors, therefore, lay the far more formidable problem of paying for Dessiou's time. The question changed from ownership of intellectual property to research costs and payment for labor. As is still familiar to most experimental physicists today, Lubbock had the methods and the observations, but now he required computer time. After a slow start, Dessiou had proved himself a competent worker, willing to spend endless hours on minute calculations early in the morning and late at night. But he needed to be paid for his services. Though this is an isolated example focusing on only one research project, the problem was not unique to the study of the tides. The more research in physical and observational astronomy relied on massive amounts of observational data—which historians have noted increased in the second quarter of the nineteenth century—the more computers needed to be found, paid, and if competent, retained. This is true of research as far-ranging as trigonometric surveys on land, astronomical surveys of the heavens, meteorological surveys of the sky, and hydrographic surveys of the sea. An understanding of the earth, air, and ocean all required the reduction and discussion of massive and unprecedented amounts of observational data. A good computer was a valuable asset, and if Lubbock was to continue to profit from Dessiou's labors, he needed to find another avenue of funding.

## Improving the *Nautical Almanac*

The *British Almanac* of the Society for the Diffusion of Useful Knowledge helped raise the level of almanac publications throughout England. Surprisingly, however, the only state-sponsored almanac, the *Nautical Almanac,* had declined in quality. Published three years in advance for navigators bound for distant regions, the nation's ephemeris had long been one of the treasures of the British navy; Britain depended on good tables to foster the wealth and strength of its global maritime operations. Now, as the editor of the first volume of the *Nautical Magazine* maintained, the *Nautical Almanac* had fallen into disrepute "arising from the numerous errors which have been found in it, to the manifest risk of navigators, and the discredit of the first maritime nation in the world."[35] Indeed, most navigators, including those in Great Britain, had turned to foreign publications.[36] To the chagrin of British mariners and astronomers alike, what had once been a bright light of British science now cast a shadow on Britain's burgeoning maritime hegemony. In March 1830 a letter to the *London Times* signed "Columbus" noted in dismay that all the lunar distances from the sun were incorrect for the year 1833, adding that "it must be a little mortifying to us to know that we are indebted to the Americans . . . for this correction of our national ephemeris."[37] Henry Brougham agreed, confidently stating that the *Nautical Almanac* was "susceptible of considerable improvement," an outrage in the age of scientific attainment. "The excellence of modern instruments, the superior education of the present race of seamen, and the zeal with which practical astronomy is now cultivated, loudly claim the interference of the government, in order to render it worthy of the national character."[38]

Nationalism could be a powerful motivating force, and as Brougham had anticipated, the government was forced into action. The Lords of the Admiralty consulted Francis Beaufort, the hydrographer to the Admiralty, who urged them to refer their requests to the Council of the Astronomical Society, "as it contained almost all the practical astronomers of the country and some seamen."[39] The Council of the Astronomical Society did what every respectable early Victorian council did: it appointed a committee.[40] The committee brought together forty of the sharpest British minds then focused on astronomy, including George

Biddell Airy, Charles Babbage, Francis Beaufort, John Herschel, John Pond, John Lubbock, and William Stratford. They submitted their report to the Admiralty listing various alterations and additions they believed were essential to reviving the *Nautical Almanac,* including a "Table of the mean time of high water at London Bridge for every day in the year, and also at the principal ports at the time of new and full moon." The tides represented an area in certain need of improvement for one important reason: there were no tide tables at all in the *Nautical Almanac.* The Lords of the Admiralty directed John Pond, astronomer royal from 1811 to 1835 and superintendent of the *Nautical Almanac,* to carry the changes into effect.[41] Since this included incorporating the latest calculations on the tides, Lubbock thought he had found a reliable source of funding for Dessiou.

While Dessiou was shackled to his desk perfecting the corrections for lunar parallax and declination, Lubbock was out politicking for funds, soliciting an official request from the Council of the Royal Astronomical Society to help with the tide tables for the *Nautical Almanac.* Lubbock immediately wrote to John Pond to explain the time-consuming calculations required to formulate tide tables. Feigning ignorance, he then asked Pond what measures his office had already taken. Lubbock knew that Pond had no such tables, and he knew he had no way of forming them without using Dessiou's work. Lubbock therefore offered to provide Pond with Dessiou's labor as long as Dessiou was paid. Lubbock was a shrewd businessman.

Dessiou's financial claim against the *Nautical Almanac* rested largely on Pond's response (or rather, on the differing interpretations of that response). "As I am directed by my instruction to insert whatever useful matter I can collect relative to the time of high water and on the tides in general," Pond responded encouragingly, "I of course cannot but consider myself authorized to defray any reasonable expense that may be incurred in procuring the same. I beg therefore you will consider the sum of £50 or any part of it as at your disposal for this purpose."[42] Lubbock, unfortunately, took "any reasonable expense" to mean that he was not limited to £50, even though Pond seemed to cap the amount. Such an interpretation of Pond's offer was a stretch at best, and Lubbock would later suggest that Pond had also intimated in a private conversation that the Admiralty would defray further costs—a conversation that was never corroborated.

Regardless of what took place behind closed doors, Lubbock certainly acted as if Pond had offered him unlimited funding. He shot off a letter to Whewell explaining his new plan for the tides. "Dessiou is going on in the formation of tables shewing the effect of the changes in the Moon's Declination on the time and Height of High water and Mr. Pond has promised me to discharge any reasonable expense that this requires from the funds at his disposal for the Nautical Almanac, so that I hope soon to have a paper on this subject for the Royal Society."[43] Encouraged by the Admiralty's orders to revise the *Nautical Almanac*, Lubbock believed he was in a position to expand his own research. No longer limiting himself to either the *British Almanac* or the London tides, Lubbock sought to expand his tidal analysis to all the major ports in England and into Britain's most prestigious scientific journal. Lubbock wrote to Beaufort shortly after Pond's enthusiastic response, saying he was persuading Airy "and some of my friends" to petition the Admiralty to have the tides observed "at all places under their control for *one given month*."[44] Beaufort did not think a private memorial from Airy, Lubbock, or any other individual would have much effect, "but I do think that a strong representation from the Royal Society and Council would succeed."[45] Beaufort also spoke with Davies Gilbert, president of the Royal Society, a conversation he referred to in a letter to Lubbock: "I assured him, as I now do you, that no one can be impressed with a stronger conviction than myself of the urgent necessity of acquiring proper data for the construction of our Tide Tables, that I considered it to be a national object, and the Government should the more readily take it in hand when they found a person qualified like you, disposed to undertake a principal part of the labour without expense to the country."[46]

Beaufort was a forceful man in a powerful position within the Admiralty, but he was careful where he used his influence. He ended his letter to Lubbock with just this notion: "But I likewise explained to Mr. G[ilbert] that, though I would give the scheme every aid in my power, I could not make the first move, and that should be done either by bringing the subject forward to the House of Commons, or by direct overtures to the Admiralty."

A direct petition to the Admiralty was tricky, but Lubbock believed he was in the perfect position to make such a request, perhaps even better than Beaufort. He had several strong points in his favor. First, observational data on the tides was important to the military and commercial interests of Britain. Second, Lubbock was a prominent member of London society. His father had been recently knighted, adding to his chances of success. Third, and perhaps most important, he held influential positions within the Royal Society. Lubbock was both treasurer and vice president, and if Beaufort believed a petition from the Royal Society would work, there was no one better situated to pull it off.

Lubbock proposed the motion to the Royal Society in August. The secretary, Peter Roget, sent the request to the Lords of the Admiralty, who forwarded it to the Navy Board, along with a copy of the council minutes.[47] The Navy Board then ordered John Barrow, secretary of the Admiralty, to have the tides measured at Sheerness, Woolwich, Portsmouth, and Plymouth, a major step in government involvement in tidal studies.[48] From that point on, observations of the tides were kept at these four dockyards by direct order of the Admiralty, a charge that continues to this day.

The sequence leading up to the Navy Board's orders is significant. Beaufort had stepped aside, pretending impotence, forcing the Royal Society to put its name on the line. That within ten days of the resolution they were rewarded with direct orders to have the tides observed at four of Her Majesty's dockyards gives reason to rethink the decline of the Royal Society during this period. A large portion of its fellows were retired naval officers, members of Parliament, and otherwise influential members of London society. Only a very few practiced science. But it was just this membership that placed the Royal Society in such an advantageous position to make requests from the government, especially relating to navigation. The Royal Society's strength resided in the amateur nature of its membership, helping sustain the link between the scientific community, the Admiralty, and the British government.

Lubbock had orchestrated another difficult maneuver, but an order to have observations made was only the beginning. The first returns from the naval dockyards were a jumbled, inconsistent mass of data.[49] What Lubbock wanted was the measure of high water from a known level, but he had failed to stipulate that in his directions. At Woolwich, the dockyard attendant, Thomas Brown, sent the "mere rise from low water."[50] Because the tide gauge had been broken, Brown had long since dismissed the person responsible for taking the observations.[51] At Plymouth no accurate tide gauge existed at all; a graduated pole and float had been placed there earlier, but it did not work "freely and efficiently."[52] At Portsmouth, the tide gauge was graduated only to feet, "and the observer must guess the rest."[53] Sheerness, it seems, was the only place that offered the potential for good observations, having erected a self-registering tide gauge. Lieutenant John Washington forwarded the curves traced by the gauge to the Admiralty, but they did not offer any information on the height of the tide from a zero point. These inconsistencies demonstrate that direct orders from the Admiralty were only the first step. Competent observers needed to be found, forms and directions lithographed, tide gauges constructed, and a calculator hired to reduce the observations as part of a standardized system.

All this work seemed to Lubbock to conform to the Admiralty's requirement that the *Nautical Almanac* include not only the tides for London, but also the tides for "the principal ports." Lubbock believed the Admiralty should shoulder the responsibility for computation, and he asked Francis Beaufort if Dessiou could use some of his time at the Hydrographic Office for this task. Dessiou's time was in short supply, however, and his work in the Hydrographic Office was already suffering from too much tidology. Though Beaufort appreciated Dessiou's financial position, he could not spare the time. Lubbock retorted that Dessiou's work, so important for the advance of physical astronomy and the renewal of the *Nautical Almanac,* could not move forward without financial assistance. "The labor which I have gone through on the subject is considerable and must be more before we have done," Lubbock despaired. "This I give most willingly but I shall think it hard to have to pay Mr. Dessiou and be a loser in money."[54]

Dessiou had been working for over a year on the corrections owing to the moon's declination and parallax, and by late 1831 he had completed another 2,136 hours. He had received his last payment of £25 from Knight but had yet to get the £50 Pond had promised. Beaufort realized the utility of Dessiou's work for the Admiralty but was reluctant to commit funds to a project that seemed to have no clear end. "If the calculations bona fide could be performed in a fortnight," Beaufort responded, "I should feel that the subject was sufficiently important to justify the sacrifice of [Dessiou's] duties for that period."[55] For Lubbock, of course, a fortnight simply would not do. Lubbock continued to profit from Dessiou's labor, presumably relying on Pond to reimburse Dessiou according to his earlier offer. Then the unexpected happened. Before Dessiou's corrections had been completed and, more important, before he had been paid the promised £50, John Pond resigned as superintendent of the *Nautical Almanac.*[56] All of Lubbock's funding strategies for Dessiou seemed to be evaporating.

## The Quarrel Begins

The tide tables published by the Society for the Diffusion of Useful Knowledge were the first step in placing the theory of the tides on a firm theoretical foundation. Letters from around England testified to their usefulness and accuracy. William Pierce, a foreman at the London Docks, compared the time of high water with the times calculated in the *British Almanac* and found them "always very near the real time, often

# TIDE-TABLE FOR MARCH, 1833.

## Mean Time.

| Week Day | Month Day | PLYMOUTH Dock Yard. Morn h. m | Aft. h. m | PORTSMH. Dock Yard. Morn h. m | Aft. h. m | SHEERNESS Dock Yard. Morn h. m | Aft. h. m | LONDON Bridge. Morn h. m | Aft. h. m | ABERDEEN. Morn h. m | Aft. h. m | LEITH. Morn h. m | Aft. h. m | Month Day | D's Age at Noon |
|---|---|---|---|---|---|---|---|---|---|---|---|---|---|---|---|
| F. | 1 | .... | 0 35 | 6 24 | 7 8 | 7 34 | 8 22 | 8 20 | 9 3 | 8 32 | 9 17 | 10 7 | 10 52 | 1 | 9·8 |
| Sat. | 2 | 1 23 | 2 7 | 7 50 | 8 31 | 9 13 | 9 54 | 9 55 | 10 38 | 9 57 | 10 32 | 11 32 | .... | 2 | 10·8 |
| S. | 3 | 2 47 | 3 26 | 9 8 | 9 41 | 10 32 | 11 4 | 11 22 | .... | 11 4 | 11 32 | 0 7 | 0 39 | 3 | 11·8 |
| M. | 4 | 3 58 | 4 26 | 10 10 | 10 37 | 11 31 | 11 57 | 0 4 | 0 39 | 11 58 | .... | 1 7 | 1 33 | 4 | 12·8 |
| Tu. | 5 | 4 52 | 5 15 | 11 1 | 11 23 | .... | 0 22 | 1 10 | 1 35 | 0 23 | 0 47 | 1 58 | 2 22 | 5 | 13·8 |
| W. | 6 | 5 39 | 6 2 | 11 46 | .... | 0 45 | 1 8 | 2 2 | 2 34 | 1 8 | 1 29 | 2 43 | 3 4 | 6 | Ful |
| Th. | 7 | 6 23 | 6 39 | 0 8 | 0 30 | 1 27 | 1 46 | 2 57 | 3 18 | 1 51 | 2 12 | 3 26 | 3 47 | 7 | 15·8 |
| F. | 8 | 7 0 | 7 18 | 0 49 | 1 9 | 2 5 | 2 23 | 3 41 | 4 4 | 2 31 | 2 51 | 4 6 | 4 26 | 8 | 16·8 |
| Sat. | 9 | 7 36 | 7 53 | 1 28 | 1 47 | 2 41 | 3 0 | 4 23 | 4 41 | 3 10 | 3 28 | 4 45 | 5 3 | 9 | 17·8 |
| S. | 10 | 8 11 | 8 29 | 2 9 | 2 27 | 3 20 | 3 41 | 4 57 | 5 14 | 3 47 | 4 7 | 5 22 | 5 42 | 10 | 18·8 |
| M. | 11 | 8 48 | 9 8 | 2 48 | 3 10 | 4 1 | 4 21 | 5 31 | 5 47 | 4 27 | 4 48 | 6 2 | 6 23 | 11 | 19·8 |
| Tu. | 12 | 9 29 | 9 53 | 3 33 | 3 56 | 4 42 | 5 6 | 6 1 | 6 18 | 5 11 | 5 37 | 6 46 | 7 12 | 12 | 20·8 |
| W. | 13 | 10 19 | 10 49 | 4 23 | 4 53 | 5 31 | 5 59 | 6 40 | 7 3 | 6 7 | 6 43 | 7 42 | 8 18 | 13 | LQ |
| Th. | 14 | 11 24 | .... | 5 25 | 6 3 | 6 33 | 7 6 | 7 28 | 7 59 | 7 23 | 8 5 | 8 58 | 9 40 | 14 | 22·8 |
| F. | 15 | 0 3 | 0 46 | 6 41 | 7 17 | 7 44 | 8 24 | 8 39 | 9 31 | 8 48 | 9 28 | 10 23 | 11 3 | 15 | 23·8 |
| Sat. | 16 | 1 26 | 2 0 | 7 52 | 8 25 | 9 5 | 9 41 | 10 24 | 11 7 | 10 4 | 10 34 | 11 39 | .... | 16 | 24·8 |
| S. | 17 | 2 33 | 3 4 | 8 55 | 9 22 | 10 11 | 10 37 | 11 43 | .... | 11 1 | 11 23 | 0 9 | 0 36 | 17 | 25·8 |
| M. | 18 | 3 31 | 3 55 | 9 46 | 10 7 | 11 1 | 11 29 | 0 16 | 0 45 | 11 44 | .... | 0 58 | 1 19 | 18 | 26·8 |
| Tu. | 19 | 4 15 | 4 35 | 10 26 | 10 45 | 11 39 | 11 59 | 1 9 | 1 29 | 0 3 | 0 21 | 1 38 | 1 56 | 19 | 27·8 |
| W. | 20 | 4 54 | 5 12 | 11 3 | 11 20 | .... | 0 18 | 1 48 | 2 4 | 0 38 | 0 53 | 2 13 | 2 28 | 20 | 28·8 |
| Th. | 21 | 5 29 | 5 47 | 11 36 | 11 54 | 0 35 | 0 54 | 2 21 | 2 40 | 1 8 | 1 23 | 2 43 | 2 58 | 21 | N M |
| F. | 22 | 6 4 | 6 22 | .... | 0 11 | 1 10 | 1 26 | 2 56 | 3 11 | 1 38 | 1 52 | 3 13 | 3 27 | 22 | 1 |
| Sat. | 23 | 6 37 | 6 53 | 0 28 | 0 45 | 1 42 | 1 58 | 3 26 | 3 40 | 2 7 | 2 22 | 3 42 | 3 57 | 23 | 2 |
| S. | 24 | 7 8 | 7 24 | 1 2 | 1 18 | 2 13 | 2 29 | 3 57 | 4 11 | 2 38 | 2 54 | 4 13 | 4 29 | 24 | 3 |
| M. | 25 | 7 40 | 7 58 | 1 35 | 1 52 | 2 46 | 3 5 | 4 26 | 4 41 | 3 11 | 3 30 | 4 46 | 5 5 | 25 | 4 |
| Tu. | 26 | 8 16 | 8 35 | 2 12 | 2 32 | 3 25 | 3 47 | 4 55 | 5 14 | 3 50 | 4 12 | 5 25 | 5 47 | 26 | 5 |
| W. | 27 | 8 56 | 9 18 | 2 55 | 3 19 | 4 9 | 4 32 | 5 26 | 5 46 | 4 36 | 5 2 | 6 11 | 6 37 | 27 | 6 |
| Th. | 28 | 9 44 | 10 11 | 3 44 | 4 13 | 4 57 | 5 24 | 6 6 | 6 29 | 5 31 | 6 6 | 7 6 | 7 41 | 28 | FQ |
| F. | 29 | 10 45 | 11 27 | 4 44 | 5 22 | 5 56 | 6 35 | 6 56 | 7 28 | 6 45 | 7 28 | 8 20 | 9 3 | 29 | 8 |
| Sat. | 30 | .... | 0 18 | 6 7 | 6 53 | 7 19 | 8 5 | 8 6 | 8 56 | 8 16 | 9 1 | 9 51 | 10 36 | 30 | 9 |
| Su. | 31 | 1 6 | 1 50 | 7 35 | 8 15 | 8 53 | 9 37 | 9 49 | 10 39 | 9 42 | 10 18 | 11 17 | 11 53 | 31 | 10 |

The times of high-water, nearly, at other places on the coast, may be found with the assistance of the above table within certain limits. Thus, the times in the Plymouth-Dock column are to be used for all places between the Land's End and Lyme Cob ; and those in the Portsmouth column, for all places between Portland Bill and Beachy Head ; by adding or subtracting the time opposite each place, according to the sign + or —.

The times of high-water at Plymouth Dock-Yard are to be used with the difference against the following places, to find the time of high-water there on the same day :—

|  | h. m. |  |  | h. m. |
|---|---|---|---|---|
| Mounts Bay and Lizard | — 1 3 | Eddystone | | — 0 18 |
| Falmouth Harbour | — 0 18 | Dartmouth and Torbay | | + 0 27 |
| Fowey Harbour | — 0 18 | Exmouth | | + 0 52 |
| Cawsand Bay | — 0 10 | Lyme Cob | | + 0 27 |

The times of high-water at Portsmouth Dock-Yard are to be used as above, for the following places :—

|  | h. m. |  |  | h. m. |
|---|---|---|---|---|
| Portland Bill | — 6 10 | Cowes | | — 0 55 |
| Weymouth Harbour | — 5 10 | Southampton | | — 1 0 |
| Christchurch and Poole Harbours | — 2 50 | Bembridge Point | | — 0 40 |
| Needles Point | — 1 55 | Selsea and Arundel Harbours | | + 0 5 |
| Hurst Chamber | — 1 40 | Shoreham Harbour | | — 0 25 |
| Lymington | — 1 25 | Beachy Head | | — 1 25 |

Figure 3.3. After working for the Society for the Diffusion of Useful Knowledge to incorporate tide tables into their *British Almanac*, Joseph Foss Dessiou worked on the first set of tide tables published by the British Admiralty. His tide table for March 1833 is reproduced here, as published in the *Nautical Magazine* 2 (1833): 121. His results were then incorporated into the *Nautical Almanac* without proper payment. University of Minnesota Libraries.

exact." He added that the tables "would be used by all the Pilots and most or all other persons who use Tide Tables, [and] particularly useful to all persons connected with Dry Docks, and bays where Vessels are laid to repair as well as at Wet Docks."[57] Lubbock and Dessiou had advanced the study of the tides to the point where minute corrections owing to the moon's declination and parallax could be detected, and though both were losers in money, they were on their way to bringing the tides to national attention.

Lubbock was without doubt quick-tempered, easily offended and quicker still to offend others. His work in physical astronomy had led to several confrontations with other astronomers in England. He often felt "ill-used" and did not hesitate to retaliate. His unsuccessful attempt to succeed John Pond as superintendent of the *Nautical Almanac* serves as an example. Lubbock badly wanted the position, a Crown appointment that he felt he deserved because of his work on the tides and his earlier work on lunar tables published in the *Proceedings* and *Philosophical Transactions* of the Royal Society. The news that the position was going to William Stratford hit Lubbock hard. The day the appointment was made, he shot off angry letters to both Beaufort and Whewell, exclaiming that Stratford "has not the mathematical knowledge of a second year man at Cambridge."[58] Beaufort apologized to Lubbock in person, suggesting that he did not realize Lubbock was so interested in the post. "With respect to the Superintendence of the Nautical Almanac," Lubbock retorted, "and in reference to our conversations yesterday, as I have for some time been engaged in investigations connected with Astronomy which have required most laborious calculations you could hardly suppose that I should have objected to become the successor of such men as Dr. Maskelyne, Dr. Young and Mr. Pond in a situation which would have enabled me to pursue the same subjects with more effort and I trust with more utility to the public."[59]

Dessiou, of course, had undertaken most of the "laborious calculations," but Lubbock was not above using that labor for his own ends. Those ends, moreover, would have helped Dessiou immensely. As superintendent, Lubbock would have inherited numerous calculators already working for the nation's ephemeris, thus ending his incessant haggling for funds. And Dessiou would have received the payment he was due.

Lubbock's resentment toward Stratford only intensified, with harsher words and deeper-seated animosity, when Stratford refused to pay Dessiou the £50 Pond had promised him for using his tide tables in the *Nautical Almanac*. Stratford had served as secretary of the Royal Astronomical Society, the same body that suggested to the Admiralty that tide tables be included. Indeed, he was a member of the committee that made the recommendations. But once he became superintendent he seemed determined, according to Lubbock, to "reduce the expense of the Nautical Almanack to the lowest possible sum." Lubbock believed he should go about it by other means than refusing "Mr. Dessiou's claim."[60]

Stratford helped place the *Nautical Almanac* back on par with the ephemerides of other nations, partly by reorganizing the computing staff. A year after his appointment as superintendent, he established the Nautical Almanac Office as a separate institution, replacing the home-based system of computers adopted by Maskelyne. No longer were calculations done piecemeal by schoolteachers or clergymen; they were now accomplished by full-time paid computers working under Stratford's direct supervision. Stratford's truck, obviously, was not with paying computers, but rather with Lubbock and the payment of calculations outside his office. But what really irked Lubbock was how Stratford supported his refusal to pay Dessiou the £50 that Pond had promised him. He suggested that Dessiou's tables were worthless. Stratford set out to compare the predicted times of all the major tide tables for the port—including those published by Bulpit, Epps, Jones, Gregory, and Dessiou—with observed times at the London Docks. "We seem to be all at Sea on the Subject," he wrote disparagingly to Lubbock; "all appear to be bad."[61] Lubbock retorted that Stratford was biased in his comparisons, unwilling to accept the merits of Dessiou's work because he was unwilling to pay for his labor. Lubbock predicted to Beaufort that the tables published in the next *British Almanac* would be far better than those in the *Nautical Almanac*. That a private society could publish tide tables for the most important port in London that were more accurate than those used by the greatest sea voyaging nation in the world was sure to catch the attention of the nation's hydrographer.

Whether or not Beaufort took this bait, he was more than ready to have the quarrel end, and he wrote a pointed letter to Stratford in October 1833 trying to head off the confrontation. "As it would give me great pain to have any thing like complaint and squabble in any of the dep[artments] with which I am connected," Beaufort warned, "you would very much oblige me by giving the matter at once your most favorable consideration and thereby stop all further misunderstanding."[62] Stratford replied to Beaufort in early November, pleading his own case against Dessiou and Lubbock. The crucial piece of evidence was the letter from Pond to Lubbock in early 1831. Stratford quite justifiably believed Pond had set aside £50 as a limiting amount, and then to be paid only if the tables proved superior to others. Based on his own comparison of observed and predicted tides, he found not a "shadow of justification for Mr. Lubbock's claim."[63] In a condescending manner, Stratford intimated that he had felt sorry for Dessiou and proposed using his tables only "with a view of rendering him a service," provided his results stood the test of comparison. "Far from deeming his tables capable of giving a degree of accuracy superior to the common methods," Stratford found only that they were "not more inaccurate than others."[64] The focus of the controversy had changed. Not only had Stratford refused to pay Dessiou, he now insisted he had never used his work in the first place.

When Beaufort sent Stratford's response to Lubbock, what immediately caught Lubbock's eye was Stratford's reference to "the common method." As Lubbock's and Dessiou's work had demonstrated, it was just this common method that was most of all uncommon. No one, not Bulpit nor White nor anyone else, had ever published such methods, and Lubbock knew that Stratford could not get access to them without considerable expense, if at all. They were industrial secrets. The methods used in the popular almanacs and navigation manuals were based on Bernoulli's work, but without the necessary corrections, they produced erroneous predictions. Lubbock had had to invent the "common method" from scratch, beginning by determining the establishment of the port. His methods were the only ones published, and thus the only ones Stratford would have had access to. He was so confident on this point that he offered a challenge: "Let Mr. Stratford say what he means by the common methods, and let him say how he could have got the Tides done for the Nautical Almanack gratuitously otherwise than by refusing to pay Dessiou. . . . Let him try. . . . We will calculate for the British Almanack for that year[,] he shall calculate by the common methods, without making use of our tables directly or indirectly and when the time comes we will compare the two."[65]

Working backward from the predicted times published in the *Nautical Almanac*, Lubbock calculated the tables Stratford must have used to acquire his tidal predictions. They corresponded exactly to the tables Dessiou had produced, proving that Stratford had quite consciously used his work. He simply would not say as much publicly, and this, according to Lubbock, amounted to theft.

Lubbock began a blistering campaign of letters to all those with a stake in Dessiou's claim, reiterating to both Whewell and Coates that Dessiou's tables "were the only ones published and they were undertaken upon the faith of Mr. Pond's letter." He offered proof that his tables were more accurate than any others through his own set of comparisons. "I have recalculated three months formerly compared by Mr. Stratford. . . . I find my results are more accurate than Mr. Bulpit's[;] Dr. Gregory's are quite out of the way. Mine are the only published Tables, for those which have been recopied from Bernoulli have never been compared with theory but by me and are quite useless."[66] That they were more correct than others was beside the point. Whether they were better or not, Stratford had used them and continued to use them without acknowledgment or payment.

Almost everyone involved seemed to think the matter should be laid to rest. "I have nothing to say for Stratford," Whewell warned, "but I would not quarrel with him if I could help it. Among other bad consequences I am afraid the cause of Tidology would suffer."[67] Whewell realized that they would need Stratford, if only for his stable of computers. Coates worried less about the science of the tides and more about the reputation of the Society for the Diffusion of Useful Knowledge. He suggested that the Society simply pay Dessiou and "for its own quiet and dignity's sake let the matter drop."[68] Lubbock, however, was unable to let things go, since the dispute had evolved beyond the payment of Dessiou's wages. It was now part of a larger controversy over the payment, in general, of calculators working for private individuals but for the good of the state. If he lost this battle with Stratford, he would certainly lose others, and he was unwilling to let his research suffer simply because Stratford insisted on "throwing [him] overboard."[69] For Lubbock, the important point was not that he demanded the government pay for his own research project. That was anathema according to prevailing opinion on government's role in funding scientific research. Rather, he couched his arguments for paying calculators in terms of the safety of British navigation in general.

The disagreement between Lubbock and Stratford had reached a stage where, according to Lubbock, British maritime interests were at stake. "Now that I have very strongly remonstrated against the injustice of employing our Tables and getting nothing for them," Lubbock wrote to Whewell, Stratford would be forced to use other tables "copied after Bernoulli, which must of course give quite erroneous results."[70] He hoped that Whewell would assist them by bringing the subject up to Beaufort, "because you are the only person who is aware of the quantity of work Dessiou has gone through, and the inaccuracy of previous published Tide Tables." As will be discussed in subsequent chapters, at just this time Whewell was becoming more involved in tidal analysis, confronting the same difficulties in acquiring adequate computers. He appreciated both Dessiou's plight and Lubbock's insistence on payment. As the study of the tides had changed into a national project, so too had Dessiou changed into an operative working for the state.

Lubbock's letter-writing campaign worked, as both Whewell and Coates wrote directly to Beaufort in support of Dessiou's claim against the *Nautical Almanac*. In the interests of the Society, Coates visited the Hydrographic Office to see Beaufort, but his conversation with the hydrographer gave him little reason to suspect that Dessiou would ever recover anything.[71] Stratford continued to insist he had not used Dessiou's tables. At Coates's and Whewell's urging, Beaufort set up a meeting with Stratford to broach the idea of paying computers outside Stratford's office. The result, unfortunately, was that Stratford staunchly defended his refusal. He would use his own computers, not Lubbock's. Beaufort felt there was little left that he could do, and though he argued strongly against any such move, he informed Lubbock that a direct appeal to the Lords of the Admiralty seemed to be the only recourse.[72] Coates gave Lubbock similar advice. Owing to Stratford's determination "to avail himself of your labors and Mr. Dessiou's without payment," he thought Lubbock could do no better than appeal directly to Parliament.[73]

Lubbock's quick temper got the best of him. Stratford, though he claimed he had not used Dessiou's tables, had kept copies of them in the Nautical Almanac Office, and Lubbock demanded their return. Since they were not yet published, they were not public property. Let Stratford work his own computers all he wished, but he would have to do without Dessiou's tables. Once he tried that, Lubbock ranted to Beaufort, "he will reprise Mr. Dessiou's claim to a very considerable recompense from the government, besides the £50 on the old score."[74] The situation had reached a boiling point, and Lubbock was now willing to take the matter as far as possible. "Should I fail in an application to the Lords of the Admiralty, I may still petition Parliament," he told Beaufort, exposing his "severe mortification" that Stratford had found a way to "retard considerably the perfection of that branch of Physical Astronomy which is now the weakest."[75]

In his third publication on the tides in the *Philosophical Transactions of the Royal Society,* Lubbock announced that Dessiou, "with undaunted perseverance," had finally completed the discussion of over six thousand more observations at the London Docks, placing the corrections for the moon's parallax and declination "upon a sure basis." These were the calculations that had gone beyond the commission of the Society, and that now, according to Lubbock's revised narrative, he had undertaken for the purposes of the state-sponsored *Nautical Almanac.* "But these cannot be published," Lubbock continued, "unless he is fortunate in meeting with more encouragement than he has hitherto experienced."[76] Lubbock was true to his word; he was barring the publication of Dessiou's results. He wrote to Spring-Rice, the chancellor of the Exchequer, in mid-November for advice on how to bring the question before the government. Informing Whewell of his letter to Spring-Rice, he reiterated that he would not publish Dessiou's new tables until he had received "something tangible for him."[77] Because the labor involved was so "prodigious," if he did not receive a positive reply he would "offer the Tables to the Bureau des Longitudes." Lubbock was treading on thin ice. France, though no longer a military enemy, was Britain's greatest commercial and military rival. The Revolutionary War and the Napoleonic Wars were still fresh in people's minds, and offering his tidal work specifically on the Thames estuary to the French Bureau instead of the British Admiralty verged on treason.

At that point, in late November 1833, correspondence between Lubbock and Stratford came to an abrupt halt. Lubbock requested that all his communications go through Beaufort, while Stratford simply skipped over Lubbock entirely and referred all his questions on the tides to Whewell.[78] And he had questions aplenty. As Lubbock had warned, Stratford was being severely pricked by the thorniest problem in physical astronomy. Without Dessiou's tables, even his battery of computers was of little use, and he implored Whewell for his recommendations in several letters in late November and early December.[79] Whewell judiciously played both sides, offering suggestions but also acting surprised that Stratford had not first consulted Lubbock. "I hoped you would have done this in the first instance not only because we owe it entirely to him and Mr. Dessiou that we are able to construct good tide tables by means of published materials and methods, but also because his directions might I think very probably be better than mine."[80] Despite the string of grammatical hedges—"might I think very probably"—this was as direct a support for Lubbock and Dessiou as Whewell could give. Whewell acted far more discreetly than Lubbock, appreciative of Dessiou's position and the progress of physical astronomy but equally aware that neither could be furthered without Stratford's help.

## The Quarrel Ends

Lubbock refrained from taking Dessiou's claim to the Lords of the Admiralty or directly to Parliament. The problem of funding computers was being solved at this very time in the most unexpected quarters. At Whewell's suggestion, the British Association for the Advancement of Science began offering research grants at its third meeting, which had taken place in Cambridge in the fall of 1833. Not surprisingly, the Association offered its first official grant to a committee composed of Francis Baily, George Peacock, John Lubbock, and William Whewell for the "*discussion* of observations of tides, and the formation of tide tables."[81] The emphasis is the committee's own; the grant was meant explicitly for funding calculators and producing tables, not for the collection of data or their theoretical analysis. Lubbock and Whewell used this first grant exclusively for Dessiou, and Lubbock then published his results in the *Philosophical Transactions*.[82] The money could not be used for past work, but Dessiou did procure ample resources for his future tidal discussions. Though most of the founding members of the Association thought they were advancing Lubbock's cause, the initial grant contained a bitter irony. The British Association was formed, at least in part, to solicit money from government to help fund scientific research. That its own deep pockets worked to release the government from further obligations to fund computers working on projects specifically for the good of the state struck Lubbock as an appropriate climax to a despicable state of affairs.

Further grants from the Association allowed Lubbock to publish Dessiou's tables, which in turn enabled Stratford to use them for the *Nautical Almanac*. Once published, they were in the public domain and quickly became "the common method." Both Stratford and Lubbock, however, felt they had been treated unjustly. Lubbock had overstepped his bounds in assuming Stratford would pay for Dessiou's calculations, and when Stratford refused, Lubbock's letters to Beaufort, Whewell, and Coates, along with only slightly veiled attacks in print, had caused Stratford needless difficulties. Speaking of himself in the third person, Stratford scribbled an angry note to Lubbock in late December 1833: "All you say about Mr. Stratford had better have been avoided. You do not of course understand him! Were I sure to give a true comedy of errors I would certainly put you in as a principal character!"[83] Stratford had worked tirelessly for the *Nautical Almanac*, rebuilding it from its dilapidated state under John Pond to a position on a par with ephemerides of other nations. His endless squabbles with Lubbock had cast an unfortunate dark shadow on his efforts.

Lubbock also had reason to be bitter. He had accomplished considerable work, without payment of course, and he was personally funding calculations that were being used by the government without recognition. Since he did not want to lose money, his choice was either to stop his research altogether or to continue and plead for funding at every turn.

That he showed little tact is to be both regretted and understood. Institutions were not yet set up for research like Lubbock's, where calculators needed to be employed full time for individual natural philosophers. Once large-scale geophysical initiatives became commonplace in British science in the second half of the century, the circumstances would be quite different. Programs like the study of the tides were then viewed as legitimate research in need of calculators funded by the government, and the Hydrographic Office and Nautical Almanac Office became notable centers of calculation. The study of the tides, however, grew out of private almanac publications, not within the halls of the Admiralty. The transition to the public sphere was no simple matter, especially in the early Victorian era, where the dawning of the age of self-help meant "improvement" came largely from private individuals and institutions. Science was certainly gaining steam as a legitimate activity of government, but this benefit was because of, not for, people like Lubbock and Dessiou.

Prodigious funding from the British Association also helped lighten the disagreement between Lubbock and Stratford. They no longer had to squabble over the funding of computers, and time evened out their strong wills. Lubbock acknowledged that Dessiou's petition did not amount to a legal claim, and within the next two years his interest in the tides waned. Stratford, however, was the one to concede the debate. "I have been looking at the tidal operations," he confessed to Lubbock in August 1835, "and there appears to be no doubt whatever that I used your Parallax and Declination Tables."[84] He then acknowledged this publicly in the *Nautical Almanac* for 1838, published in late December 1835. Stratford, it seems, was attempting to mend bridges. "I wish to aid you," he wrote kindly to Lubbock, "and therefore propose that you and Mr. Whewell cut out a good quarter of a Year's Work. You shall then have my whole force for one Week for its accomplishment."[85] In addition, he offered his two best computers, Mr. Russell and Mr. Jones, for Lubbock's more delicate operations.[86] But the wounds were many and had yet to heal. Lubbock found the paragraph in the *Nautical Almanac* "unsatisfactory" even though Stratford had attempted to follow Lubbock's very words.[87] Since he was already losing steam in tidal analysis, he offered to give Stratford's entire "battalion" of computers to Whewell if he wanted them.[88] The dispute, moreover, had grown from a squabble over £50 to a debate over the funding of calculators in general. A battalion of calculators working for one week was a good start, but it had come only when the race was already well under way.

Lubbock experienced similar disappointment over the tide observations made at the Naval Dockyards. The Royal Society had set up a "Tide Committee" to examine the observations received from the Admiralty and had resolved that the best of the observations from each port should be reduced and printed.[89] The Royal Society lacked the funds, however, and the committee asked the government to take up the affair. Roget's request to John Barrow, the secretary of the Admiralty, stressed the importance of having the "observations of the Tides classified and reduced from time to time by proper calculators."[90] Since this was an extremely time-consuming and important part of the work, Roget believed these expenses "might with peculiar propriety be defrayed by the government." At present, he argued, the tides were "calculated by unpublished methods upon unknown data, and without any progressive improvement in their accuracy"—a disgrace in this age of scientific attainment.

The Lords of the Admiralty sent the request to Beaufort for his evaluation. He suggested that the observations of each of the stations be printed for two successive years in order to fix with certainty the establishment of that station, but he thought they should refuse to pay for reductions. Though the expense would be minimal, argued Beaufort, "yet government seems to do enough in furnishing the raw material; and may surely expect that Philosophers and Societies will do the rest for their own fame and from the love of science." The Admiralty would collect and print the publications, but "they declined furnishing the calculations."[91] Lubbock again had been only partially successful. He had used his position within the Royal Society to have observations of tides collected and now printed, but the means to have the data discussed was still wanting.

## Conclusion

Because Dessiou lived and worked in London, he and Lubbock often met to confer about their tidal analysis. Unlike the voluminous correspondence that exists between Whewell and his calculators, Thomas Gamlen Bunt and Daniel Ross, little remains of Dessiou's correspondence concerning his own achievements in tidology. Small clues suggest, however, that Dessiou moved far beyond the tedious tabulations usually associated with the work of an underling. In a letter to Charles Knight in November 1832, for instance, Dessiou apologized for his tardiness, explaining that he had an "anxious desire that the Tides should be calculated from New Tables."[92] Even Lubbock was surprised at this, an initiative that obviously had not originated with him, and though Knight thought Dessiou "not the best even if he were punctual," Lubbock insisted that he be kept on as the Society's computer.[93] He was obviously capable, and if he had not been paid for his services, he would have received much more credit for his work. Indeed, David Cartwright has recently suggested that at least one of the papers published under Lubbock's name in the *Philosophical Transactions* was written by Dessiou himself.[94] Moreover, the tables he produced were technically his, a point Lubbock consistently acknowledged in print. "By the permission of Mr. Dessiou," Lubbock began his third paper on the tides published in the *Philosophical Transactions,* "I am enabled to communicate to the Society some results which he has obtained." Similar statements can be found throughout Lubbock's published papers and personal correspondence. They represent more than a passing recognition of a hired hand; they embody professional respect and a sincere debt to Dessiou's skill and expertise.

Beyond these indirect references to Dessiou's "undaunted perseverance" and "indefatigable labor," what is known with some certainty is the time Dessiou devoted to calculations for the London Docks. From June 1829 to April 1832, Dessiou had tallied an astounding 6,300 hours of work solely on tidal calculations for the one port. At only a shilling an hour, this amounted to £315.[95] And £100 pound a year would just about double the salary he received from the Hydrographic Office. By the following April 1833, Dessiou had earned another £150, some of which went to assistants he hired and boarded.[96] Whewell, in his first publication on the tides read the following month, acknowledged Lubbock's "very important" theoretical investigations, but he praised equally the foundation that work rested on: "The calculations by which they were obtained were performed by Mr. Dessiou; and the task which he has thus executed, is, perhaps, in the amount of labour, and in the judicious and systematic mode of its application, not inferior to any of the most remarkable discussion of large masses of astronomical or meteorological observations by other modern calculators."[97]

Because Dessiou counted on a wide variety of people and institutions for payment, he did not always receive his due. The bulk of the early payments came from Charles Knight and the *British Almanac,* with further funding from Beaufort and the Hydrographic Office, Coates and the Society for the Diffusion of Useful Knowledge, and Lubbock's private funds. About £50 was in arrears, the amount Pond promised but Stratford refused to pay.

Such an amount, though significant, fell far short of the better-known controversies then erupting in science. It was no vicious priority dispute or heated theoretical debate such as could be found at the time in geology

XXI. *On the Tides in the Port of London. By* J. W. Lubbock, *Esq.,* V.P. and Treas. R.S.

Read June 16, 1831.

I HAVE the honour to present to the Society a discussion of observations of the tides made at the London Docks, in the form of various Tables, which show the time and height of high water, not only at different points of the moon's age, but also for the different months of the year, for every minute of the moon's parallax, and for every three degrees of her declination. This work has been accomplished by Mr. Dessiou of the Admiralty; but for the arrangement of the Tables, and the methods employed, I alone am responsible.

The tides take place in the river Thames with extreme regularity, and, as the rise is considerable*, the observations are made with facility. Those upon which the annexed Tables are founded are made at the entrance to the London Docks; the time of high water, or the time when the water has just made its mark, is there noted on a slate by the watchman on the pier-head, generally only to the nearest five minutes; this is afterwards copied in a book kept for the purpose by Mr. Peirse. I am enabled, through the kindness of Mr. Solly, the worthy Chairman of the London Dock Company, to present to the Society the books containing the observations which serve as the foundation of Mr. Dessiou's Tables. These observations are not made with sufficient care; but they are valuable from the extent of time during which they have been carried on, as they were instituted soon after the opening of the Docks in 1804, and have been continued without interruption to the present time. I am not aware of any series of observations of the tides so extensive, except that made at Brest by order of the French Government. Mr. Peirse informs me that the observations of the night are generally more correct than those of

* I believe about nineteen feet. I am not however able to speak with precision, not having yet been able to examine the observations of low water.

Figure 3.4. The title page to John William Lubbock's first publication on the tides in the prestigious *Philosophical Transactions of the Royal Society of London*. Note that he acknowledges Dessiou's contributions in the first paragraph and calls the tables "Dessiou's" in the second paragraph. Note also that, as he admits in the footnote, Lubbock had not observed the tides in the port of London. J. W. Lubbock, "On the Tides in the Port of London," *Philosophical Transactions* 121 (1831): 379.

and physics, so it has remained lost to the historical record. But with it lie concealed the significant contributions calculators made to early Victorian physical astronomy. Quick tempers and strong personalities certainly added fuel to an otherwise manageable flame, but no amount of tact or diplomacy could have prevented the conflagration that broke out once research begun in the private sector spread to the public sphere. As the amount of data in the geophysical sciences exploded toward midcentury, so too did problems associated with funding human computers. Save for Stratford and one or two others, very few held paid positions in science in England outside the universities. Lubbock, for one, was practicing science on his own account, supporting himself through his lucrative work in his family bank combined with shrewd investments, such as the St. Katherine Docks. Scientists—whether in word or profession—had yet to be created, and the mechanisms to support their work also lay in the future. Once Lubbock's research moved beyond the *British Almanac,* he found himself with a ready-made research program based on Bernoulli's equilibrium theory and years of observations to work with, but no apparatus for paying calculators to combine theory and observation. Because he was advancing a research field appropriate for the state, he did not believe he should have to shoulder the financial burden. His time was his own concern, but paying others for work that was the state's responsibility seemed opposed to the progress of scientific knowledge. What if he had not been so financially secure? Though an isolated incident in a relatively small field of study, the problem with calculating the tides is representative of the geophysical sciences in general, including meteorology, hydrography, terrestrial magnetism, and others. The problem of funding computers materially hampered the advance of all of physical astronomy in the early nineteenth century.

Computers in astronomy and navigation, like their counterparts in countinghouses throughout Britain, had matured along with industrialization. They were, in fact, going through a transformation in the first half of the 1830s. Just as Stratford had centralized his computers in the Nautical Almanac Office, George Biddell Airy was revamping the organization of computers in observational astronomy. As director of the Cambridge Observatory in the late 1820s, he ruled with a strong hand and ensured that observations were not only made but also promptly reduced.[98] He continued his dictatorial governance of staff and observatory when he left Cambridge to succeed John Pond as astronomer royal in 1835. One example was his reorganization of the computers at the Royal Observatory. Thus, both the Nautical Almanac Office and the Royal Greenwich Observatory had reorganized their calculators in the

1830s, incorporating them more fully into the practice and process of science.

What became a matter of contention was the lack of such centers of calculation for work done outside these official institutions. As research in the geophysical sciences became more and more dependent on massive numbers of observations, the tabulation, organization, and reduction of those data required that computers be hired, whether or not the research was physically located within a state institution. This was what riled Lubbock. His research, and by extension the safety of British shipping, was being unduly halted because of the lack of funding for computers, a problem he thought Stratford could have fixed. "It is very mortifying to me that Mr. Stratford gives us so little help," Lubbock complained, "with his apparatus of calculators he might dispatch in a week or two, what we shall not wade through, by the hours taken from poor Dessiou's natural rest, in months."[99] Lubbock was forced to use clerks from his father's bank, a solution that solved neither Dessiou's financial problems nor the problem of calculators in physical astronomy more generally. "It has taken one of our clerks three weeks doing nothing else," he wrote to Whewell, "to make the additions requisite for my recent investigation."[100] Lubbock could afford the inconvenience; Dessiou could not.

Laboring ten to twelve hours a day at the Hydrographic Office and an additional six to eight hours a day on tidal computations before and after that, Dessiou was working himself to death. Of the £50 still owed him, he had been sent £20 by Lubbock and another £20 by Coates, but this still left £10 unaccounted for. Years passed and Dessiou had yet to receive payment. He was forced to keep writing to Lubbock, Coates, and Beaufort to ask for funds. "I would not trouble you about it after what has already passed," he wrote to Lubbock in August 1837, "if I were not in absolute want of it, in consequence of waiting so long for what I have received namely the £40. . . . I was served with a Copy of a Writ because I could not pay a small account. . . . It really is sad to be spoken of. It is painful to my feeling thus to address you, but imperious necessity compels me so to do."[101]

One can still feel Dessiou's embarrassment, and similar letters quickly followed. "I fear you have forgotten me," he wrote to Coates, but still no recompense. "Lest my case should have escaped your memory," he reminded him, "I am now compelled to repeat, that this morning I have received another threatening letter, which I doubt not will be put in execution, and I be arrested, if I do not pay £10 of the sum that I am indebted to the party."[102] When Coates finally did reply, Dessiou's heart sank. Rather than sending him news of the missing £10 and sparing him from arrest, Coates asked how Dessiou was planning to repay the £40 already advanced to him. Dessiou expressed his disbelief to Lubbock. He was never going to see any money from Stratford, he rejoined, "and therefore how could I repay the £40."[103] Such was the life of a computer working outside the main centers of calculation.

Lubbock too remained bitter to the end. He continued to jab at Stratford in print, writing in the *Philosophical Transactions* of 1837 that Stratford's limited alterations to Dessiou's tables "cannot be important." Stratford in turn used the only power he had over Lubbock, pulling all his calculators from the tides. "I have no hesitation in saying that it is utterly out of my power to spare any of my present force for any purpose whatever at present."[104] This was in direct response to Lubbock's latest attack in the *Philosophical Transactions,* the response Whewell had feared all along. By this point, in 1837, the study of the tides had ample observations and a well-articulated theory. The problem—one that increasingly became central to the practice of physical astronomy—was a lack of well-funded computers. Dessiou had lost his claim.

# FOUR

# "Tidology"

So much for philosophy. When I talk of giving the rest of my life to *it*, I always reserve to myself the *tides*, as a corner of physics which I shall go on and work at till all is done that I can do. I have made so much progress that I have great hopes I shall have that subject in a condition very different from that in which I found it. **WILLIAM WHEWELL** *TO JOHN HERSCHEL,* 3 **APRIL** 1836

When the Royal Society of London was formed in the late seventeenth century, its first members tried to ascertain a correct theory of the tides. Robert Moray and Henry Oldenburg distributed forms to as many people as possible, but the results were modest at best. Isaac Newton offered a general theory that Daniel Bernoulli and others refined in the mid-eighteenth century, but theory and observation were still considerably at odds. One could not use the Newton-Bernoulli equilibrium theory to predict the tides on the coasts of Britain; it was a general theory that explained the astronomical forces acting on the tides but was not easily applied to particular coasts. While the practical significance of tidal investigation had grown throughout the eighteenth century, its academic study had waned. When George Innes, the author of the Aberdeen tide tables, tried in 1830 to incorporate a more scientific approach in constructing his tables, he searched in vain for a scientific treatment published in England. The science of the tides simply was not discussed in any scholarly text. "In conversing with several people of Science on the subject," complained Innes, "the regular answer is: I have not directed my attention to the subject of the Tides."[1] Natural philosophers in Britain had failed to take up the study.

John William Lubbock resurrected the theoretical study of the tides in Britain from relative neglect. Like Robert Moray and Henry Oldenburg, he began by sending out blank forms, which Francis Beaufort in the Hydrographic Office and Henry Brougham in London distributed throughout the British Isles. The difference was that the early nineteenth century was beginning to see the rudiments of a scientific culture that could sustain such an endeavor. Specialized scientific societies such as the Geological Society of London (1807), the Royal Astronomical Society (1830), and the Royal Geographical Society (1830) complemented the national organizations such as the Royal Society and the newly created British Association for the Advancement of Science (1831). Local Literary and Philosophical Societies and Mechanics' Institutes also extended the call of science to the middle and, to a lesser extent, working classes. Unlike the early *Philosophical Transactions,* moreover, popular publications such as the *British Almanac* reached a broad audience that was receptive and commercially conscious. Tidologists took advantage of science's growing popularity. Lubbock was the first to advance Bernoulli's work and publish the methods needed to produce accurate tide tables, and Dessiou was the only computer to have compared predicted results with a long series of observations.

As vice president and treasurer of the Royal Society, Lubbock also used his influence to have observations collected from naval dockyards around the southeast coast of Britain. The complaint within the nascent scientific community that the halls of the Royal Society were filled not with active researchers but with former navy officers was precisely why the Society remained so influential within government, especially the Admiralty. Lubbock introduced the topic of the tides as a valuable research project for natural philosophers intent on demonstrating the practical significance of physical astronomy for the good of the state. One researcher in particular took up the study with a vengeance.

When William Whewell first used the term "tidology" in the early 1830s, the study of the tides languished as a field "still to be begun."[2] Lubbock had confined his research to the Thames estuary and refused to generalize his results either geographically to other areas or theoretically beyond his empirical findings. He was hampered by a lack of funding for the reductions such an expanded study would require. Though Francis Beaufort referred to the tides as a "national object" and argued the "urgent necessity of acquiring proper data for the construction of our Tide Tables," the Admiralty had yet to make allowances for the necessary calculations.[3] According to Whewell, there was no system in the gathering of tidal observations, no method of paying for the laborious calculations, and no laws of the phenomenon to guide the expansion of the subject beyond London. The observations, their reductions, and the Newton-Bernoulli equilibrium theory had yet to be "extended and generalized."[4]

While tide table makers and almanac publishers studied the tides in their local context, Whewell saw the tides as a state would—as a major research project to be furthered by the Admiralty, the British Association, and the Royal Society of London. As the subject's most influential theorist, Whewell worked within these elite institutions to envision and shape it throughout the 1830s. From his entry into the field, Whewell acknowledged his debt to Lubbock but also differentiated his research and his methods from his former student's. He wanted to establish tidology as a viable research frontier based on adequate funding, the necessary equipment, and a worldwide network of observers. Whewell's interests differed from those of the mariner, banker, insurer, and shipowner; his aims were not aligned with those of Brougham, Knight, or even Lubbock. The way he viewed his sphere was far more synoptic, a legacy that reached back to Edmond Halley and Alexander von Humboldt.

Whewell turned to the study of the tides because he believed it was there that he could garner quick results.[5] He was guided in his approach through a detailed study of the history of physical astronomy, and his early researches in tidology helped him formulate his mature views on scientific discovery. After a short but intense fascination with meteorology, Whewell chose a topic that was in every way suited to profit from his studies of the history and philosophy of science, one in which data had been gathered but laws had yet to be discovered. Tidology offered Whewell an opportunity to further Newton's work, produce practical results for the British Admiralty, and enhance his own scientific stature.[6] Understanding Whewell's promotion of tidology as a legitimate field of research not only contributes to the history of the organization, funding, and practice of physical astronomy in the Victorian era but also corrects a long-standing misinterpretation concerning one of the premier advocates of science. To fully understand Whewell's broad achievements, a study of his science is crucial.

In sharp contrast to Lubbock's approach, Whewell's methodology incorporated the rank and file of British observers. As Whewell repeatedly stressed in his papers presented to the Royal Society, in *Reports of the British Association*, and in correspondence with members of the Admiralty, tidology was a shamefully underdeveloped subject connected with the most fully developed science—physical astronomy—and accessible to even the humblest researchers. Whewell turned to a wide array of sources, including missionary societies, Admiralty surveyors, and members of the British Association, offering popular evening lectures at British Association meetings and publishing queries throughout the pages of the *Nautical Magazine*. He did not have to search far for help. The cultural fascination with science was growing, and the members of these organizations were already collecting data and commenting on theory. Indeed, as I have tried to show, in many respects the associate laborers were leading the way. Whewell's position within the scientific community, combined with his unique method of advancing a burgeoning field of research, enabled him to organize their efforts on a grand scale.

## William Whewell in History

Astronomer John Herschel once wrote of Whewell that "a more wonderful variety and amount of knowledge in almost every department of human inquiry was perhaps never in the same interval of time accumulated by any man."[7] Some subjects, such as Whewell's philosophy of science, have been studied extensively, especially in the past two decades, but his work on the tides has yet to be integrated with his other metascientific pursuits. He is best known for his multivolume *History of the Inductive Sciences* and his equally imposing *Philosophy of the Inductive Sciences*, works that helped define what "science" was in the early Victorian era. His *History* focused on the great revolutions in the sciences, work that he used in his *Philosophy* to determine how science advanced, the importance of nomenclature, and perhaps most important, how further advances could be made. Though Whewell's reputation rests largely on these two works, they only scratch the surface of his dauntingly broad intellectual pursuits. He published works in experimental physics, geology, astronomy, tidology, architecture, poetry, and religion.[8] He invented the self-registering anemometer and originated many scientific terms including ion, anode, cathode, Eocene, Miocene, scientist, and physicist. He spent his entire intellectual career at Trinity College, Cambridge. After graduating second wrangler in 1816, he received his MA degree in 1819 and held professorships first in mineralogy, then in moral philosophy. In 1841, Queen Victoria offered Whewell the mastership of Trinity College, a position he held for twenty-five years, until his death in 1866.[9] He was one of the most powerful figures in British science in the mid-nineteenth century.[10]

For most, Whewell was a philosopher, a historian, an educator, and an administrator of science, all wrapped into one.[11] He was thus able to command respect in evaluating science, a growing but still undefined force in the 1830s. Along with John Herschel, Whewell became the premier evaluator of science in Britain, practicing what Richard Yeo has termed "metascience." Indeed, the parallel with John Herschel is instructive. Both graduated with distinction from Cambridge University and played integral roles in bringing French analytical methods to England in their early postgraduate years. In the 1830s, both viewed science from above, trying to define its parameters and outline its methods. Herschel's *Discourse on Science* (1831) inspired Whewell, who dedicated his *History of the Inductive Sciences* (1837) to Herschel. Whewell pursued a research project on the tides, while Herschel followed a decade later with a similar initiative in meteorology. In his later years Herschel translated Homer's *Iliad;* Whewell translated Plato's *Dialogues.* Their interests were similar and wide ranging.[12]

Yet John Herschel is acknowledged as one of the leading practitioners of science in England during the Victorian era, while the accepted view of Whewell is quite different. His secure position as an evaluator of science has arisen, at least in part, from his *lack* of involvement in any one scientific discipline. Past scholarship has referred to the importance of Whewell's engagement in science but has continually turned to a wide variety of subjects other than his work on the tides to exemplify that importance, including geology, biology, electricity and magnetism, architecture, mineralogy, the wave theory of light, and meteorology.[13] "Whewell desired to assimilate and organize knowledge, to get the big picture," argues Harvey Becher. "He had no urge to investigate every nook and cranny not yet explored."[14] By treating Whewell as a "critic, adjudicator, and legislator of science without being a major scientific practitioner or discoverer," Richard Yeo has likewise concluded that Whewell's role differed from that of Herschel.[15] Menachem Fisch took this view one step further, to conclude that "unlike Herschel, at no time did he ever fully devote himself to any one scientific project."[16]

Whewell, however, was not a geologist or a biologist, he was not a meteorologist or a mineralogist, nor was he a political economist or a wave theorist. For more than two decades, he was the premier tidologist in Britain. He devoted himself to the subject for over twenty years, published fourteen papers on it in the *Philosophical Transactions,* and was awarded the Royal Society's gold medal for his efforts. It outshone his work in other fields of science. Reviewing recent scholarship on Whewell, Joan Richards commented, "Whewell dabbled in [science]; he dropped in occasionally but on balance was more an observer than a participant in their enterprise."[17] This is a fair representation of past research on

Figure 4.1. William Whewell at the height of his tidal studies about 1835. Portrait housed in Trinity College, Cambridge.

Whewell but an unfair portrayal of his systematic work in tidology. As the opening quotation suggests, Whewell always reserved the tides for himself as a corner of physics he pursued throughout his intellectual career.

## Whewell's Search for a Scientific Subject

Whewell was productive from the beginning of his postgraduate days, but 1819 marked his entry into science. That year Whewell received his MA degree, helped form the Cambridge Philosophical Society, and first appeared as an author with the publication of his *Elementary Treatise on Mechanics*, a text on applied mathematics with a heavy dose of

Continental symbolism.[18] Whewell's *Treatise* became the standard text at Cambridge, and he followed it with a second textbook, *A Treatise on Dynamics* (1823), which openly used the new analytical methods of the French.[19] Whewell was not a radical reformer in mathematics like Herschel and Babbage; he stayed clear of their abstract approach.[20] At once both innovative and conservative, Whewell's two textbooks, followed by many new and substantially modified editions, brought him recognition as a leading agent for bringing the French analytical methods into English science.

Along with his other pursuits, this recognition led to his election as a fellow of the Royal Society of London. But Whewell was not content with standing on the sidelines, merely translating science and not taking part in its development. He craved an active role. In the fall of 1823 he wrote to his Cambridge colleague and lifelong confidant John Herschel: "When I was admitted into the Royal Society I intended, if possible, to avoid belonging to the class of absolutely inactive members, and I have since been on the look out to find among the speculations in my way some one which might possible be worth presenting to it."[21]

Whewell initially chose crystallography as his subject. He was drawn to the "extremely imperfect and anomalous state of crystallography as a mathematical science"—its unsystematic methods and its lack of an adequate and general notation. His work in crystallography foreshadowed the approach he would take in all his scientific research: a concern with nomenclature, methodology, and utility.

Whewell's first paper, "A General Method of Calculating the Angles Made by Any Planes of Crystals, and the Laws according to Which They Are Formed," was read to the Royal Society in November 1824 and published in the *Philosophical Transactions* for 1825.[22] Whewell's aim was to rid crystallography of its arbitrary symbols, to relate its notation to its method of analysis, and to "reduce the mathematical portion of crystallography to a small number of simple formulae of universal application."[23] H. Deas, in his survey of the history of crystallography, claimed that this approach "laid the foundation of mathematical crystallography."[24]

Through the 1820s, Whewell expanded his interest in crystallography to all of mineralogy. He set off for Germany in 1825, hoping to obtain a good collection of mineral specimens, and above all, to work under Professor Friedrich Mohs in Freiberg, one of the leading practitioners in the field. On his return he accepted the professorship in mineralogy at Cambridge and was intent on treating it as more than a sinecure.[25] In his "Essay on Mineralogical Classification and Nomenclature," published in

1828, Whewell used a natural as opposed to an artificial mode of classifying minerals, a revision meant to replace Mohs's system.[26] His teaching and research in mineralogy, his intense study of the mathematics underlying its laws, and his tour of the Continent to collect specimens and study under the field's leading lights went far beyond the duties expected of a professor at that time.[27] Whewell studied the field's history, how its laws were created, and what it could hope to achieve. He was formulating a scientific methodology, one geared not only toward how science worked but also toward how one could develop a field from the ground up. At this stage in his own career, he was particularly interested in how *he* should practice and advance a burgeoning field of research.

From relatively humble beginnings—he was the son of a master carpenter—Whewell now held a professorship at the University of Cambridge, a position he retained until 1832, when he lectured on the state of mineralogy to the British Association. Yet he had lost interest in the topic as a scientific pursuit, confiding to his friend Richard Jones that it would "take a long time to turn me into a good mineralogist."[28] His interests changed under the influence of two younger researchers, the first a new acquaintance from Scotland, the second a friend and former student at Cambridge.

James David Forbes in Edinburgh viewed Whewell as his mentor, excited by Whewell's insistence on using French analytical methods.[29] The liberal education offered in Scottish universities differed greatly from that at Cambridge; students received only one year of natural philosophy as part of their four-year curriculum.[30] Whewell and Forbes began to exchange letters after Forbes visited Cambridge in May 1831.[31] Along with pedagogy, their favorite topic was meteorology. It was "so important and so interesting" that Whewell had "read a good deal on [it] and related subjects."[32] He rejoiced to hear that Forbes was to give a paper on the subject at the next British Association meeting in 1832, and he offered at length his own thoughts on advancing the field. "All the subjects on which it most depends," Whewell noted, "are in that rapid state of change and progress which makes them very attractive topics."

The topic was attractive to Whewell because progress seemed imminent. He wrote to Richard Jones in July 1831, excited about his own prospects on entering such a promising field of research. "Among other things I have got hold of a new science, which is altogether admirable both for my theology and for my induction. . . . What do you think my new pet is?—Meteorology. The people have been collecting facts for a very long time,—(ever since Noah) and are now just beginning to get a notion of the general laws and proportions in to which the mass is to be resolved. I do not know any subject which is at present in so instructive a condition."[33]

In meteorology facts had long been collected but had not yet been formed into laws. The field's young, undeveloped nature, combined with its prospects for relatively quick advancement, also made it exceedingly instructive for the process of discovery itself. Forbes responded to Whewell's suggestions with similar excitement. He told Whewell that he desperately wanted to move beyond "the meagre and too often fruitless display of columns of a Register" and seek out the organizing principles that would set the science on a fruitful course. Forbes and Whewell were a sharing a reaction against naive empiricism that was pervasive at that time; they also acknowledged that this meant "someone of sufficient talent taking up the theory."[34]

Forbes took up meteorology with fervor; Whewell did not. What Whewell did hit on as a subject was in many respects similar, but even more promising of rapid advancement: the tides. And the influence this time was from another young researcher, John William Lubbock, whom Whewell had worked closely with at Trinity College, Cambridge. Whewell had been Lubbock's tutor at Cambridge for only a year, and though Lubbock was ten years younger, their scientific work overlapped enough that they formed a rather close bond. Lubbock's insistence on studying French mathematics, his interest in probability theory, and his application of mathematical techniques to life insurance and political economy closely matched Whewell's own interests. Whewell suggested topics for Lubbock to follow and looked to him for help in his own research, a mutual influence and admiration that lasted throughout their lives.

The earliest correspondence between the two concerning the tides was in the fall of 1829, when Lubbock told Whewell he had been discussing observations made at the London Docks and planned to publish his results in the *Companion to the Almanac*.[35] He sent the paper to Whewell for his comments.[36] Whewell's only major complaint was that Lubbock's essay was not "more general in its status."[37] Since Whewell was to take exactly this approach—generalizing Lubbock's highly localized and empirical conclusions—the seeds of his own interest were perhaps planted at this time.

By the summer of 1830 Whewell was persuading Lubbock to reach further than the *British Almanac*, perhaps even to compose an entire treatise. "I think it would be useful here if you could do it," Whewell prodded Lubbock, "for we have nothing readable on the subject."[38] Their correspondence throughout 1830 and 1831 centered mostly on extant sources of observations and on corrections and additions to Lubbock's papers in the *Companion to the Almanac* and the *Philosophical Transactions;* they made very little mention of theory or future research. But Whewell's interest in the tides was on the rise. He pondered the subject's similarities with meteorology and began to amass materials in his rooms at Trinity College, providing a constant source of new material for Lubbock's (and his own) future research. They also discussed nomenclature. "A book called Tydology was in 1810 published by a M. Sade," Whewell informed Lubbock, adding rather contemptuously, "I suppose he thought the y made the word look Greek."[39]

Early in 1831, Whewell offered all the materials he had "in hand" for a paper Lubbock proposed to submit to the Royal Society that year. "With care and labour," Whewell confided, "I dare say something may be generalized out of them."[40] This advice should be read as an implicit admission that Whewell had himself begun to take up the subject. "I think the best thing to do would be to make but such broken and conjectural lines of contemporaneous tide as one can with such material and then represent strongly how imperfect they are and how good a deed it would be to improve them." More explicitly, Whewell admitted that he "should really like to have the rummaging of the subject and the bringing together of the facts." His focus on graphical representation and his "bringing together" of diverse observations were entirely different avenues of research than Lubbock's current program. Whewell was differentiating his own efforts from Lubbock's, marking his own territory. Whewell added that at present he had no time for such matters and was happy that Lubbock had "made the subject peculiarly yours." In a postscript, Whewell asked Lubbock if "cotidal Lines" would not be the proper term to represent the contemporaneous tides.

As with previous subjects that he had seemed content to jump in and out of, Whewell was thinking of the nomenclature to be used and how to produce quick results. By the beginning of 1832, however, he had become engrossed in the new subject and intended to publish. In just over a year, tidology had eclipsed his interest in meteorology, partly owing to the differences in the two subjects. His switch was based on his textbook writing, specifically his interest in the history of mechanics and dynamics. Since the publication of his two textbooks, *Dynamics* and *Mechanics,* Whewell had issued new editions, sometimes revised beyond recognition. Two new editions undertaken in the latter half of 1831 and published in 1832 contained histories of Newtonian mechanics, subsumed years later in his *History* (1837). Menachem Fisch has referred to these "formative years" as Whewell's search for "a theory of excellent science."[41] The historical sketches convinced Whewell of the prescriptive role of the inductive method. He formulated a theory of excellent science based on the model of physical astronomy that taught him how to reach general propositions from a mass of unconnected data.

Whewell's study of the history of physical astronomy convinced him that a study of the tides would be especially fruitful. Tidology, classed under the mechanical sciences, could draw on the revolutionary findings of Newton and follow the lessons of physical astronomy. It thereby came with guidelines, something meteorology did not have. Furthermore, Lubbock refused to generalize beyond his empirical findings, leaving tidology open for just such an approach. Masses of data had been collected, but the laws explaining the phenomena had yet to be discovered.[42] In Whewell's own formulation, he could serve as the Kepler of tidology, uncovering the phenomenological laws of the tides. On the lookout for a scientific subject since his election to the Royal Society, Whewell had finally found it.

In April 1832 Whewell admitted to Lubbock that he intended to take up the subject as part of a larger research agenda.[43] He asked if Lubbock had any objection "to my mixing my labours with yours and supposing that I could satisfy myself so far as to write a paper on the subject?" Lubbock was honored, proud that the tutor was now following the student. Whewell's letters, however, retained their furtive tone. Lubbock had to rely on reports from others, such as Mary Somerville, suggesting that Whewell was indeed hard at work.[44] Lubbock asked Whewell to keep him abreast of his advances, but Whewell continually responded that much more needed to be done and that the work so far was taking much more "time and trouble" than he had anticipated. At this point Whewell was working without a computer, and the data were overwhelming. He admitted to Lubbock that he had "borrowed largely" from Lubbock's own article on the tides that had appeared in the *British Almanac,* "thinking you would allow this."[45] Two months later Whewell was still rather quiet on particulars, but he announced that he had begun work on a map of cotidal lines that could be used to study the tides in the world's oceans. "I had rather set off doing any thing about cotidal lines till I can collect more materials and give more time to the subject than I can now; but if it appears that the only way to set people in distant parts to work upon the subject is to make some conjectural assertions, (not concealing that they are so), I should have no objection to take this method."[46]

Whewell, with one eye toward a unifying law of the tides as they move through the oceans, had his other eye open for intersting observers "in distant parts to work upon the subject," an active part of his methodology. Lubbock had by then shrugged off any attempt at organizing observations and contented himself with studying the tides from observations made at the port of London for many consecutive years. Whewell wanted to expand his own approach far beyond Britain.

By the fall of 1832, Whewell had begun a sustained research project on the tides. He followed many of Lubbock's leads, including his method of data analysis and his insistence on comparing extant observations with the Bernoulli equilibrium theory. But Whewell also forged his own path. His method was influenced by his view of how a science should proceed, based on his study of the history of physical astronomy. His views on the philosophy of science, he believed, offered the most promising path for advancing a science such as that of the tides. By the end of 1832, Whewell was actively blazing just such a path to further his own research in tidology.


## A Collaborative Enterprise

From Whewell's introduction into tidology, he was intent on collecting as much data as possible from sources as disparate as he could imagine, confident of the contribution to be made by rank-and-file British observers. Whewell's first circular, a two-page "Suggestions for Persons Who Have Opportunities to Make or Collect Observations of the Tides," printed in 1832, outlined all the important information a mariner could gather for the theorists back in London and Cambridge. It was concerned almost exclusively with instituting long-term observations so as to find the corrected establishment of various ports. The last line, however, written it seems almost in passing, suggested that "Comparative Observations of the Height at different places in the same seas . . . may also be of great value." He included the rhetorical hedge "may" because no one had ever attempted such an analysis. It required an army of observers stationed along the coast, organized and systematized through a central authority. Whewell set out to make himself that authority, establishing relationships with the Admiralty, missionary societies, and the British Association for the Advancement of Science.

*Missionary Societies*

Whewell began his data collection by soliciting help from missionaries serving abroad. Even before traders embarked on their quest for reliable sources of raw material to support British manufacturers, and before the navy's impulse to acquire accurate charts and surveys to support that trade, missionaries had begun to sail around the globe. Francis Beaufort, the leading cartographer in London, was often astounded at their penetrating power. "The activity of those zealous proselytizers is astonishing," Beaufort wrote to Thomas Coates, "and being always the forerunners, not only of settlers and colonists, but of military expeditions and commercial travelers, all our discoveries and all our recent geographical knowledge has been derived from them."[47] These "zealous proselytizers" also were educated, attaining the rudiments of a liberal education that included at least an exposure to science.[48]

Most important for Whewell's research, the missionaries also were widespread. The nineteenth century was "the great century of missions," with Protestant missionary societies, such as the Church Missionary Society founded in London in 1799, reaching into Africa, South Asia, Latin America, and the islands of the Atlantic, Pacific, and Indian Oceans.[49] By the second quarter of the nineteenth century, missionary work had grown into a mature organization, with a hierarchical power structure and highly successful publication outlets. It was advanced enough to help Whewell in his search for observational data that were otherwise difficult, if not impossible, to obtain.

Though Whewell's family is not mentioned in most scholarly work on Whewell, he did keep up a continuous correspondence with his sister, informing her of his many lines of academic scholarship. His sister, in turn, often sent her older and more learned brother queries on scientific subjects. In March 1832, at about the time Whewell began to study the tides, she wrote to Whewell concerning peculiar tides in the Pacific Ocean. She had read in a journal published by the Church Missionary Society of London that the tides of Tahiti ebbed and flowed at the same time every day.[50] Whewell assured her that such tides could not occur and that the missionary unknowingly had reported the effect of the wind and weather. "But if you do find in your missionary books, or any other books, any notices about the tides," Whewell added, "let me know, for I believe I shall have to write something about them, and all information on the subject will be useful."[51] Written before Whewell's announcement to Lubbock, this is the first letter in which Whewell wrote that he intended to pursue tidal research. Significantly, the missionary's observations concerning Tahitian tides were accurate and Whewell's response was incorrect.

It is difficult to determine exactly what missionary Whewell's sister was referring to, but it could have been William Ellis, who had visited Tahiti and noted that the tides there were so regular that the Tahitians marked the shore to determine the time of the day. Ellis followed the same reasoning as earlier observers who had registered tides inexplicable by Newton's theory. Though Ellis hesitated to oppose the "established opinions of great and learned men," he concluded that the tides could not possibly be caused by the force of the moon, the basis of the Newton-Bernoulli equilibrium theory.[52] This did not stop Whewell from asking for his help. He wrote directly to Ellis, enclosing instructions for him to pass along to the missionaries of the London Missionary Society. Ellis informed Whewell that John Williams, a missionary from Scotland, had made observations in the Society and Navigator Islands and would shortly return home from his voyages. Because the tides at the Society Islands, according to Ellis, obeyed only the sun, while those in the Navigator Islands, according to Williams, obeyed both the sun and moon, Ellis suggested that observations at Easter Island, midway between the two, would be highly instructive for any theory of the tides. Ellis added that he had not observed the tides there because "the inhabitants are jealous and hostile."[53] Evidently scientific curiosity and missionary zeal had limits, and hostile natives was one. Williams duly sent Whewell his observations when he returned to Britain, then returned to the islands of the southern ocean in the late 1830s, where he was killed by natives—and supposedly eaten![54]

Whewell wrote to his sister again seeking more observations of the tides on the shores of the Pacific islands. "And I think the persons most likely to be able and willing to make such observations," wrote Whewell, "are the missionaries."[55] Whewell, far from apologizing for using the missionaries this way, suggested that they "will like very much to be so employed."[56] Whewell made this remark after having read John Williams's *Narrative of Missionary Enterprises,* a popular account of missionary work in the South Seas. Whewell found the material of such interest that he transcribed a particularly seductive passage: "It is to me a matter of regret that scientific men, when writing upon these subjects, do not avail themselves of the facts which missionaries might supply, for while we make no pretensions to great scientific attainment, we do not hesitate to assert, that it is in our power to furnish more substantial data on which to philosophize, than could be obtained by any transient visitor, however profound in knowledge, or diligent in research."[57]

Whewell could not have agreed more. He asked his sister to examine all her books and magazines "in which accounts are given of the various missionary establishments all over the world, and put down the stations, the names of the missionaries or teachers at them, the societies by which they are sent or supported, and the address of the official persons of those societies." In return, Whewell offered as a "reasonable proposal" to subscribe to any society that would help him in his task.

His sister duly sent the information, which led Whewell to correspond directly with William Jowett. Jowett was educated at St. John's College, Cambridge, graduated twelfth wrangler in 1810, and afterward became a fellow of St. John's. In 1813 he left college life and volunteered for the foreign service of the Church Missionary Society. Most important for Whewell, he was now the Society's clerical secretary.[58] Jowett promised to contribute to Whewell's research if Whewell would send him detailed instructions.[59] Whewell again asked his sister to forward anything she came across concerning tides in the Pacific, even though the information might be unsuitable for scientific publication. "This is of no consequence," Whewell assured her, "for I am pursuing my inquiries for my own satisfaction, and only want to know how things go on in the great oceans, and especially in the Pacific." Though the registers received from missionaries were not great, they did supply Whewell with a wealth of information on anomalous tides from isolated islands and distant outposts that otherwise would have been impossible to obtain. Once Whewell turned to the Pacific Ocean in the 1840s, these accounts proved especially fruitful, for far fewer British vessels were plying those distant waters. Whewell's continued correspondence with missionary societies, moreover, demonstrates one important aspect of his tidal research: he was willing to search every nook and cranny.

*The British Association for the Advancement of Science*

In 1830 Charles Babbage published *Reflections on the Decline of Science in England*, a scathing critique of the lack of direct government support for scientific research in Britain. Its publication initiated a reform movement within the still inchoate scientific community, intent on raising the status of science and its practitioners. It was in the midst of this debate, one year after Babbage's publication, that the British Association for the Advancement of Science was formed, a peripatetic organization that used its meetings to extend science from London into the provinces and eventually to Australia and Canada. Its wide geographic leanings matched the type of science it fostered, in which researchers from all classes of society and all areas of Britain and its colonies could contribute. Furthering the particular interests of the British scientific elite, the Association advanced a political agenda as well. It attempted to foster an image of science as above politics, class, and religion. The organizers believed that in this way science could ease the political and religious strife endemic to

Britain's early Victorian culture. The annual meetings were enormously successful, with local pride and favorable publicity assuring that the activities of the band of natural philosophers reached a wide public.[60] The leading lights within the British Association stressed that science could become an agent of national prosperity and international harmony, a resource that could not help succeeding in its aim to "promote the intercourse of those who cultivate Science with one another and with foreign philosophers." It should be no surprise that the initial members of the Association found the tides an especially appealing topic of research.

The British Association held its first meeting in York during the summer of 1831. Neither Lubbock nor Whewell attended, but this did not stop the Association from entering into tidal studies. It was founded to offer direction and assistance to branches of science that required collaboration and data collection from around Britain, and tidology seemed an obvious choice. Vernon Harcourt, who was behind the first constitution, wrote to Lubbock after the first meeting telling him that "it was suggested in the committee as a proper project for corresponding observations on different parts of the coast, to ascertain the height of the Tides for 24 hours."[61] That the leading lights in tidology were not present was obvious from the method the Association proposed to study the tides. Twenty-four-hour observations had never been suggested by Lubbock or anyone else.

One other significant development in the name of tidology came out of the first meeting. William Scoresby, a famous whaling captain who was an active early member, told Harcourt there were accurate observations of the tides at Liverpool spanning almost thirty years, and he surmised that similar observations might exist throughout Great Britain.[62] Harcourt mentioned this to Lubbock and formally invited him to address the next meeting at Oxford "with a Report on the actual state of our information upon this subject and on the data remaining to be obtained, in order that the Association may exert itself to procure them." Harcourt proposed a bargain: if Lubbock would give such a report, the officers would give him "any assistance which [he] may desire in collecting the requisite information."[63]

Lubbock was eager to receive further information on the observations at Liverpool but was far less enthusiastic about amateurs' spreading out through Britain to gather information. As Jack Morrell and Arnold Thackray aptly put it: "Lubbock was decidedly unimpressed by the enthusiasm of the new Association."[64] Lubbock had yet to receive observations good enough to use other than those kept by the dockmaster at the London

Docks, and those were useful only because of the length of time they had been kept (twenty-five years). It is not surprising that Lubbock was extremely interested in thirty years of observations made at Liverpool, observations that overlapped with those at London, and far less keen about amateurs' sending information that he had no way to judge. And for Lubbock's interests—empirical corrections using long-term observations to construct tide tables—short-term observations were worthless.

Though Lubbock did write a paper on the state of tidology in Britain, he refused to attend the second meeting of the Association, held in Oxford in the fall of 1832, shortly after Whewell had announced his own intention to embrace tidology. In his stead, Whewell presented Lubbock's paper, initiating a fruitful relationship between the Association and the study of tides. Hearing that Whewell would attend, Vernon Harcourt asked him to include a map of cotidal lines by way of demonstration. Whewell informed Harcourt that "Lubbock will not meddle with any thing of the kind," but he was also skeptical that he could pull the material together on such short notice.[65] Since he had no computer and had not had time to sift through the data, he did not want to "dissatisfy people" by making assertions he could not support. By "people," he meant Lubbock, since they disagreed about making inferences from the data at hand. Whewell therefore confined his remarks to Lubbock's written report, but Lubbock had failed to highlight the value of provincial observers in collecting data, which Harcourt had outlined to Lubbock as important for his paper. On the backward state of observations, Lubbock wrote, "I trust that the influence of the British Association will be exerted to remove in some degree this national reproach."[66] This remark was not directed toward ordinary British observers but referred to the British Association's "influence" on government. At this time Lubbock was trying to persuade the British Admiralty to initiate tide observations at the naval dockyards.

Lubbock's indifference, of course, did not stop the Association, which heaped money on tidal studies. Since attendance at its meetings reached the thousands within a few years, the Association found itself embarrassingly prosperous. Under Whewell's recommendation, it offered a scheme of grants to individuals with established research programs. At its third meeting, held in Cambridge with Whewell serving as vice president, the Association offered its first official grant of £200 to a committee composed of Francis Baily, George Peacock, Lubbock, and Whewell for the reduction of tidal data. Because the funds were allocated to a committee and not to him alone, Lubbock grew agitated. He complained to the secretary, John Phillips, that "no one has suggested any improvement upon the methods of discussion I have pursued all along, and which I was the first to indicate. If I am allowed to proceed my own way I know what I am about."[67] In the years that followed it was Whewell, not Lubbock, who incorporated the British Association into tidology.

At a well-attended Thursday evening session during the Cambridge meeting, Whewell gave a popular lecture on how those living near the coast could help by making observations that would be useful as ground-work for general laws. Published as an appendix to the third report of the Association and similar to the separate sheet he published for Beaufort to distribute among the officers in the navy, it included directions for mak-ing observations and defined the nomenclature used by tidal theorists. Throughout the 1830s, Whewell continued to attend the British Associ-ation meetings, to keep the Association abreast of his latest researches, and to give popular lectures in the evenings. Most important, he also drew grants for reducing the incoming data. Whewell did not have to battle the astronomer royal or the Hydrographic Office for computer time; he found all he needed in the deep pockets of the Association. At the next meeting, held in Edinburgh in 1834, £50 was added to the £200 already offered, and in 1835, an additional £250 was granted, again ex-pressly for the discussion of observations. In the end, Whewell received over £1,000 for reducing tide observations. His emphasis on the tides as a branch of physical astronomy that even the humblest researchers could contribute to made it especially appealing. Using this line of argument to great effect, Whewell garnered funds from the British Association to keep the reduction of data abreast with theoretical and observational advances.

*The British Admiralty*

After the 1832 British Association meeting, observational data that had previously passed from Whewell to Lubbock began to flow in the other direction. When Whewell first said he intended to investigate the tides, he asked Lubbock for "the references to the sources of some of your facts."[68] Lubbock offered all he had, largely published materials from sailing directions, books on astronomy, and manuals on navigation. Sharing extended from observational data to introductions to people Lubbock had interacted with during his own investigations. In January 1833, while Whewell was visiting London, he wanted to ask Lubbock for a letter of introduction to the dockmaster at Portsmouth.[69] Whewell intended to visit the royal dockyards within the next few days and dis-cuss the actual process of making tide observations. But Lubbock was not at either his home or his bank, and Whewell returned to Cambridge empty-handed. Lubbock regretted their miscommunication and asked Dessiou to write to Whewell about the materials on the shelves of the Hydrographic Office. "It seemed to promise so much at the Admiralty," Whewell reported back to Lubbock, "that I returned to London instead of going to Portsmouth as I intended; and I was glad that I did so."[70] Whewell found considerable stores of materials that he could "examine and compare," and for a time they sated his empiricist appetite.

Whewell received a warm reception at the Admiralty, especially from Francis Beaufort and the scientific servicemen connected with the Hydrographic Office. In practical terms, the Admiralty was the institution that would profit most from Whewell's labors. It increasingly embraced projects that demanded familiarity with the latest mathematical, scientific, and technical knowledge. Indeed, such ventures had been on the rise since the turn of the century, and because of the growing scientific and technical character of many of the problems surfacing within the Admiralty, the government formed a Committee on Science in 1828 to advise on "all questions of Discoveries, Inventions, Calculations, and other Scientific Subjects."[71] The next year, Francis Beaufort was appointed hydrographer to the Admiralty, a position he held for twenty-five years. After his appointment, the Hydrographic Office became the research and development wing of the British Admiralty.

Known for his accurate charts and for the wind scale that still bears his name, throughout the second quarter of the nineteenth century Beaufort was also responsible for organizing scientific expeditions to span the world's oceans and coastlines. He drafted instructions and chose the scientists for each voyage and was in charge of equipping each expedition.[72] Largely owing to Beaufort's scientific interests, the Admiralty placed him in charge of the budget for the scientific branch within the Admiralty, formed in 1831 and composed of the Hydrographic Office, the astronomical observatories at Greenwich and the Cape of Good Hope, the Nautical Almanac Office, and the Chronometer Office. Thus Beaufort had control over all offices within the Admiralty that dealt with scientific and technological topics, and he took a leading role in the Admiralty's collaboration with the foremost scientific societies of the day, including the Royal Society, the Royal Astronomical Society, and the British Association for the Advancement of Science.[73] Nowhere is Beaufort's valuable role more evident than in the study of the ocean tides. If Lubbock and Whewell provided the mathematical techniques used in tidal theory, Beaufort's tireless efforts on their behalf made the research possible.

The Admiralty's reliance on science filtered down to the technical training for its military engineers, dockyard apprentices, and the professional military men associated with the trigonometric and coastal surveys. As David McLean recently noted, "even common seamen and dockyard workers needed training and elementary knowledge, and if such was not available in the coastal towns then the Admiralty was reconciled to making its own provisions."[74] While scientific servicemen often received a university education, these other men increasingly obtained highly technical training at the Royal Naval College at Portsmouth, the Royal Military Academy at Woolwich, the Royal Engineer's Institution at Chatham, or other schools founded for apprentices at the royal dockyards.[75] Members of the Admiralty, including civil engineers and dockyard officials, were therefore in a position to further accurate observations and add to the theoretical investigation of the tides. This included the invention of self-registering tide gauges and their proliferation to the major ports in Britain.

The construction of self-registering tide gauges in the early 1830s allowed for the improved accuracy needed in tidal observations. Only then were short-term observations of the tides dependable enough for Lubbock and Whewell to use them to perfect theory. Henry Palmer, a civil engineer at the London Docks, deserves credit for inventing the first self-registering tide gauge. Palmer had directed a survey for the railroad between London and Rochester and thus was trained as a surveyor; he was quick with numbers and at home with instruments. When he was appointed civil engineer for the London Docks Company, Palmer needed to study the tides, especially the effects of the removal of the old London Bridge and the erection of the new bridge. His constant attention to observing the tides, however, left little time for his other duties. He needed a method of observation free from "those inaccuracies and doubts which the frequent and long-continued observations of individuals, through nights as well as days, must be liable to."[76] His mechanical contrivance, complete with a wind gauge, was close to completion when Palmer sent its description to Lubbock to place before the Royal Society.[77] The Society published the description in the *Philosophical Transactions,* one of the few entries by a civil engineer in this period.

Palmer's gauge was copied by other artisans and dockyard officials interested in tabulating accurate tidal data, including J. Mitchell, the civil engineer at the Sheerness naval dockyard. The first issue of the *Nautical Magazine* gave a description along with a large plate illustrating the tide gauge and the house it was placed in. A primitive tide gauge existed at the dockyard in Sheerness, but because the attendant was not on

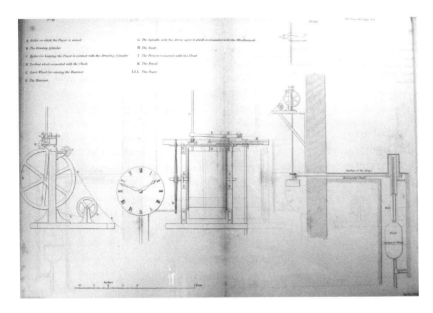

Figure 4.2. Henry Palmer presented an account of his self-registering tide gauge to the Royal Society of London, and it was subsequently published in the *Philosophical Transactions,* one of the few entries by an artisan in the early Victorian era. Henry Palmer, "Description of a Graphical Register of Tides and Winds," *Philosophical Transactions* 121 (1831): 209–13.

duty at night and was often absent during the day, Mitchell redesigned the machine to make it self-registering.[78] Because his machine traced the entire tide, the observer could detect "the most trifling irregularity which may have occurred at any time during the lunation."[79] The editor of the *Nautical Magazine* hoped that "ere long we may see such establishments formed at all our principal sea ports."[80] Such gauges, if used systematically, would provide information pertinent to the local causes of the tides "and shortly enable us to explain all the laws which influence and regulate this most interesting phenomenon."[81] Both Palmer and Mitchell were civil engineers working at the Admiralty dockyards or in the service of the British navy, seeking to update the Admiralty's method of tide observation through innovative mechanical inventions.

John Washington, a surveyor working under Beaufort for the Admiralty, sent Beaufort the diagrams from the Sheerness dockyard, suggesting that they construct similar gauges at Portsmouth and at Plymouth, an expense he believed would be "trifling compared with the importance of the results."[82] Washington exemplified the scientific interests of the surveyors responsible for gathering material for the British Admiralty,

Figure 4.3. Close-up view of the graphic register on Palmer's self-registering tide gauge. Henry Palmer, "Description of a Graphical Register of Tides and Winds," *Philosophical Transactions* 121 (1831): 209–13.

material that increasingly included scientific observations of the world's oceans. "Surely it is high time," he added, "we should make as accurate observations of one of the most striking phenomena of the ocean." Surveyors used their interest in science to advance professionally within the Admiralty. Washington, for one, was particularly interested in the tides, and he eventually succeeded Beaufort as hydrographer to the Admiralty.

Beaufort issued detailed instructions to the officers in charge of each survey from 1829 onward. The most important aspect was the sounding of the ocean or channel floor. To make correct soundings, however, the depths needed to be reduced to low water. As Beaufort noted,

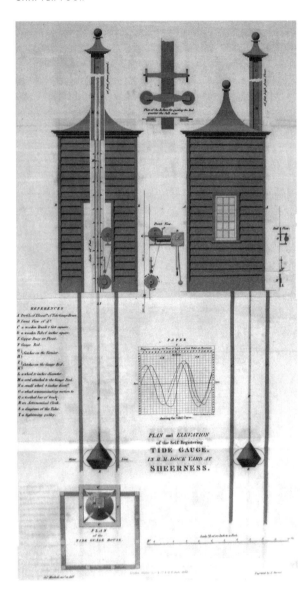

Figure 4.4. The first self-registering tide gauge actually used to take observational data. It was constructed by J. Mitchell, civil engineer at the Sheerness Dockyards. A description of the gauge, along with this plate, was published in the first volume of the *Nautical Magazine* (1832).

"those depths can be obtained only by carefully comparing the work in the boats with the simultaneous depths marked by the self-registering tide gauge, or shown on the tide pole; the judicious place of which will therefore be a preliminary step of prime importance."[83] After the efforts of Lubbock and Whewell had begun, Beaufort also included a

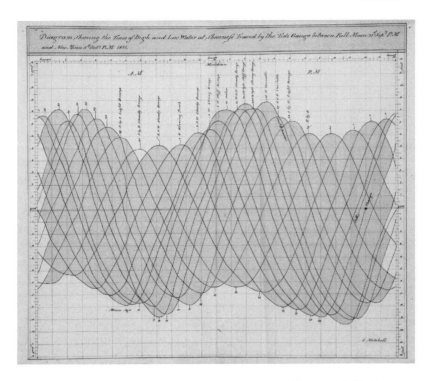

Figure 4.5. Diagram showing high and low tide at Sheerness, traced from J. Mitchell's tide gauge between 27 September 1831 and 5 October 1831. Through these diagrams, researchers could determine "by eye" the semimenstrual inequality, the diurnal inequality, and other laws of the tides. This tracing is reproduced from the original housed in the Royal Society Papers, Meteorological Manuscripts 108, Royal Society of London.

special section on the tides in each surveyor's instructions: "A minute examination of the tides, including all the data by which the establishment for each port may be accurately computed, their local set, and the extent to which they are influenced by periodic winds and by sea currents, is so prominent a part of every survey that it need scarcely be dwelt upon here. But as the subject of the tides has been lately taken up in such a large and masterly manner by Mr. Lubbock and Dr. Whewell, so it should be the willing duty of all officers who have it in their power to contribute fresh facts for their investigation." Beaufort suggested that the captain of each survey assign "two or three steady persons, perhaps quarter masters" to devote their time exclusively to the tides and "who should be taught to feel a pride in its correctness and completeness." Beaufort ended his instructions by noting that tidal readings were being taken at the major naval ports in Britain, and "nothing is wanting but additional series of correct observation, to extend indefinitely the number of those ports."

After Lubbock and Beaufort had persuaded the Lords of the Admiralty to have tide measurements initiated at the major naval dockyards, Beaufort tried to extend the observations throughout the globe.[84] He asked Captain Woolmore to begin tide measurements at Ramsgate[85] and asked Commander William Mudge to make very delicate observations on the western shores of Wales.[86] If the surveyors themselves could not take tidal observations, Beaufort asked that they try to get others to do so. Beaufort suggested that Lieutenant Manuel John Johnson at St. Helena persuade the "very sensible and liberal governor to establish an accurate tide gauge in such a situation as not to be affected by the swell, and to have it constantly watched and registered for a series of years."[87] St. Helena was significant because it was the farthest from the mainland of any island in the Atlantic.

Beaufort also was in close contact with Captain James Horsburgh, the hydrographer of the East India Company. Owing to the extensive trade that Britain maintained with the East Indies, the two hydrographers had collaborated to keep the surveys, soundings, and charts up to date. Indeed, that the charts of India were more complete than the charts of Britain's own shores reflected the importance of India to British trade. In November 1832, Beaufort wrote to Horsburgh detailing the Admiralty's work on the tides, explaining the additional benefit that would be derived from tidal data from a few ports in the East Indies. "I want to know from you whether the Directors, whose minds appear to be ever open to the cause of science, would order their servants to cooperate." Beaufort suggested Bombay, Prince of Wales Island, Singapore and Macao, "and at any other place that the Court may think conducive to the object of forming a general theory of Tides."[88] Horsburgh assured Beaufort that the Court of Directors would do everything in its power to have the tides registered. As with the observations initially made at the London Docks, tide observations already had been taken at many ports in the East Indies but had lain dormant because no one knew what to do with them. The East India Company was more than happy to have these observations put to some constructive use, especially if they could be bartered for charts from the Hydrographic Office of the Admiralty.

By using contacts such as Captain Horsburgh, Beaufort did all he could to have measurements of the tides made outside Britain. By 1833, tide data was coming in from Malta, the Cape of Good Hope, Port Royal, Jamaica, Halifax, and Bermuda, all by orders from the Admiralty on

Beaufort's recommendation. Beaufort also used the Admiralty's involvement and the interest of Lubbock and Whewell to persuade others not directly under his supervision to take up the subject. To Sir Edward Parry, the autocrat of Port Stephens, he wrote, "I think you will find the tides to be an excellent object of quiet pensant. I am happy to tell you that two of our best men Whewell and Lubbock have taken up the subject with great energy. The Admiralty has ordered day and night observations to be recorded for their use at several ports in England Wales and East Indies, Gibraltar, Cape of Good Hope etc, and it would be a happy adaptation of your present resources as Governor to your old devotion to science if you established a tide gauge at Port Stephens. I will send you a couple of forms for that purpose."[89] The high expectations conveyed in Beaufort's letter should not be missed. If scientists achieved quick advances in research programs in physical astronomy, Beaufort was the person to truly feel its pulse. He was at the center of an expanding network radiating from his staff at the Hydrographic Office to surveyors and port officials, harbor masters and dockyard attendants, and commissioned and noncommissioned officers stationed at home and abroad.

Francis Beaufort's appointment as hydrographer helped strengthen the relationship between the Admiralty and the scientific community and led to increased funding of new, empirically based projects in physical astronomy. Beaufort's encouragement also heightened the enthusiasm of the natural philosophers themselves. In November 1833 Whewell told Lubbock there was little doubt that they could eventually move the tides to the level of other parts of astronomy "and have observers and calculators employed upon it in all parts of Europe." First, however, they needed to demonstrate what could be done. "This is the way in which science generally begins, and we may as well make up our minds to it."[90] Thanks to Beaufort's wholehearted participation, Whewell had gathered reams of data. He was now poised to publish his first paper on the tides, work he hoped would lead the government to "set up a tide department in the Hydrographer's office.[91]

## Whewell's Research Program

By the summer of 1833, Whewell had amassed enough data to read his first paper on the tides to the Royal Society of London.[92] It outlined past advances in tidology, defined the correct terminology, and delineated the research he hoped to do in the field, a program vastly different from Lubbock's earlier work. Lubbock's research was overtly empirical, highly mathematical, and though of practical significance for tide table calculators and navigators, addressed to at most a handful of physical astronomers across England and the Continent. Whewell aimed for a much broader audience. Though practicality might have been Whewell's ultimate goal, his first publication was an addition to tidal theory. Whewell began by establishing his niche in tidal research, distancing himself from Lubbock while acknowledging his great debt to his former pupil. He introduced the tides as a problem of physical astronomy, governed by Newton's law of universal gravitation and thereby the positions of the sun and moon. In attempting to perfect Newton's theory, both Daniel Bernoulli and Pierre-Simon Laplace had resorted to unrealistic assumptions, such as a perfectly spherical earth completely covered with water, which made their analysis highly questionable. According to Whewell, continents, islands, and narrow channels had to be accounted for rather than assumed away.

The first section of Whewell's paper deduced the laws of the motion of fluids in bodies of water separated by landmasses; he determined theoretically what the tides should look like once islands were introduced. This included completely derivative tides found in rivers and channels where the tide was not affected by the direct action of the sun and moon.[93] Whewell's method entailed comparing the actual paths of the tides from observations with those deduced from theory to uncover the differences, and reducing those differences to mathematical laws. He could then use these laws to find the establishments of ports around the world. But Whewell reserved this discussion for the third section of the paper; I likewise will reserve this use of global observations to form a tidal map of the world's oceans for the next chapter. Following his earlier approach to scientific subjects such as mineralogy, Whewell interjected a section on the precise terminology needed to systematize the accumulation of observations. He then defined the important quantities necessary to determine the establishment of the port, including the right ascension of the moon from the sun, the semimenstrual inequality, and the age of the tide.[94]

The last section of Whewell's paper, "Suggestions for Future Tide-Observations," outlined his research program, which he would follow intensively for several years and then work on intermittently for the next two decades. It emphasized the need for systematic observations based on theoretical considerations and echoed a paper he gave at the third meeting of the British Association that same year. "It has appeared in the course of the preceding discussion," Whewell reflected, "how

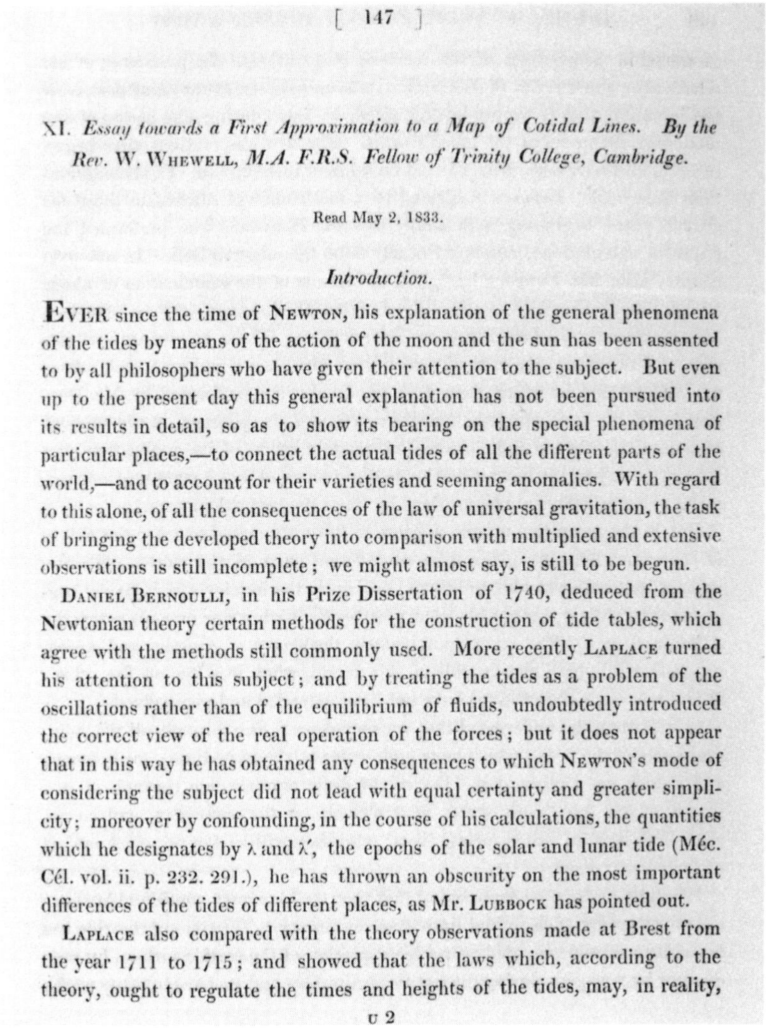

[ 147 ]

XI. *Essay towards a First Approximation to a Map of Cotidal Lines. By the Rev.* W. WHEWELL, *M.A. F.R.S. Fellow of Trinity College, Cambridge.*

Read May 2, 1833.

*Introduction.*

EVER since the time of NEWTON, his explanation of the general phenomena of the tides by means of the action of the moon and the sun has been assented to by all philosophers who have given their attention to the subject. But even up to the present day this general explanation has not been pursued into its results in detail, so as to show its bearing on the special phenomena of particular places,—to connect the actual tides of all the different parts of the world,—and to account for their varieties and seeming anomalies. With regard to this alone, of all the consequences of the law of universal gravitation, the task of bringing the developed theory into comparison with multiplied and extensive observations is still incomplete; we might almost say, is still to be begun.

DANIEL BERNOULLI, in his Prize Dissertation of 1740, deduced from the Newtonian theory certain methods for the construction of tide tables, which agree with the methods still commonly used. More recently LAPLACE turned his attention to this subject; and by treating the tides as a problem of the oscillations rather than of the equilibrium of fluids, undoubtedly introduced the correct view of the real operation of the forces; but it does not appear that in this way he has obtained any consequences to which NEWTON's mode of considering the subject did not lead with equal certainty and greater simplicity; moreover by confounding, in the course of his calculations, the quantities which he designates by $\lambda$ and $\lambda'$, the epochs of the solar and lunar tide (Méc. Cél. vol. ii. p. 232. 291.), he has thrown an obscurity on the most important differences of the tides of different places, as Mr. LUBBOCK has pointed out.

LAPLACE also compared with the theory observations made at Brest from the year 1711 to 1715; and showed that the laws which, according to the theory, ought to regulate the times and heights of the tides, may, in reality,

υ 2

Figure 4.6. The opening page of William Whewell's first publication on the tides in the *Philosophical Transactions.* William Whewell, "Essay towards a First Approximation to a Map of Cotidal Lines," *Philosophical Transactions* 123 (1833): 147.

extremely imperfect, and in many cases contradictory, are the statements which we at present possess concerning the establishments or tide-hours at different places. This has arisen in a great measure from the circumstance, that these observations were made without any settled rule or any definite object."[95] Whewell was restating views expressed parts of the ocean it is in advance of that place. The same is the case with the IV o'clock line, but the advance being greater, the two convex portions at the two ends of the island are turned towards each other; in the V o'clock hour line these portions touch; and thus the line may be considered as formed of two, which meet at the point of contact just mentioned, one line having its two ends on the shores of the island; the other line running across the ocean like the uninterrupted lines, but with an indentation towards the island. After this time these two lines give rise to two separate waves, 6 and VI; the former moving in a retrograde direction towards the island; the latter moving forwards, and gradually obliterating the indentation produced by the island.

It appears in this way that there is *a point of divergence of cotidal lines* on the side of the island which is towards the coming tide-wave, and *a point of convergence* on the opposite side.

| Fig. 3. | Fig. 4. |
|---|---|

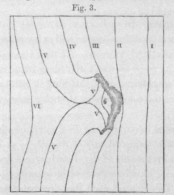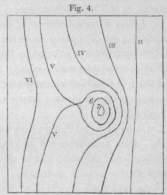

And if there be shallower parts of the ocean, not connected with any land, or connected only with small islands, the effect upon the form of the cotidal lines will be of the same nature, but may go still further. See fig. 4. In advancing upon such a part of the ocean, the cotidal lines immediately behind it will be brought closer together, while to the right and left of the place they will proceed without a corresponding thronging. Hence the cotidal curve on the two sides will advance beyond the islands, while it cannot pass directly over the islands themselves. The undulation will be propagated from the right and left

Figure 4.7. Whewell's diagrams demonstrating how islands, bays, and estuaries affect oceanic tides. From his "Essay towards a First Approximation to a Map of Cotidal Lines," *Philosophical Transactions* 123 (1833): 153.

in his correspondence with James David Forbes a year earlier. Like meteorology, tidology needed someone to take up the field and advance it, to organize its data collection according to unifying principles, systematize its nomenclature, and mathematize its principles—in short, to define the discipline.

Whewell presented two "circumstances of real importance" for the furtherance of tidology. His second circumstance, "comparative simultaneous tide observations," will be discussed in the next chapter; it represents Whewell's spatial approach to the tides. The first circumstance, "good and long-continued observations" at any single port enabled the researcher to determine the establishment of the port, along with the semidiurnal equality, the age of the tide, and the effects of the parallax and declination of the moon and sun. The history of physical astronomy suggested this approach as the correct way to advance a young and undeveloped field. History was instructive, and it could guide the modern researcher on the most promising course of action:

It is, perhaps, remarkable, considering all the experience which astronomy had furnished, that men should have expected to reach the completion of this branch of science by improving the mathematical theory, without, at the same time, ascertaining the laws of the facts. In all other departments of astronomy, as, for instance, in the cases of the moon and the planets, the leading features of the phenomena had been made out empirically, before the theory explained them. The course which analogy would have recommended for the cultivation of our knowledge of the tides, would have been, to ascertain, by an analysis of long series of observations, the effect of changes in the time of transit, parallax, and declination of the moon, and thus to obtain the laws of phenomena; and then proceed to investigate the laws of causation.[96]

Beginning in 1833, four years before the publication of his *History* but near the time he started writing it, Whewell embarked on this exact route. Lubbock, with Dessiou's help, had produced tables for the effect of the moon's hour of transit, declination, and parallax from long series of observations. The next step, a lesson from the analogy with the history of physical astronomy, was to obtain the general laws that accounted for these results, and only afterward to investigate the laws of causation.

## Long-Term Observations

After Whewell published his first essay on the tides and outlined his research path, he also turned more faithfully to what he referred to as his "Induction," meaning his history and philosophy of science. He was thus pursuing both studies at the same time. "Between my Tides and my Induction," he wrote to Richard Jones in mid-November 1833, "I have given myself a holiday of three days to write letters." In a passage that demonstrates Whewell's own views on the close relationship between his work on the tides and his *History of the Inductive Sciences,* he continued: "I am meditating the returning forthwith and in earnest to my beloved Induction. I have been employed all the term hitherto upon a thumping paper on the Tides, which I intend to be a step of some consequence in the theory. I wish I could explain to you how useful my philosophy is in shewing me how to set about a matter like this, and how good a subject this one of the Tides is to exemplify it."[97]

Whewell wrote this letter to Jones the day after he sent the Royal Society of London his second paper on the tides, "On the Empirical Laws of the Tides in the Port of London." Even its title built on Lubbock's earlier "Tides in the Port of London." Whewell was extending Lubbock's work, generalizing his empirical tables through an analysis of phenomenological laws. Newton, Bernoulli, and Laplace had all offered mathematical solutions of the tides based on suppositions remote from reality. None had found the empirical laws of the phenomena and compared those with their admittedly simplified theories. "In short, the mathematicians who have treated this subject have not completed their task by giving rules for the calculations of tide tables," Whewell wrote, "and showing that the tables so produced agree with the general course of the observations in all essential circumstances."[98]

Whewell began with the laws of the times of high water. He first confirmed Lubbock's earlier work on the corrections owing to the declination and parallax of the moon, using the tables Dessiou produced, then moved into uncharted territory, extending the analysis first to the solar corrections and then to the empirical laws of the *heights* of high water. With Dessiou's help he found the correction for both the lunar and solar parallax and the declination for height and times of high water, which he then had Dessiou form into tables to compare with theory. Reflecting on the comparison of theory and observation, Whewell admitted it would be imprudent to deduce any general views, since the discussion of observations at any one place would probably not give the true laws of the tides. Furthermore, the tides at London were extraordinary, the result of two tide waves traveling around the island by different routes. Deducing the true laws of the tides, Whewell confessed, would require the accumulation and discussion of masses of observations at various places. His method, Whewell acknowledged, might not be the most practical way of arriving at the true theory, but it was "at least that to which, founding our expectation on the past history of science, we may look with most hope."[99] Whewell had made some progress. By comparing the formulas found from observations with those deduced from theory, he saw that the different configurations of the sun and moon determined the tide's lagging behind the equilibrium tide, what he termed the age of the tide. Whewell concluded by noting that the empirical formulas he obtained matched the observations with "tolerable exactness" and could be used to calculate tide tables "as readily as any other empirical rules."[100] Lubbock had doubted that Whewell could obtain the laws of the inequalities, and Whewell could now reply that "I have now not only done this, but also got nearer the theory than I expected to do."[101] The problem, of course, was that he had done this for only one port, a port admittedly peculiar in its tidal action. Philosophically, this would not do.

Whewell once again turned for help to the British Association for the Advancement of Science. He had hired Dessiou to work on the tides for the port of London using monetary grants that he and Lubbock had received from the Association. What especially excited Lubbock, and now Whewell, was the discovery at the Association's first meeting of nineteen years of observations made at Liverpool by William Hutchinson, dockmaster and author of the well-known *Practical Navigator*. Whewell was especially eager to have these Liverpool observations reduced, "for it will be infinitely interesting to see whether the inequalities due to parallax and declination assume the same form there as at London." Whewell took charge of the application of the materials, as well as asking the British Association for funds to have them reduced, and used the grant it provided to put Dessiou to work.[102] From Dessiou's reductions of the Liverpool observations, Lubbock published a paper in 1835 similar to the one he had written on the tides for the London Docks. Consisting of three pages of text and nineteen pages of tables, Lubbock and Dessiou's paper contained the empirical tables needed to produce tide tables for the port of Liverpool.[103]

Whewell again sought to generalize these results, to find the phenomenological laws that followed from Lubbock and Dessiou's empirical tables. His results, he believed, would have the added benefit of "testing and improving the formulae to which I was led by the London observations."[104] His approach therefore was very similar to his previous paper on the phenomenological laws for the port of London, and he was again elated about his results. His researches on the Liverpool tides, Whewell boasted, "have both confirmed, in general, my formulae, and have given me the means of very much improving them. The corrections for lunar parallax and declination, which, as far as they depended on the former investigation [on the port of London], might be considered as in some measure doubtful, and probably only locally applicable, have been so fully verified as to their general form, that I do not conceive any doubt now remains on that subject."[105]

Whewell acknowledged that the constants used for Liverpool—the mean interval of the tide and transit of the moon; the age of the tide; and the amount of the solar and lunar inequalities—were different from those needed for the London tides. But once these different constants were found, they agreed with the London tides "with a precision not far below that of other astronomical phenomena." He concluded that it "is impossible to doubt, under these circumstances, that the theoretical formula truly represents the observed facts."[106] Whewell would use similar reasoning (and language) in his *Philosophy* when discussing the way hypotheses should be tested. In his famous discussion of the "consilience of induction," Whewell offered guidelines by which researchers could assess the truth of their hypotheses. One of those guidelines posited that a hypothesis should *"foretel* phenomena which have not yet been observed;—at least all phenomena of the same kind as those which the hypothesis was invented to explain."[107] That his formulas, derived from the port of London, fit so well with the observations of the same kind at Liverpool was for Whewell a sure sign that he had advanced the field of research.

## Conclusion

The tides had long been of significant practical interest to the maritime community, advanced by tide table makers and almanac publishers before the Society for the Diffusion of Useful Knowledge forced it on the British scientific elite. Lubbock, in turn, brought the study of the tides to the attention of the larger scientific community, and Whewell, for one, quickly made it his own. Through his papers read before the Royal Society and published in the *Philosophical Transactions,* in his reports read before the British Association for the Advancement of Science, and in his correspondence with the leading natural philosophers and scientific servicemen in Britain, Whewell transformed the field's aims and methods. By the mid-1830s, before the Magnetic Crusade had gotten under way and before similar work was done in meteorology, Whewell had established tidology as a legitimate geophysical frontier.

By the time Whewell published his second paper on the Liverpool tides in 1835, he was already numbering his publications—fourth series, fifth series, and so on—carving out a research space that would end only with his fourteenth series in 1850.[108] His work as the leading tidologist in Britain allowed him to comment on the nature of science more generally, making him a successful critic and evaluator of science. He was more than comfortable applying his own research to demonstrate what science was and how it could be advanced, using his tidal research extensively in both his *History* and his *Philosophy* to comment on methodology. His twenty-year research project on the tides belies the claim that "at no time did he ever fully devote himself to any one scientific project."[109]

At least since his election to the Royal Society of London in 1823, Whewell had been "on the look out" for a scientific subject that he could actively contribute to. His writing of textbooks and his early excursions into crystallography and mineralogy helped him create a unique methodology geared toward fostering a burgeoning field of research. He went to the foundation of the subject, studying its nomenclature, its history, and the certainty of its mathematical laws. In its mature form, this approach was published in 1840 as his *Philosophy of the Inductive Sciences* and explained both the kind of certainty science could attain and the best methods for achieving it. In the early 1830s, however, Whewell used this heuristic in his own research to advance the particular field of tidology.

In doing so, Whewell made the tides visible. One important way his approach differed from Lubbock's was his reliance on associate laborers. While Lubbock limited his own investigations to long-term observations and largely scoffed at attempts to include amateur observers, Whewell worked assiduously in the halls of the Admiralty and the meetings of the British Association to incorporate observers from all areas of the globe. As with the inception of the study on the coasts and in the estuaries of Britain, the theoretical revival of the tides was hierarchical and collaborative. Whewell's involvement in the tides demonstrates that research in physical astronomy relied on a broad base of interest, activity, and support. From Admiralty surveyors to Christian missionaries, diverse groups were marshaled from disparate areas. In the study of the tides, natural philosophers advanced the theory from observations they obtained from the Admiralty, observations reduced by calculators like Dessiou, paid with funds from the deep pockets of the British Association. The invention of self-registering tide gauges, built by royal dockyards engineers, improved the accuracy of observational data. The rise of physical astronomy in the second quarter of the nineteenth century necessitated the involvement of the Admiralty, and even philosophers as adamant as Whewell concerning the autonomy of scientific investigation increasingly accepted financial support from the British government.

That the Admiralty was so involved is not surprising. Naval historians widely acknowledge that the British Admiralty was at the forefront of technological and scientific advance.[110] As trade expanded, and with it the need to protect that trade, the Admiralty had to grapple with problems of a highly scientific nature, such as terrestrial magnetism and the theory of the tides and waves, and was forced to rely more heavily on scientific specialists. The result was a stronger relationship between the Admiralty and the scientific community and increasing government patronage of science. Indeed, as the next chapter demonstrates, the most significant aspect of Whewell's methodology was extending the study of the tides from its limited confines in the Thames estuary to all the world's oceans, a geographical turn that nicely complemented the Admiralty's desire to rule the waves.

FIVE

# The Tide Crusade

Setting boundaries with respect to space, time, scale, and environment then
becomes a major strategic consideration in the development of concepts, abstractions, and theories. It is usually the case that any substantial change in these
boundaries will radically change the nature of the concepts, abstractions, and
theories.

DAVID HARVEY, *JUSTICE, NATURE AND THE GEOGRAPHY OF DIFFERENCE* (1996)

When William Whewell wrote to John Lubbock in early
April 1832 announcing that he meant to enter the field
of tidology, two other young and adventurous gentlemen
were on the other side of the Atlantic, viewing the lush
tropical forests of Brazil for the first time.[1] Charles Darwin
and Robert Fitzroy had embarked three months earlier on a
voyage of exploration that eventually circumnavigated the
globe. John Henslow, Darwin's former professor in Cambridge, had given Darwin a copy of Charles Lyell's *Principles
of Geology* shortly before his departure, and the aspiring geologist pored over it between bouts of seasickness while on
board HMS *Beagle*. He convinced himself of the validity of
Lyell's reasoning while traveling through South America,
and like Lyell, he used the immensity of geological time
to uncover the timeless laws of nature. When he entered
the tropics of South America, however, it was not Lyell but
Humboldt who first captivated Darwin's romantic imagination. "I formerly admired Humboldt," Darwin gushed, "I
now almost adore him."[2] Darwin memorized large sections
of Humboldt's *Voyage to the Equinoctial Regions of the New
World*, enthralled by the animated descriptions of practicing science, especially the need to travel to the interior of

**157**

distant and exotic regions. Like so many other British surveyors, naturalists, and collectors, Darwin was following in Humboldt's footsteps.[3]

Darwin, as his adoration of Humboldt suggests, contemplated much more than the expansiveness of geological time during his months of journeying inland through South America. Geographical space proved equally important. "When on board H.M.S. *Beagle*, as naturalist," Darwin wrote to open his *On the Origin of Species*, "I was much struck with certain facts in the distribution of the inhabitants of South America, and in the geological relations of the present to the past inhabitants of that continent."[4] Relating the present to the past enabled Darwin to trace the changes in flora and fauna through geological epochs. But following Humboldt, the distribution of species through space—within a continent, between continents, or among islands in the Galápagos—provided Darwin with the intellectual framework to comprehend the overarching patterns of the world's diversity.

Darwin was not alone in his reliance on geographical space. As Darwin traveled inland to study South America's abundant biota, the captain of the *Beagle*, Robert Fitzroy, was stuck on board taking soundings and collecting tide observations for the Admiralty as part of a similarly expansive endeavor in physical astronomy. While Darwin had received the second volume of Lyell's *Principles* while on board the *Beagle*, Francis Beaufort had sent Fitzroy a copy of Whewell's paper on cotidal lines. Fitzroy was practicing a similar type of science, but with a slight twist: he looked outward over the ocean rather than inward through the continent.

Analogous to the role of geological time in natural history, astronomers had attempted to answer questions in physical astronomy through the expansiveness of astronomical time. Tidologists based their analysis on the positions of the earth, moon, and sun, a heavenly three-dimensional dance that brought a beautiful regularity to its laws. The relations of these celestial bodies, however, differed each year to produce exceedingly different tides on the world's coasts. Thus Lubbock tried to acquire enough observations to cover the system's entire nineteen-year cycle. He secured twenty-five years of observations from Isaac Solly, chairman of the London Docks Company, and nineteen years of observations from the Athenaeum in Liverpool. These long-term observations enabled him to determine the major tidal constants for the most important ports in England and cancel out the variables due to wind, weather, and human error. From the beginning, Lubbock studied the tides temporally, as he would do throughout his career.

Like Darwin and Fitzroy, however, Whewell viewed his science spatially; he believed the science of the tides should be explored over geographical areas rather than astronomical time. He relied on short-term but connected observations, extending those observations throughout the oceans just as Darwin and other naturalists extended their comparisons inland through continents. Throughout the 1830s, Whewell helped transform physical astronomy into a geographical science by combining Humboldt's spatial approach with the abundant resources of the British Admiralty. Whewell's was a horizontal science, one that covered vast distances and involved a conglomeration of people and institutions.[5] The Admiralty excelled at such extensive undertakings, for it was the most wide-ranging and powerful military institution in existence. British expansion made the spatial approach to science possible, and spatial science, in turn, helped extend Britain's imperial reach.

The advantage to the natural philosophers from this mutually beneficial relationship was often simply one of numbers: hundreds of scientific servicemen willing to undertake research that required the accumulation of observations from around the globe. As an anonymous entry in the first volume (1832) of the *Nautical Magazine* suggested, "There is, probably, no class of society which has more frequent opportunities of adding to the general stock of scientific knowledge than that composed of persons in the royal and mercantile navy. Their various avocations necessarily carry them to distant regions many seldom visited; while they not unfrequently have at their disposal various portions of time, which, added together, become considerable, and which might be most beneficially employed in the advancement of general science."[6]

The Admiralty had control over scores of mariners stationed at far-flung outposts of the British trading routes, from the chill waters of the Arctic to isolated islands in the Atlantic and Pacific. These servicemen, moreover, brought their efforts to the attention of both the government and the scientific elite.[7] In the mid-1830s Lubbock and Whewell received numerous letters from members of the Royal Sappers and Miners, the Preventive Coast Guard, and the Lighthouse Board.[8] Indeed, Whewell would use one such letter to great advantage to help transform tidology into a spatial science. But the Admiralty's participation involved more than numbers and expansion to "distant regions many seldom visited." The Admiralty was a secure source of patronage, willing and able to advance science in ways no other institution possibly could.

The initial creation and execution of the multinational tidal experiments of 1834 and 1835 highlights the collaborative nature of research in physical astronomy, particularly among the elite theorists in Britain and the many scientific servicemen and commissioned and noncommissioned officers of the Royal Navy. Whewell's global tidal experiments included institutions ranging from the East India Company to the East Liverpool Docks Committee, from the Bombay Geographical Society to the South African Literary and Philosophical Society. The people involved ranged from Joseph Foss Dessiou, an expert calculator, to the Duke of Wellington, the foreign secretary. The organization of global geophysical initiatives, including the oceangoing vessels and the observational equipment involved, vastly exceeded the scientific budgets of individual nations, and their execution and coordination crossed the arbitrary lines of nation-states. The result was an international scientific collaboration among France, Germany, Russia, the United States, and Britain.

The need to reduce shipwrecks and protect their economic interests underpinned other nations' willingness to participate in the global tide experiments of the mid-1830s. The loss of ship, cargo, and crew was not solely a British preoccupation. As Helen Rozwadowski has recently argued in her study of deep-sea exploration, the vast oceanic environment made cooperation among nations essential.[9] Only through the massive accumulation and sharing of observational data among nations could questions such as the patterns of the tides be answered. Scientists and politicians stressed how the purely scientific aspects of such research would lead to international harmony and goodwill, a rhetorical strategy that only slightly veiled the more political, if not outright militaristic, nature of such endeavors. Even though Whewell's research promised to lead to better tide predictions on the shores of Britain's rivals, other nations believed it was worth the price to gain knowledge of the tides on their own shores. In the study of the tides at least, an international scientific ethic transcended nationalist sentiments.

The transnational complexion of tidal studies mirrored a world that was increasingly held together through European imperial and colonial trade. The internationalism apparent in tidal studies was linked to economic and political developments in the late eighteenth and early nineteenth centuries.[10] It depended chiefly on the changing nature of worldwide economic relations and the internationalization of the economy that was gaining force at the time. During the 1820s and 1830s, a small group of political economists revived Adam Smith's arguments in favor of free trade. David Ricardo, a member of Parliament in the 1820s, argued fervently against outdated mercantilist policies, stressing that if all countries traded freely in the products that gave them a comparative advantage, the total quantity of goods would be raised and all countries would benefit. In the hands of a cogent thinker and lively rhetorician like Richard Cobden, leader of the Anti–Corn Law League, it was a small step to argue that adopting free trade would also help end international conflict. The Anti–Corn Law League transformed Adam Smith's promotion of the merits of free trade into a means for social unity and international peace.

Cobden advocated free trade on moral and religious grounds throughout the late 1830s and early 1840s. The earliest exposition of his views came in two pamphlets—*England, Ireland and America,* published in 1835, and *Russia,* published in 1836—overlapping the global tide experiments of the mid-1830s. Rather than waging war in defense of commerce, Cobden argued that society should look to commerce itself as "the grand panacea, which, like a beneficent medical discovery, will serve to inoculate with the healthy and saving taste for civilization all nations of the world."[11] Freer trade would produce more jobs, higher wages, and cheaper food.[12] But his most cogent argument was geopolitical. Increased international trade would bring governments together and lead to international harmony and goodwill.[13]

In the midst of the turmoil and domestic strife running through Britain in the 1830s, both political economists and natural philosophers fostered an ideology of internationalism to persuade the British government to change its outmoded policies. In the economic realm, the effect was a steady move away from mercantilist policies and toward freer trade and greater interdependence between nations. In ocean science, the effect was a more open dialogue among scientists from different nations and a move toward standardizing nomenclature, instruments, and methods. Yet there is little doubt that an acute nationalism pervaded the international tide experiments from the beginning, a mix of interests and motivations that brought both complexity and urgency to the science. While sailors, surveyors, military men, and interested amateurs took tidal readings, they also sketched bays and estuaries and measured currents and streams to determine how best to maneuver their vessels safely in and out of distant ports. The practice of global, spatial science was a means to control areas from afar, in this case from the halls of Trinity College, Cambridge, and the Hydrographic Office in London. The sheer mass of activity reinforced the notion of the superiority of science and the ability of the new scientists to control the shoals, bays, and estuaries of Britain's expanding possessions.

The Admiralty was most interested in the safety of its sea-lanes, but as its maritime interests expanded, this objective increasingly required a global view of the ocean as an organized and connected whole. Researchers realized, for instance, that the tides in the middle of the South Atlantic produced the tides that ran through the rivers in Britain, and

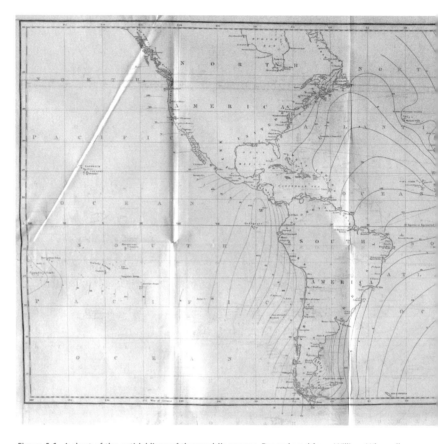

Figure 5.1. A chart of the cotidal lines of the world's oceans. Reproduced from William Whewell, "Essay towards a First Approximation to a Map of Cotidal Lines," *Philosophical Transactions* 123 (1833): 147–236.

that the earth's weather patterns and changing magnetic field were likewise global. Researchers used Humboldt's isolines, a graphical technique, as the filaments to construct an organized vision of the seas. Whewell, adapting the work of Edmond Halley in the seventeenth century and Thomas Young and Alexander von Humboldt in the early nineteenth, used what he termed "cotidal maps," filled with lines connecting places on the globe, including those far in the uncharted oceans, that experienced high tides at the same time. For instance, he drew lines connecting all the places that experienced high water at noon, at one o'clock, at two o'clock, and so on, to determine how the tide progressed from the deep ocean to shores of Europe.[14] As the science expanded geographically and

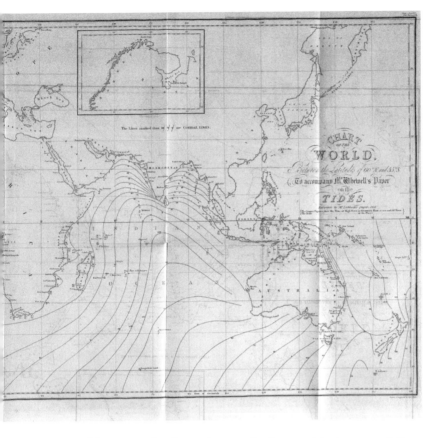

the myriad forces involved grew in complexity, graphical representation became the preferred mode of exhibiting scientific results. Spatial science and graphical representation matured apace.[15]

These advances occurred simultaneously with Britain's move from being the world's most industrialized country to being its largest imperial power. The product of Whewell's global tidal experiments, the cotidal map, was a document laced with economic and political significance. The isomap was part of a larger project of ordering nature's diversity according to rational categories of scientific analysis. It was also part of a broader economic and political goal of making the world's oceans safer for navigation and trade.[16] Recreating the collaboration behind such maps helps uncover the hidden practices that gave them such power. In assembling the isomap, Whewell fashioned a view of the ocean as knowable and thus controllable, a space now within the purview of science. Amid this confluence of political, economic, and scientific forces,

Whewell also formulated his views on scientific discovery and his definition of the modern scientist. Science and the new scientist ordered the world's oceans and contributed to Britain's global ascendancy. In the mid-1830s, as Whewell's cotidal lines graphically connected the coasts where he and Darwin practiced their science, his definition of the scientist also linked the two philosophers methodologically in a common endeavor set squarely within the politics of imperialism and the economics of worldwide trade.

## Whewell's Spatial Turn

The type of data early tidologists had at their disposal necessarily guided their research methods. Lubbock used observations over long periods to find the major constants for the ports of London and Liverpool, including the inequalities owing to the declination and parallax of the sun and moon. Whewell initially followed his former student in relying on long-term observations, but he fretted over the problems associated with it. Beyond London and Liverpool, no such continuing observations existed, and if they did, they were beyond the philosophers' reach. Prized as personal property, observations of tides were a valuable commodity. The nineteen years of observations Mr. Corbett made at Plymouth from 1811 to 1830 would have been exceedingly useful to Lubbock and Whewell to compare with observations made during roughly the same years at the London Docks and the port of Liverpool. But neither Whewell nor Lubbock could ever acquire them. Corbett had made the observations as part of his duties for the Breakwater Service in Plymouth, and they were still in his possession. "Unfortunately," wrote William Snow Harris, "I rather think Mr. Corbett is very jealous of parting with them as I could not after he left the Breakwater Service get him to let me copy them."[17] It turns out he was hoarding them to produce his own tide tables. Whewell amassed large numbers of local tables during his career as a tidologist, but he had trouble procuring the observations or the mathematical rules professional calculators used to produce them.

For Whewell, however, the problems associated with theory far outweighed the lack of observational data. According to the Newton-Bernoulli equilibrium theory, tide-producing forces caused by the sun and moon acted on the tides in the open ocean, which then advanced onto the shores of Britain and Europe. The position of the sun and moon not on the day of those tides, but at some (unknown) time preceding it, produced the tides in most ports. The tides in the port of London, for instance, were derivative tides that traveled from the South Atlantic up the Thames estuary. To determine the forces that produced the tides on these coasts Lubbock and Whewell used the position of the sun and moon at some previous time or "epoch." The equilibrium theory did not offer any rules for determining the correct epoch; it had to be calculated for each port directly from observation. Lubbock usually began with the epoch contemporaneous with the tide as it hit the coast of England and worked backward until he hit on the numbers most consistent with actual observations, a roundabout and exhausting way to save the phenomenon. Such labor-intensive empirical investigations constituted the main occupation of tidal calculators. Lubbock had Dessiou create tables for the declination and parallax of the sun and moon at different epochs of twelve hours. The closest fit with observational data determined the correct epoch. Combined with the paucity of extant observations, this gap in theory made tidology look exceedingly backward compared with the advanced state of other parts of astronomy.

Whewell came up with an alternative way to close these gaps in both observation and theory. In the last section of his first publication on the tides in 1833, he outlined his two approaches.[18] The first was an extension of Lubbock's work to find the phenomenological laws (mathematical rules) to account for the data through long-term observations. Following Lubbock, Whewell's next several publications on the tides concerned the phenomenological laws of the tides at the ports of London and Liverpool. "But in the meantime," Whewell argued, "no one appears to have attempted to trace the nature of the connexion among the tides of different parts of the world."[19] Whereas Lubbock had focused on the tides at certain strategic ports using only long-term observations, Whewell created his own niche in tidology through this second approach, by tracing the connections among the tides over the entire surface of the earth.

Extending the boundaries of tidology this way was uncharted territory, both geographically and methodologically. Consistent with his philosophy of science, Whewell began with a bold unifying principle. By drawing cotidal lines connecting all parts of the ocean where high water occurred at the same time, he could determine the correct epochs directly from his chart of cotidal lines without years of observations at every port. "For the age of the original tide in any part of the open ocean being known," Whewell explained, "the age of the tide derived from the original tide in any other part would be known from the number of intervening cotidal lines."[20] All he had to do was count the lines. Furthermore, owing to the difficulty of taking tidal readings in the open ocean, his graphical method actually created data points in "parts of the ocean where no tide observations have been, or perhaps can ever be made."[21] It extended both the number of his original observations and their geographical area.

Whewell's sources for his initial map of cotidal lines published in the *Philosophical Transactions* for 1833 were by no means accurate. As I described in the previous chapter, Whewell used the work of missionaries, local observers, and Admiralty surveyors as well as all the observations Beaufort could acquire through his position as hydrographer. Whewell also used the best printed sources then available: Joseph-Jérôme Lalande's *Astronomy*, J. W. Norie's *Epitome of Navigation*, sailing directions of all kinds, and the charts and manuscripts in the dusty archives of the Hydrographic Office of the British Admiralty. The graphical method allowed Whewell to test the accuracy of this diverse data, to determine which he could use and which he should reject. It brought "as much regularity and similarity in their form and intervals" as the data would allow, an early example of the bold unifying process that he championed so heartily in both his *History* and his *Philosophy*. This way of using the graphical representation of data, Whewell argued, provided the best means of approaching a science plagued by inconsistent, incomplete, and haphazard data.[22]

Whewell's use of cotidal lines to bring regularity to geographically diverse observations represented a novel direction in tidal research. Newton, Bernoulli, and Laplace had never imagined such an approach, and though both Young and Lubbock included diagrams of cotidal lines in their work (Young called them "contemporaneous lines"), neither used the graphical method for anything more than presenting the data. Whewell used his cotidal maps as an organizing principle to determine further laws, to find the appropriate epochs, and ultimately to guide further research. For Young and Lubbock the cotidal map was a visual display; for Whewell it was a unifying tool. Convinced of his method and confident in his approach, Whewell wrote expectantly that "we may in a very few years be able to draw a map of cotidal lines with certainty and accuracy; and thus to give, upon a single sheet, a tide table for all ports of the earth."[23] Whewell's goal was to create a synoptic view of the world's tides on a single sheet of paper.

Lubbock's published works were highly mathematical and aimed at a select audience who had access to the *Philosophical Transactions;* only those adept at the mathematics involved in physical astronomy could follow his research. Whewell's goals were broader: to reach the maritime community, not necessarily for its own sake, but to advance his geographical approach to the science of tidology. He described his new approach in several circulars, through popular evening lectures at the British Association meetings, and in numerous entries in the *Nautical Magazine*, a monthly journal published out of the Hydrographic Office for mariners stationed throughout Britain's vast possessions. In the first volume of the *Nautical Magazine*, Whewell noted that besides the irregularities caused by wind and weather, "there are many permanent anomalies, which no theory has yet explained, and to which we do not know how to adapt our tide-tables."[24] That is, tide tables did not travel well. Those that were exact for one place did not work for another part of the coast, even close by, "and mathematicians have not yet learned to make accurate and trustworthy tables for any place, without having a long and careful series of observations *made at that very place.*"[25] To solve this problem, Whewell outlined his new approach: "Continued observations at the same place are connected by relations of time; comparative observations at different places are connected by relations of space. The former relations have been made the subject of theory, however imperfectly; the latter have not."[26]

This was Whewell's spatial turn. And it necessitated a radically different way to exhibit the data. Comparative observations, Whewell noted, were almost impossible to reduce to calculation owing to the extreme complexity of the forces involved. "But, though the connexion of the tides in different places cannot be calculated," Whewell argued, "it can be expressed." While long-term observations had traditionally been placed in tables, comparative observations demanded the modern graphical method.

Whewell's first method was based on time—that is, observations at strategic ports over long periods. His second approach was organized around space, conceptualizing the science geographically and representing it graphically. By constructing a theory of the tides as they progressed in the ocean and then expressing that theory through a synoptic map, he could extrapolate from one port to the next and eventually to all the ports in Europe. To develop such a theory of the progression of the ocean tides, he required short-term observations at a large number of places along the coast rather than years of observations at every port. But there was a catch. The observations had to be simultaneous. They needed to be carried out by trustworthy observers with calibrated instruments stationed in strategic areas around the globe. This meant that the observations had to be organized beyond what a single researcher could do.

Whewell thought he was just the person to oversee this approach. He combined scientific status and political influence with a scientific methodology based on a study of the history of science geared toward

# SUGGESTIONS

## FOR PERSONS WHO HAVE OPPORTUNITIES TO MAKE OR COLLECT

## Observations of the Tides.

It was shewn by Newton, nearly 150 years ago, that the fact of the Tides and several of their circumstances, resulted from the law of the Universal Gravitation of matter. But in this interval of time scarcely any thing has been done which might enable us to combine into a general view the phenomena of the Tides as they take place in all the different parts of the world; and at very few places have good and continued observations been made and published. It is conceived that by collecting such observations as have been made, or may easily be made, the connexion and relation of the Tides of all the parts of the Ocean may be in a short time clearly made out ; and that persons may be induced to make such careful observations as may serve to be compared with the theory. In this hope the present paper is circulated.

The most useful Observations with reference to our general knowledge of the Tides are the following, beginning with those which are most easily made :

1. The Observation of the Time of High water at a known place, on any day, and especially at new and full moon.

2. The Observation of the Time of High water on several days in succession at the same place.

3. The Observation of the Height of several successive Tides at the same place.

4. Observations of the comparative Time of High water on the same day at different places in the same seas.

1. An observation of the Time of High water at a given place on *any* known day may be useful.

If the Time of the *Moon's southing* on the same day be noted, this will facilitate the use of the observation, and will furnish an additional evidence of the correctness of the date.

The Time of High water on the days of *New and Full Moon* is more particularly useful than on other days.

Observations of the Time of High water may be made with sufficient accuracy without a tide-post. A place ought to be selected where the water is tolerably smooth.

2. If there be opportunity at any place, it is desirable to observe the Time of High water *every day for a fortnight.*

If it be ascertained that the two tides on the same day occur at regular intervals, *one* of them only need be observed.

But there are often irregularities in the relative Times of the morning and evening Tide; and these irregularities are different for different ages of the moon. In this case *both daily Tides* should be observed.

Figure 5.2. Circular sent out by William Whewell in 1833. Notice his spatial move in the second sentence, arguing that "scarcely any thing has been done which might enable us to combine into a general view the phenomena of the Tides as they take place in all the different parts of the world." Reproduced from Whewell Papers, R.6.20/8.

advancing a developing field of research. He had successfully adapted the tides to his heuristic to advance tidal theory in years rather than the centuries needed for the perfection of other branches of physical astronomy. Now all he needed were organized, simultaneous observations from around the globe. He didn't have any. But he did have a promising prospect.

## Science and the Preventive Coast Guard

Whewell's work on the tides brought him into continual correspondence with men in the Royal Navy. He was gaining some notice for his research, mostly through short entries in the *Nautical Magazine* that summarized and popularized his lengthy essays in the prestigious *Philosophical Transactions*. Through these publications and his correspondence with Beaufort and the staff at the Hydrographic Office, Whewell was in contact with eminent scientific servicemen of the British Admiralty as well as cartographers, dockyard officials, and tide table calculators. But of all the communications Whewell received from navy personnel and private individuals concerning the tides, one letter in particular stands out. Richard Spencer, a captain in the Royal Navy, reminded Whewell late in 1832 that naval officers of the Preventive Coast Guard were stationed within a few miles of each other around the entire coast of Great Britain, under the immediate direction of Captain William Bowles, the comptroller general. Spencer had no doubt that a request from Whewell would spur Captain Bowles to take action, perhaps even to order the chief officers to observe the tides at their stations.[27]

Reforms within the Coast Guard made Spencer's suggestion an exciting possibility. After the Napoleonic Wars, mass unemployment—what historians refer to as the "Great Slump"—hit the Royal Navy. After war ended in 1814, the Admiralty cut the ships in commission from 713 to 121 over the next four years, and the number of men fell from 140,000 to a mere 20,000.[28] Most of these men simply drifted back into their civilian lives, while some found employment outside the Royal Navy, though still on half pay, by either transferring directly to the merchant navy or enlisting with other European states such as Portugal, Russia, and Greece.[29] The Royal Navy tried many tactics to find continuous work for its men on half pay, one of the most successful being through the expanded and reformed Coast Guard.[30] "In 1829, and still more thoroughly in 1831," writes naval historian Michael Lewis, "the whole Coast Guard was welded into an even closer unit, and enlarged so that, by 1832, there appeared in the Navy List quite a formidable number of officers grouped under the heading 'employed in the Service of the Coast Guard.'"[31] By the fall of 1832, when Whewell received the letter from Richard Spencer, an expanded and reformed Coast Guard encircled Great Britain.

So many trained officers stationed around the coasts excited Whewell. He immediately told Beaufort about the idea, and Beaufort set out to recruit Bowles. After briefly recounting the history of tidal theory, Beaufort boasted to Bowles of the Admiralty's accomplishments in acquiring tidal observations and the advances in theory made by the nation's top natural philosophers. Beaufort told Bowles, however, that one of those philosophers, "Mr. Whewell, wants something more: a consecutive line of observations along the coast of Great Britain made simultaneously on the same tidal wave and continued for a fortnight. Is there a possibility that you could accomplish such a grand operation for us at the whole series of your stations at one and the same time?"[32] A positive reply would lead to a renewed attack on the theory of the tides, allowing Whewell to begin the second phase of his research: short-term observations taken simultaneously at a large number of places along the coast.

Bowles admired Beaufort and the work produced out of the Hydrographic Office. His staff at the Preventive Coast Guard sailed in and out of ports through dangerous waters along the entire coast every day. He relied heavily on the charts and tables published in Beaufort's office. The tide experiment was his chance to repay Beaufort while participating in the latest scientific research by a distinguished philosopher of the Royal Society. Bowles wrote to the deputy comptroller general, Samuel Sparshott, to feel out the chances for such a large undertaking. Sparshott responded enthusiastically but thought that not all the stations needed to contribute data, "many of them not being more than three or four miles apart."[33] Quite to the contrary, Beaufort shot back. He wanted the experiment to be as thorough as possible, representative of the Admiralty's resolve to contribute to the science of the day.

Bowles sent a copy of Beaufort's response to all the officers in each district, ordering them to carry the observations "most fully and carefully into effect."[34] He also returned a list of the Coast Guard districts with the names of the commanding officers, telling Beaufort he should send the requisite forms and instructions directly to them.[35] Whewell was asked to write out a set of directions for the officers to follow. "Forward them to us," wrote Beaufort's assistant, "and we will suit them to Sailor's vocabularies and have plenty of them printed and issued ready to commence on all parts of the coast."[36] Once Whewell drafted the instructions and

Beaufort issued them to the commanding officers, these officers in turn wrote letters explaining in layman's terms even the most minute details of rather mundane directions. Every Coast Guard station in Britain, over four hundred, was ordered to have observations of the tides recorded and returned directly to Beaufort at the Hydrographic Office.

The enthusiastic participation of the British Coast Guard had direct and immediate consequences. Both Lubbock and Whewell had previously used their personal contacts to acquire tidal observations, but they encountered similar problems: even if their contacts agreed to make observations, the reports often proved unreliable.[37] Once members of the Coast Guard received an official order, however, it became their duty to take careful observations and provide a full report. Beaufort's assistant at the Hydrographic Office, Alexander Becher, traveled to several of the stations in preparation for the observations and found yet another value in enlisting their help. "The Coast Guard people, it appears are so regular in all their proceedings," Becher informed Whewell, "that time to them is essential and they are therefore tolerably well provided with watches etc."[38] The cooperation of the Preventive Coast Guard also ensured that the requisite instruments were on hand, in this case accurate timekeepers. Guaranteeing accurate time had always been a problem in making tide observations far from major ports or cities. On several occasions, when observations looked anomalous, Whewell found that the observer had used a sundial, a method far from accurate, questionable during cloudy weather, and completely useless at night.

Having Beaufort and his staff so involved also allowed Whewell to freely question the data coming into the Hydrographic Office. Whewell was skeptical, for instance, of Commander William Mudge's observations on the coast of Ireland, and he queried Beaufort about it in rather harsh terms. Beaufort assured Whewell that Mudge would not "feel in the least sore at what you have said," adding that "the object of the surveyor is to observe, of the philosopher to reason, and of both Truth."[39] He then wrote to Mudge directly, noting that Mudge's recent observations were "making a great noise among our Philosophers here. They say such a difference of time in so short a distance is contrary to the laws of hydrostatics."[40] Mudge was a salty seaman who knew a thing or two about tides, and he assured Beaufort that his observations were sound, none too happy that a professor in land-locked Cambridge was questioning his work as an expert surveyor. "However it may turn out," wrote Beaufort in an attempt to calm the waters, "you will be delighted to find that men of real talent are now zealously taking up the neglected subject of tides."[41] Not surprisingly, Mudge's observations proved exceptionally accurate.

Beaufort did not stop with Commander Mudge. He wrote letters and instructions to all his surveyors, making certain that those stationed at the tide gauges would be particularly vigilant during the two weeks in June.[42] He also sent out "Hydrographic Notices" to all the commanding officers of Her Majesty's vessels. Rear Admiral William Parker, the commander in the Pacific, distributed Beaufort's instructions to numerous ships in his squadron and a few copies to the British consuls at Oporto and Lisbon, promising all the "further publicity" in his power.[43] Beaufort's connections expanded far beyond the coasts of Great Britain, and even before the 1834 British experiment began, he realized it could function as a dry run for a much larger undertaking.

In each of the letters to the officers of the Coast Guard, Beaufort urged the commanding officers to communicate freely with Whewell, "to make the results of this great experiment more satisfactory to men of science and consequently more agreeable to the Admiralty."[44] Beaufort realized the significance of the tidal experiment as a way to create tide tables and thus help control the seas. Whewell wanted to visit several of the Coast Guard stations, and Beaufort wrote to the officers introducing him, suggesting that "perhaps by some slight alterations in your process you might be able to better meet his views."[45] Beaufort was a man of action and was elated that Whewell was taking such an active role in the experiment. "It is delightful," he remarked to Lubbock as Whewell set off to visit some of the stations, "to see a man following up his speculation with such active enthusiasm."[46] In preparation for the experiment, Whewell checked the placement of the instruments, inspected the clocks and tide poles, and for the first time, made a few tide observations himself.

## Calculating the Tides

Every Coast Guard station in England, Ireland, Scotland, and Wales registered the tides every fifteen minutes for a fortnight in June 1834 and sent its observations directly to Beaufort. Once the returns starting coming in, the Hydrographic Office continued its administrative role. Beaufort charged Dessiou with overseeing the reduction of data. "The returns of last June are more consistent and accurate than I could have anticipated," Whewell noted in his paper in the *Philosophical Transactions* the following year, "made in many instances with ingenious and suitable contrivances."[47] In addition to taking the observations, the Coast Guard men included models for proposed tide gauges, plans for the best way to make observations out of sight of land, and offers to continue the observations, often on their own time and at their own expense. Beaufort received so many models, plans, and suggestions for building tide gauges that he did not know what to do with them all. In response to a model sent by Thomas Wright, a commissioned boatman of the Coast Guard Station at Glynn, Ireland, Beaufort replied: "The minds of so many people having been lately turned to that subject we have received several models and plans, for effecting the same object, and I have no doubt that by carefully combining the best parts of each of them something very simple and very satisfactory might be produced."[48] The problem was not with the observers on the coastlines of Britain but with the theorists back in London and Cambridge.

Lubbock and Whewell, now following different tracks of research and utilizing different methods with different aims, were at odds over how to discuss the data. Lubbock asked Dessiou to calculate the establishment of the tide at each station throughout the coast of Great Britain by the same methods he had followed for the long-term observations at London and Liverpool. Whewell, now convinced of his spatial approach and confident in his graphical method, hoped to do much more. He wanted to determine whether the establishment progressed in a regular manner along the coasts of Great Britain, and he asked Dessiou to represent all the observations "in curves" to see if such a progression existed.[49] Dessiou found himself caught in the middle. Beaufort suggested that he make samples of each method and send them to both Whewell and Lubbock. "As it would be idle to go through such a voluminous mass in two ways," Beaufort wrote to Lubbock with a tinge of frustration, "I wish you would take the trouble of settling the point with Mr. Whewell."[50] In the end Lubbock acquiesced; this was Whewell's experiment.

Whewell was confident he could get what he wanted out of the Coast Guard data, but he feared that slow results would hamper the chance of making further observations. Dessiou's work was much too sluggish for Whewell. He needed the reductions, even if this meant applying for possession of the original Coast Guard observations and having them discussed himself.[51] He repeatedly badgered Dessiou to send him the calculations and graphs, but Dessiou was already overburdened. He was still working through the Liverpool observations for Lubbock, reductions he performed at home before and after work, and other duties at the Hydrographic Office often took precedence. "Some sailing directions for Bristol Channel must be written for immediate publication," he wrote apologetically to Whewell, "I must drop these tides, until the directions are completed."[52] Dessiou's busy schedule forced Whewell to undertake

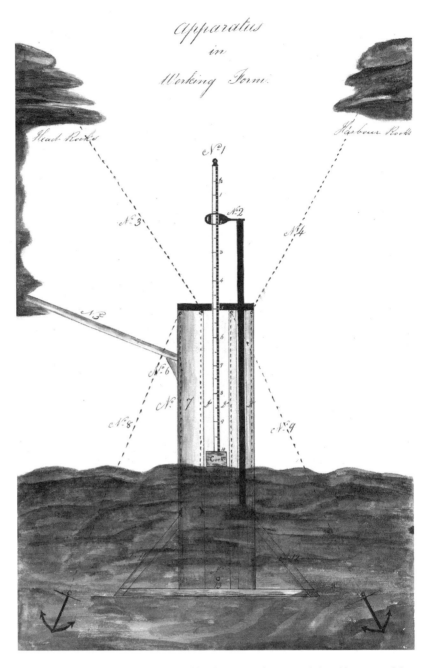

Figure 5.3. Model of a tide gauge invented by Thomas Wright, a commissioned boatsman of the Coast Guard Station at Glynn, Ireland. Wright used the gauge during the 1834 tide experiment, and it demonstrates the active involvement of the men of the Coast Guard not only in observing the tides but in building instruments and advancing observational techniques. The picture is enclosed in Captain Sparshott to Francis Beaufort, 14 August 1834, RHO, S395.

some of the reductions himself. He applied to the Hydrographic Office for the original observations and asked "more distinctly" that Dessiou discuss the rest. "It would not be doing justice to the subject of the observations," he wrote Beaufort late in November, "to publish my results derived from a portion of the observations, when they may be compromised or contradicted by the rest."[53] As in many similar instances, the burden of reducing the data, not their acquisition, halted progress.

What Whewell did get from Dessiou and from his own reductions tantalized him. First, the graphs suggested that no general irregularities affected the tides along the coast, and therefore the irregularities that did exist could be ascribed to local circumstances.[54] Second, they also demonstrated that the range of the tide was different for different places on the coast, sometimes for stations right next to each another. Whewell, with both Newton and Laplace in his sights, noted that this put a definitive end to any attempt to deduce the mass of the moon from the tides. Third, Whewell could show that the major tidal constant (the semimenstrual inequality caused by the position of the moon) had a common form at different places but was different in amount. This, along with other factors, demonstrated conclusively that tide tables for one port could not be used to determine the tides at another by simply adding or subtracting a constant interval. This conclusion was a warning for mariners, who often simply added or subtracted times from the major ports in Britain to determine the tides for adjacent areas on the coast. And finally, the observations allowed Whewell to return to his favorite topic, the progress of the tide wave throughout the ocean, and modify the map of cotidal lines published in his "first approximation" in 1833. The cotidal lines ran more closely together at the main promontories in Great Britain and nearly parallel to the shore at other parts of the coast, modifications that researchers could generalize to all such promontories and coasts around the world.[55] Whewell's spatial approach combined with his method of graphical representation had made these conclusions possible.

## The Multinational Tide Experiment of 1835

The results from the 1834 observations proved so useful for furthering the theory of tides and for constructing tide tables that Beaufort and Whewell agreed the experiment should be repeated the next year. Whewell hastened to publish his results; he wanted to place a finished product in the hands of people who would be repeating the work.[56]

He was in some sense contented: he had scores of observations to work with—almost too many—and a calculator, albeit a bit slow, to help with the reductions. By repeating the experiment, he would be able to test the accuracy of the earlier data, correct for meteorological anomalies and human errors, and ultimately quantify the multiple variables that account for the mathematical laws of the tides on the coast of Britain.

Beaufort had other designs. He wrote to Whewell in February of 1835 with a query: "Would it not be a delightful appendage to the batch of Coast Guard Tides which are to be observed this year if we were to procure simultaneous observations along the shores of Holland and France, Newfoundland, Nova Scotia and North America? If so, no time should be lost in determining on the periods at which the operations should take place. Suppose from the 9th to the 27 of June?"[57] Whewell, it appears, had not fully realized the extent of the network he was now connected to. "I should have attempted to obtain something of the kind," Whewell replied, excited by such a grand undertaking, "though perhaps I should hardly have ventured to extend my hopes so widely as your plan proposes."[58] Whewell added that he would make the necessary applications to the foreign countries and draft the instructions for the observers. But again, Whewell had underestimated Beaufort. Whewell would not be the one to apply to foreign governments or write out the instructions. The Admiralty, for better or worse, had taken charge, with Beaufort at its helm. What began as Whewell's experiment to advance the theoretical understanding of the tides was rapidly turning into Beaufort's project to amass the knowledge needed to control the world's coastlines and oceans.

Beaufort was always on guard against overextending himself or his surveyors, especially if it appeared to tax the Admiralty's budget. Through his personal connections with Bowles and the surveyors within his own department, he had been able to have the 1834 observations made without the intervention of the Lords of the Admiralty or direct overtures to the House of Commons. Beaufort realized, however, that for the grand undertaking he now had in mind, he would need to extend his influence a bit further. When he informed Bowles that he planned to repeat the tide experiment, the response was far from enthusiastic. Tide poles and other materials came at a price, and Bowles wanted to know whether the Admiralty would be willing to defray the expenses.[59] Money changes things, and Beaufort recognized that the Lords of the Admiralty would have to become officially involved. He also knew that once the experiment extended beyond Great Britain, he would need a dispatch signed by the foreign secretary. And that would entail a direct petition to government.

In mid-February 1835, Beaufort wrote to the Lords of the Admiralty stating his plans to extend the simultaneous observations of the tides from the coasts of Great Britain and Ireland to the opposite coasts of the English Channel and to all of the Atlantic.[60] A sharp increase in trade among maritime nations had required that hydrographers from European countries share their charts, and the realities of empire encouraged extending this cooperation to other continents.[61] Because Beaufort had spent so much time and expense fostering a relationship with hydrographers or their equivalents in other countries, he could tell the Lords Commissioners he had high hopes for full cooperation from these foreign governments.

The secretary to the Lords Commissioners wrote to the Foreign Office at the beginning of March enclosing Beaufort's letter, requesting in the "most urgent manner" the cooperation of the foreign secretary. The Duke of Wellington then applied to the foreign governments through the respective British ministers. The letter, written by Beaufort and sent by the foreign secretary, focused not on the practical benefits of the observations, but on the scientific aspects of the "great tidal experiment."[62] It assumed the foreign governments would cooperate based not on their naval interests but on their past enthusiasm for scientific projects. The experiment was scientific, Beaufort argued, and would enhance international relations. Everyone knew what knowledge of the tides meant: access to coastlines, ports, estuaries, rivers, homes. In the competitive environment of the nineteenth century, an argument based on science rather than seamanship, on international goodwill rather than national interests, seemed less antagonistic.

Once the Admiralty became fully integrated into the endeavor, the project took on a different form. The government, if correctly nudged, had deep pockets. Beaufort could now communicate to Bowles that he would "cheerfully pay" all expenses accrued the preceding year and defray all the "petty expenses" of the ensuing campaign.[63] Government involvement also allowed Beaufort to fill some of the holes in the observations. Whereas in the previous year Beaufort and Whewell had depended on the personal interests of naval men outside the Coast Guard to initiate tide measurements, this year they could rely on direct government pressure.

The case of the Channel Islands is instructive. Observations from these islands between England and France could add considerably to the knowledge of the tides as they progressed from the Atlantic up the English Channel and into the North Sea. But because there were no Coast Guard officers or Admiralty surveyors stationed there, Beaufort needed

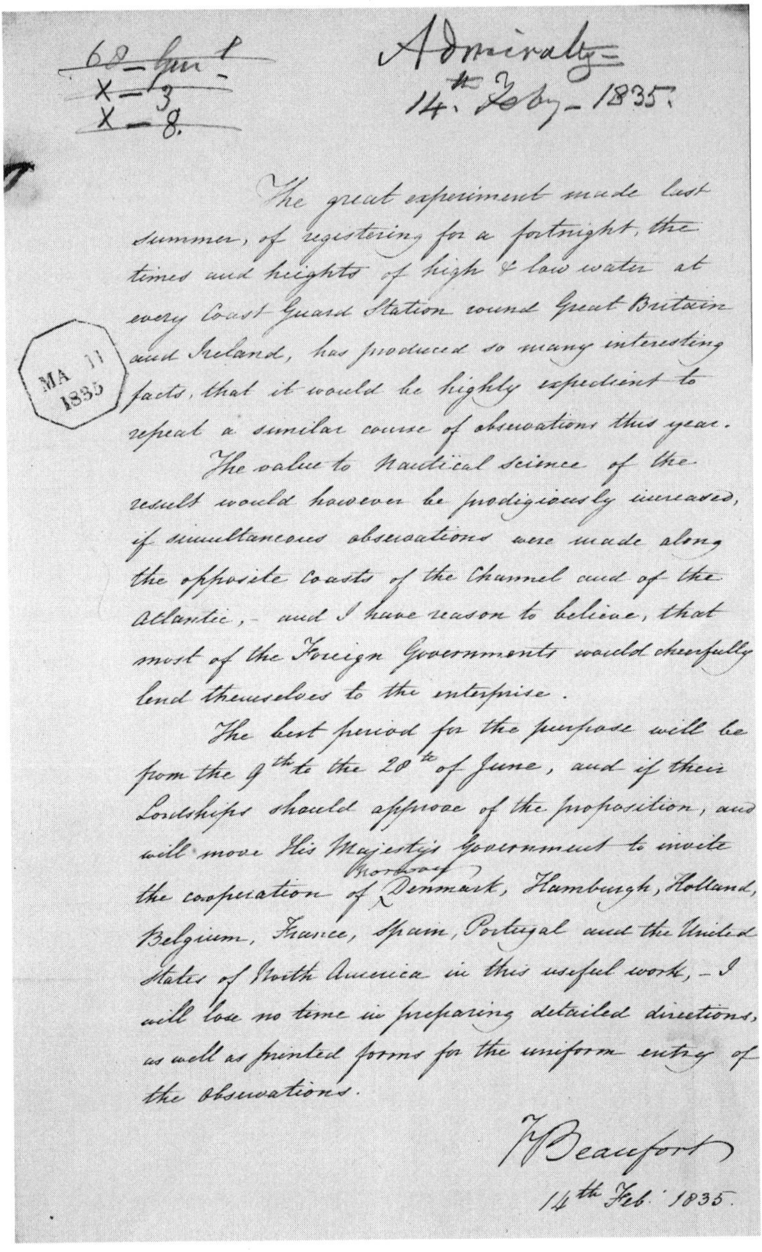

~~68 Gen 1~~
~~X – 3~~
~~X – 8.~~

Admiralty
14th Feby – 1835.

MA 11
1835

The great experiment made last summer, of registering for a fortnight, the times and heights of high & low water at every Coast Guard Station round Great Britain and Ireland, has produced so many interesting facts, that it would be highly expedient to repeat a similar course of observations this year.

The value to Nautical science of the result would however be prodigiously increased, if simultaneous observations were made along the opposite coasts of the Channel and of the Atlantic, – and I have reason to believe, that most of the Foreign Governments would cheerfully lend themselves to the enterprise.

The best period for the purpose will be from the 9th to the 28th of June, and if their Lordships should approve of the proposition, and will move His Majesty's Government to invite the cooperation of Norway, Denmark, Hamburgh, Holland, Belgium, France, Spain, Portugal and the United States of North America in this useful work, – I will lose no time in preparing detailed directions, as well as printed forms for the uniform entry of the observations.

F Beaufort
14th Feb. 1835.

Figure 5.4. Draft of a letter from Francis Beaufort to the Lords Commissioners of the Admiralty, 14 February 1835, suggesting that the tide experiment of 1834 be extended to other nations. Reproduced from PRO, ADM.1.3485.

another way to collect them. Competition with France made this even more urgent. "The French are going about the business [on the Channel Islands] with considerable spirit," wrote Thomas Spark, a lieutenant in the Royal Navy, "and we must not be outdone by them."[64] But the spirit of the Union Jack paled in comparison with the weight of the British pound. Spark warned that finding someone to measure the extreme rise and fall of the tide would be exceedingly difficult unless "pretty well paid for it."[65] To Beaufort this meant that the observations would not be made correctly, so here as elsewhere, he demonstrated his astute political maneuvering. Realizing that the foreign secretary's name would not be enough, he asked the Lords Commissioners of the Admiralty if they also could persuade Lord John Russell, the home secretary, to transmit the directions for tide observations to prominent political figures in those Islands.[66] A letter from the future prime minister eventually did the trick. The correspondence concerning the observations at the Channel Islands demonstrates both the technical difficulties with tidal studies at a station outside Great Britain and the difficulty of finding someone dependable enough to take the requisite readings.[67]

Beaufort had taken great care to ensure the success of the multinational experiment. As the letters to the foreign dignitaries went out, Beaufort shot a letter off to Whewell, proud of how "warmly the Lords Commissioners of the Admiralty have taken up the affair."[68] Beaufort also corresponded with all the captains of the surveying vessels, ordering them to make tide observations on the requisite days in June. Moreover, it was probably no coincidence that on the same day the letters went out to the foreign dignitaries, Beaufort corresponded with each of the foreign hydrographers, offering them charts that had been printed in the Hydrographic Office during the preceding year. The responses from foreign governments were all enthusiastic. The Norwegian government lost no time in setting up the means for the observations, transmitting the necessary orders to the northernmost points of Norway. The Senate of Hamburg likewise reported its willingness "to contribute in any way to the encouragement of so useful and scientific an attempt."[69] Some of the responses stressed the way relations between the countries had been strengthened through similar scientific endeavors, though the military importance of the experiment was not lost on the foreign heads of state. The Dutch government, while viewing the experiment as a new stage in the improved relations between the two countries, also insisted that it be furnished with the observations made on the British coast.[70]

The French needed very little inducement to contribute. They had long been interested in the tides, owing particularly to the recent efforts of the French hydrographer, Pierre Daussy. Daussy had worked closely with the French *ingénieurs hydrographes* under the command of the French minister of marine, Charles François Beautemps-Beaupré, to determine the establishment of the principal ports throughout France, focusing especially on the effects of atmospheric pressure on the height of the tides. He was particularly eager to extend his observations beyond the coasts of France. François Arago, an eminent French scientist and a foreign member of the British Association, brought Beaufort's proposal before the Chamber of Deputies "in the strongest terms" and published an account of the experiment in several Paris newspapers. He then read extracts of Whewell's papers on the tides at the French Academy before meeting with Beautemps-Beaupré, confessing that the efforts of the British had "put him to the blush."[71] Beautemps-Beaupré immediately directed observers to measure the tides at sixteen stations in France and the Channel Islands. In the Netherlands, likewise, no expense of energy or funds was spared. Eighteen officers, each furnished with a sextant and a chronometer, took observations of the tides for the entire month of June. Gunboats were dispatched to several of the stations owing to their precarious position on the rocky coast. The observations were then reduced and tabulated at the expense of government and sent to Beaufort.[72]

The still inchoate scientific community in the United States was especially eager to contribute to such an international enterprise. Orders traveled up and down the command chain. The Duke of Wellington wrote to Sir Charles Vaughan, the British consul in Washington, who in turn addressed a note to the U.S. secretary of state.[73] The secretary of state then wrote to the secretary of the navy. Since it was not convenient to send naval officers to all the stations, the secretary of the navy solicited the aid of the secretary of the treasury, who ordered officers of the army and officers attached to the revenue service to make the observations. Alexander Dallas Bache, at that time an aspiring young officer in the topographical engineers, reproduced several of Whewell's and Lubbock's papers on the tides for inclusion in the *Journal of the Franklin Institute* and distributed copies to all the stations furnishing observations.[74] Officers posted at twenty-eight stations extending two thousand miles from Maine to the mouth of the Mississippi were instructed to make tidal measurements for the entire month of June, representing the first systematic attempt to observe the tides on the coasts of the United States.[75]

For twenty days in mid-June 1835, nine countries—the United States, France, Spain, Portugal, Belgium, Denmark, Norway, the Netherlands, and Great Britain and Ireland—participated in a multinational venture to simultaneously observe the ocean tides bordering their countries and their possessions. Over 650 tidal stations participated, covering both sides of the Atlantic and at least 20 stations along the American Pacific and Gulf coasts. The observations reached from the Gulf of Mexico to Nova Scotia, from the Straits of Gibraltar to the North Cape of Norway. It was a massive global research project that covered not only Europe and America but numerous islands under British protection, including Scilly, the Isle of Man, Mauritius, Malta, Ceylon, and three of the Channel Islands. Table Bay and Simon's Bay in South Africa, along with selected ports in Australia and New Zealand, also contributed data. Observers registered the tides every fifteen minutes, day and night, often at places inaccessible by land. Observations of the tides across the boundaries of national space at the same moments in time were tabulated, graphed, mapped, and charted. The state of the weather at all the stations, along with the barometric pressure and force of the wind, filled thousands of tables that eventually made their way back to their center of calculation, the Hydrographic Office in London.

## Graphing the Ocean

In a letter to James David Forbes written over twenty years later, Whewell referred to the 1835 tide experiment as his "crowning achievement in Tidology."[76] He aimed to create a map that showed how the ocean tide wave progressed through the Atlantic and onto the shores of the major maritime nations and their possessions. The observations, however, turned out to be a confused jumble. On the coast of America, for instance, the observations suggested a convergence and divergence of the tide wave at very close intervals, quite contrary to Whewell's previous tidal map and incompatible with his theory. The incomplete registers from South Africa, Whewell admitted, were inconsistent and entirely useless. Taken as a whole, the data suggested a turbulent, even violent, ocean. As the returns began rolling in, Whewell's job was to extract and stabilize, transforming this jumbled mass into a meaningful assemblage for European scientists. "It became me," as Whewell put it, "to turn to the best advantage the large mass of materials thus collected."[77] It was his task to graph rationality onto a seemingly chaotic region of the globe, to infuse the ocean with order and stability. In the process, Whewell quantified and qualified the ocean in ways that helped modify Western conceptions.

The most pressing concern was simply reducing the data, the job electronic computers do today and human calculators accomplished in

1835. For high tides alone, forty thousand data points needed to be calibrated to a constant zero point, arranged into tables, and correlated with wind, weather, and their position on the globe. As Whewell noted, reducing the data went far beyond "the powers of an individual," and the Admiralty again stepped in.[78] Captain Beaufort agreed to put Dessiou in charge of the calculations "as far as the business of the office left him time," but that was far too little to get the job done. The first lord of the Admiralty, Lord Auckland, allowed Beaufort to assign two additional calculators, Daniel Ross and H. Boddy, to work under Dessiou's guidance.[79] In addition to the tedious tasks of examining and tabulating the data, these three calculators arranged the data into curves following Whewell's graphical method. At each tide station, they laid down in curves the difference between the time of high water and the time of the preceding transit of the moon, with all the various errors and peculiarities of the observations. "The inspection of these curves," Whewell wrote, "afforded me the means of judging of the best mode of combining them so as to get rid of local and casual anomalies" and thus find the major constants for the tides on each coast.

Whewell later explained this graphical method more fully in his 1840 *Philosophy of the Inductive Sciences,* in a section titled "Special Methods of Induction Applicable to Quantity."[80] Whewell listed these special methods as the method of curves, the method of means, the method of least squares, and the method of residues. Taken together, they represent the modern graphical method, and though commonplace for researchers in the sciences today, its systematic use was novel in Whewell's time.

The graphical method, Whewell recognized, was based on the premise that "order and regularity are more readily and clearly recognized, when thus exhibited to the eye in a picture, than they are when presented to the mind in any other manner." In his *Philosophy* he used the example of trying to make sense of locally published tide tables. Graphical representation, however, accomplished much more than bringing order to "obscurity and complexity."[81] First, it could be used to invent data points in the open ocean in essentially the same way as having those data points measured. Second, it allowed the researcher to determine which observations were accurate and to correct those that were not. By drawing a curve "not through the points given by our observations, but *among* them," the researcher could "obtain data which are *more true than the* individual *facts themselves.*"[82] The graphical method, Whewell argued, actually invented observations far out in the ocean that were *more real* than the actual observations on his desk.

Table X. (Continued.)

### Coast of America.

Honourable MAHLON DICKERSON, Secretary of the Navy, United States.

| Station. | Observers. | Latitude N. | Longitude W. | Date. | Greatest Range. | Date. | Least Range. |
|---|---|---|---|---|---|---|---|
|  |  | ° ′ ″ | ° ′ ″ |  | ft. in. |  | ft. in. |
| Eastport (Maine) .... | Jery Burgin, Inspector. | 44 54 0 | 66 56 0 | 11 P | 22 10 | 21 A | 14 8 |
| Mount Desert Island .. | Henry S. Jones. | 44 9 0 | 68 31 0 | .... | 13 4 | 22 P | 8 1 |
| Portland........... | John Williams. | 44 39 16 | 70 20 30 | .... | 12 2 | 21 A | 7 0 |
| Portsmouth Navy Yard | { Jos. R. Jarvis, Lieut. United States Navy. | 43 4 44 | 70 45 0 | 10 P | 10 4 | 20 A | 6 1 |
| Gloucester......... | John Webber. | 42 36 0 | 70 42 0 | 10 P | 12 8 | 21 P | 6 9 |
| Boston Navy Yard.... | Commodore John Downes, Duncan Bradford, Professor of Mathematics, Henry French, passed Midshipman. | 42 20 0 | 71 4 9 | .... | 14 8 | 22 A | 10 11 |
| Cape Cod ......... | Richard Ainsworth. | 42 2 6 | 70 4 0 | .... | 12 6 | 21 A | 7 3 |
| Province Town ...... | Major James D. Graham, United States Corps of Topographical Engineers. | 42 2 45 | 70 13 0 | .... | 12 6 | 22 A | 7 1 |
| Nantucket ......... | William Coffin. | 41 16 12 | 70 7 42 | 12 A | 2 6 | .... | 0 11 |
| Newport........... | Col. J. G. Totten, Engineers, assisted by Lieut. Child, Artillery. | 41 29 0 | 71 21 14 | 10 P | 6 0 | 21 A | 2 6 |
| Warren ........... | Lieut. Joel Abbot, United States Navy. | 41 44 0 | 71 15 15 | .... | 6 8 | 20 A | 2 7 |
| Gardiner's Bay ...... | M'Perry, Master Commander, United States Navy. | 41 4 0 | 72 5 0 | .... | 3 5 | 21 A | 1 5 |
| New York Navy Yard.. | Commodore C. G. Rigeby, Commander M. F. Mix. | 40 42 40 | 74 1 8 | .... | 6 6 | 20 A | 1 6 |
| Sandy Hook ........ | Josiah Tattnall, Lieut. U.S. Navy. | 40 28 0 | 74 1 0 | .... | 7 1 | .... | 2 7 |
| Delaware (Breakwater) | A. R. Hetzel, 2nd Infantry. | 38 57 0 | 75 10 0 | 10 P | 6 4 | 20 P | 3 0 |
| Old Point Comfort.... | C. H. Kennedy, Lieut. United States Navy. | 37 0 0 | 76 22 10 | 10 P | 3 9 | 21 A | 1 10 |
| Gosport Navy Yard .. | William P. S. Sanger, Engineer. | 36 50 50 | 76 18 47 | 11 P | 4 5 | 21 A | 2 1 |
| Cape Hatteras ...... | Isaac S. Farrow, and Joseph C. Jennett. | 35 14 0 | 75 30 0 | 9 P | 5 6 | 19 A | 2 0 |
| Cape Fear River .... | J. Dimeck, Capt. Artillery. | 33 48 0 | 78 9 0 | 10 P | 6 11 | 20 A | 2 7 |
| Charleston ......... | W. H. Pettes, Lieut. Artillery. | 32 44 0 | 80 1 0 | 11 P | 7 11 | .... | 3 6 |
| Savannah ......... | C. S. Merchant, Capt. Artillery. | 32 2 0 | 81 3 0 | 10 P | 8 5 | .... | 1 5 |
| St. Augustine....... | F. L. Dancy, Lieut. Artillery. | 29 48 30 | 81 35 0 | 10 P | 6 7 | 21 A | 3 1 |
| Key West ......... | F. L. Dade, Brevet Major, United States Army. | 24 29 0 | 81 55 0 | 13 A | 2 6 | 21 P | 1 6 |
| Tampa Bay......... | R. A. Lantzinger, Major, United States Army. | 28 5 0 | 83 18 0 | 15 P | 3 3 | 17 P | 0 8 |
| Pensacola Navy Yard.. | W. Chauncey, commanding Navy Yard, W. K. Latimer, Master Commandant, and Nahum Warren, Sailing Master. | 30 32 0 | 87 12 0 | 13 A | 2 3 | 20 A | 0 10 |
| Mobile Point ....... | F. S. Belton. | 30 13 0 | 88 21 0 | 11 A | 2 1 | .... | 0 8 |
| Fort Wood......... | John M. Creylar, Assistant Surg., United States Army. | 29 15 0 | 89 35 0 | 13 P | 2 7 | 20 P | 0 2 |
| Fort Pike ......... | John Mountfort, Major, Artillery. | 28 0 0 | 89 0 0 | .... | 1 8 | 21 A | 0 0 |

Figure 5.5. An example of the tabular representation of the range of the tides on the east coast of the United States. The actual numbers are not as important in this table as are the places of observation and the names of observers. To convey the information that numbers and tables used to give, Whewell used to the graphical method (fig. 5.6). Reproduced from William Whewell, "Researches on the Tides," 6th ser., "On the Results of an Extensive System of Tide Observations Made on the Coast of Europe and America in June 1835," *Philosophical Transactions* 126 (1836): 326.

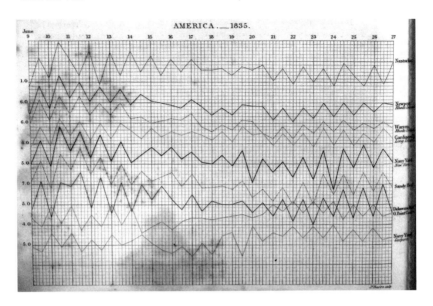

Figure 5.6. An example of graphical representation of tide observations made on the east coast of the United States. The graphical method offered a quick representation of the range of the tide, which Whewell could then compare with other observations, both on the coasts of the United States and on coasts of other continents. Reproduced from William Whewell, "Researches on the Tides," 6th ser., "On the Results of an Extensive System of Tide Observations Made on the Coast of Europe and America in June 1835," *Philosophical Transactions* 126 (1836): 306.

Whewell's use of the graphical method allowed him to speak with some authority on how the tides progressed around the coasts of Great Britain and Ireland. Sailors and scientists had long known that the tides at the London Docks were extraordinary. They resulted from the confluence of two distinct tide waves, one traveling around the west coast and down from the North Sea, the other straight up the English Channel. At the London Docks, moreover, these two tide waves met exactly in conjunction. Though extraordinary in their production, the tides in the Thames appeared quite ordinary, though extremely large. This was not the case, however, where the two tide waves were not exactly in conjunction—for instance, in places farther up the English Channel. In these regions, the tides acted in a peculiar manner, since the combination of the two tides reaching the shore at different times caused several high and low tides each day, yielding entirely different heights and times very close together. Recall that Whewell had harshly criticized Captain Mudge's seemingly anomalous observation of the tides on the coast of Ireland. The results of the 1835 experiment caused Whewell to recant:

"Knowing the anomalies which prevail in this neighbourhood, I do not now doubt that Captain Mudge's statements are all entirely correct."[83] This was not the first time, nor would it be the last, that a land-locked philosopher found himself retracting criticism of a surveyor's observations in the field.

But Whewell was not interested in individual tides on individual coasts. His principal object was to fix with precision the cotidal lines as they traversed the earth's seas and oceans, and his results were numerous enough to include a section entitled "On a Second Approximation to a Map of Cotidal Lines, and Especially of Those of the German Ocean." He admitted that the map he included in his first paper on the tides in 1833 required considerable revision, especially the cotidal lines of the North Sea. Rather than a wave progressing up the English Channel, as Whewell had assumed, "it appears that we may best combine all the facts into a consistent scheme, by dividing this ocean into two rotatory systems of tide-waves." A peculiar consequence followed: there should be a place in the middle of the North Sea where no tide existed at all.

Observing the tides offshore proved exceedingly difficult, and no data existed for the central parts of the North Sea or any other large body of water. Before Whewell's use of cotidal maps, observers could measure the tides in the open ocean only where there were islands, such as St. Helena in the Atlantic, making those observations especially desirable. This was one reason Beaufort was so adamant about obtaining observations from the Channel Islands. The use of cotidal lines allowed Whewell to create from theory the rise and fall of the tide in the deep ocean, but he still had no way to measure those tides, and thus no way to test his theory. Whewell had thought hard about how to make such observations, but a surveyor of the Royal Navy would be the one who eventually came up with the method, made the observations, and published them in the *Nautical Magazine*.

Lieutenant Becher wrote to Whewell shortly after the results of the tide experiment had been published, asking if it had "ever occurred to [him] to determine the state of the tide out of sight of the land [by means of] soundings."[84] Becher was stumped as to how he could accomplish this, and he suggested that first "some theoretical mode must be adopted." Whewell, ever the theoretician, offered his own suggestion, accompanied by a sketch. It included a buoy moored a few feet below the surface by three anchors resting on the ocean floor. A staff that could slide up and down with the rise and fall of the tide was then fixed to the buoy, and the observer could make the observations through a telescope from a vessel stationed nearby. Whewell was not sure such a scheme

Figure 5.7. William Whewell's map of cotidal lines of the British Isles, showing the progress of the tide wave as it travels up the English Channel and onto the shores of Britain, France, the Netherlands, the Germanic states, and Norway. Reproduced from "Researches on the Tides," 6th ser., "On the Results of an Extensive System of Tide Observations Made on the Coasts of Europe and America in June 1835," *Philosophical Transactions* 126 (1836): 289–341.

would be practical, but he had "no doubt your nautical friends would devise [a plan] better than I can."[85] Whewell was right to be skeptical of his own invention and equally correct in his valuation of Becher's nautical friends. Commander William Hewett devised a workable scheme to measure Whewell's no-tide zone and then set out to the middle of the North Sea to test it.

Hewett was a seasoned veteran who knew the North Sea perhaps better than anyone in the British navy. He received orders to survey the North Sea early in the 1830s, including directions for observing the tides.[86] He was a well-respected officer with an interest in scientific subjects, and Beaufort often suggested that he collect tides at special places for the "Tide philosophers."[87] After the publication of Whewell's paper, Hewett

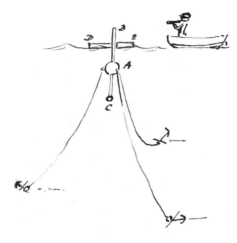

Figure 5.8. William Whewell's sketch explaining his suggestion for measuring the rise and fall of the tide out of sight of land. Reproduced from William Whewell to Alexander Becher, 19 November 1836, RHO, W547.

was eager to determine if such a no-tide zone really existed. Several years passed, however, before he sailed his vessel, HMS *Fairy*, near Whewell's proposed rotary zone. His first attempt in the summer of 1839 failed because the waters were too turbulent, and Hewett returned in the fall of 1840 to try again.[88] The absence of a tide did not translate into the absence of a strong tide stream, and he was confronted with a fierce flow of water while conducting the delicate observations. He published a description of his methods in the *Nautical Magazine* for 1841 as a letter addressed to Beaufort, including a diagram of the way he made the observations and a table of the results.[89] Though his diagram looks similar to Whewell's, the two methods differed considerably. Hewett moored to the bottom of the ocean not a tide gauge but a vessel from which he made soundings, following Becher's initial suggestion. Hewett took soundings every thirty minutes for two days, reporting to Beaufort that his results confirmed the existence of Whewell's no-tide zone. Beaufort then congratulated Whewell on "the singular sagacity with which you had unravelled the mysterious laws that guide the moments of the great deep."[90]

Hewett ended his article with hearty but ultimately posthumous congratulations to Whewell. Though the voyage suggested a successful ending to Whewell's cotidal analysis of the North Sea, it proved a disaster for the British navy. As the results were going to press, the *Fairy* was lost at sea. It foundered in a storm on 17 November 1840, and the entire crew was lost. Beaufort took the news especially hard. Hewett had been a surveyor under him for a decade, beginning when he first became hydrographer in 1829, and he was sick with grief for Hewett's family.

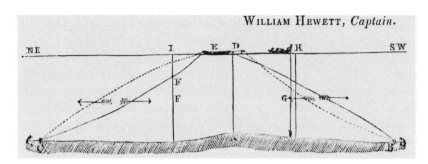

Figure 5.9. A diagram depicting Captain William Hewett's method of measuring the rise and fall of the tide out of sight of land. He performed this experiment in 1841 while surveying the North Sea on board HMS *Fairy*. Reproduced from William Hewett, "Tide Observations in the North Sea—Verification of Professor Whewell's Theory," *Nautical Magazine* 9 (1841): 183.

Hewett left eight children from nine months to fourteen years old, and his widow lost not only her husband but also her eldest son (a midshipman) and her brother (the master).[91]

Despite Hewett's confirmation of Whewell's theory and its further verification the following year by Captain John Washington of HMS *Shearwater*, Whewell's theory of a no-tide zone produced some dissent among scientists.[92] George Biddell Airy, in his monumental "Tides and Waves," published in 1845 in the *Encyclopaedia Metropolitana*, opposed the idea violently, having "little hesitation in pronouncing this to be impossible."[93] Airy offered a cotidal chart of his own that illustrated two systems of waves traveling along opposite coasts. The debate concerning the North Sea tides has since enjoyed a notable history reaching into the twentieth century. T. H. Tizard included a no-tide zone in the North Sea in a cotidal map published by the Hydrographic Department of the Admiralty in 1909.[94] However, two years later in an article on tides in the *Encyclopaedia Britannica*, George Darwin offered a cotidal map that followed Airy's system of two waves. Indeed, most of the leading figures in tidology in the first half of the twentieth century, including Rollin A. Harris in 1904, Albert Defant and Robert von Sterneck in 1920, and Joseph Proudman and Arthur T. Doodson in 1924, used either Whewell's or Airy's system of cotidal lines to chart and rechart the North Sea. Whewell's 1836 cotidal map and the Admiralty's chart no. 301, published one hundred years later, are reproduced in figures 5.10 and 5.11. The close correspondence between Whewell's chart, composed using an incorrect theory, and the Admiralty chart, which included observations taken under the auspices of the International Council for the Exploration of the Sea using radar from submarines, is striking.[95]

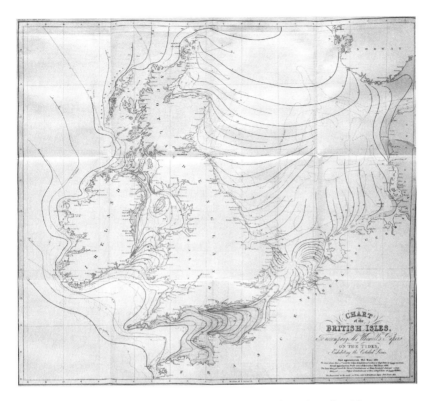

Figure 5.10. William Whewell's map of cotidal lines for the North Sea. Reproduced from "Researches on the Tides," 6th ser., "On the Results of an Extensive System of Tide Observations Made on the Coasts of Europe and America in June 1835," *Philosophical Transactions* 126 (1836): 289–341. Notice its close parallel to figure 5.11.

Though Airy had been exceedingly critical of Whewell's results, especially the map of the North Sea, he had only praise for his methods. In his article in the *Encyclopaedia Metropolitana*, Airy "confidently" referred the reader to Whewell's investigations as "one of the best specimens of the arrangement of numbers given by observation under a mathematical form."[96] He was dismayed that Whewell had clung so relentlessly to the Bernoulli equilibrium theory, but he acknowledged that "viewing the two independent methods introduced by Mr. Whewell, of reducing the tabular numbers to law by a process of mathematical calculation, and of exhibiting the law to the eye without any mathematical operation by the use of curves, we must characterize them as the best specimens of reduction of new observations that we have ever seen."[97] Whewell's conclusions might be wrong, Airy argued, and he might have applied an

Figure 5.11. Cotidal map of the North Sea, prepared and published by the Hydrographic Office in 1931 as Admiralty chart no. 301. Notice how closely it resembles Whewell's map published almost one hundred years earlier. From A. T. Doodson and H. D., Worburg, *Admiralty Manual of Tides* (London: Her Majesty's Stationer, 1941). Reproduced by permission of the Royal Hydrographic Office.

incorrect theory, but the methods he used to reduce and represent the ocean were exemplary.

Beaufort agreed. Whewell's efficient and rational ordering of the ocean fit perfectly with the Admiralty's own designs. The ocean and coastlines now looked regimented and ordered, reduced to a single visual that could be printed, reproduced, and dispersed throughout the sea-lanes of the world.[98] Science was helping to control the ocean. Whewell requested twenty-five additional copies of his isomaps, which he then sent to his fellow theorists, naval officers preparing for voyages, and interested amateurs living abroad.[99] Beaufort also sent copies of Whewell's maps to the commanding officers of the Coast Guard and to all his surveyors stationed around the world. The ability to graph the ocean conferred remarkable power on the British Admiralty while heightening the status of the scientist.

The transition in the study of the tides from a scientific project to an Admiralty initiative highlights the importance of the isomap. In many respects the Admiralty felt the multinational tide experiment of 1835 was entirely its own. Beaufort was the one who first suggested the international collaboration, then put it into action; the Preventive Coast Guard gathered the data; the government funded the venture, furnished the calculators, published the maps, produced tide tables from the results, and ultimately used the results. Beaufort and the men in the Hydrographic Office had access to the technical resources, the men, the international connections, and most important, the financial resources to make global research possible. The isomap was theirs, and they used it as an imperial tool to keep their vessels safe.

## Unfolding the Isomap

As science expanded geographically and the object became to find the relationships among increasingly complex variables, the visual graph emerged as a means of accurate and useful representation. The graph was ideally suited to represent massive amounts of data at a single synoptic glance; visual trends appeared in the data that were difficult if not impossible to discern through tables. Graphs also proved ideal for comparing large amounts of data, especially to find similarities or differences in data sets from separate continents, countries, or coasts. Since graphs could be used to suppress errors and create facts "more true than the individual facts themselves," they provided the best means of organizing data acquired from around the globe by "persons of various ranks and countries."[100] They stripped the observations of their inaccuracies (human or otherwise) to reveal the most apparent regularities, irrespective of who gathered the data or where. Moreover, for natural philosophers in the 1830s, these graphs increasingly took on an explanatory burden. Whewell's prose centered on the data rather than the data collectors, on the steady hand of the scientist rather than the turbulent surface of the ocean.[101]

Graphs, whether line graphs or isomaps, are intended to reveal the relations between changing variables. Whewell's graphs related the variables of time and space, specifically the time of high water throughout the space of the globe, but they hid much more than they revealed. They masked the embedded authority relations between the individuals assembling the maps. Indeed, this is the object of the graph: not to exemplify but to simplify, not to make real but to make useful. In this sense graphs do not depict nature, they create it. Hidden under Whewell's isotidal maps are a myriad of calculations and equations piled up on sheets and sheets of graph paper. These in turn depend on the different types of data amassed (numbers, tables, and curves) and the instruments used to gather them (tide poles and tide gauges, barometers and wind gauges, chronometers and military vessels).[102] Connected with these instruments are the innumerable individuals who used them, persons of "various ranks and countries," connected through the imperial trade routes of the British Empire. This was not limited to the Coast Guard and the surveyors and military men associated with the Hydrographic Office and British Admiralty. Science was riding the coattails of British expansion, just as empire was propelled by the indefatigable efforts of the scientists.

The institutions, people, and scientific devices combined to give authority to the map.[103] Behind the graph paper and equations, however, lies a less tangible, less obvious array of diverse interests and intents, both national and international, as much personal as professional. Beaufort and the Hydrographic Office wanted to obtain correct tide tables to safeguard Britain's vessels. But much more was at stake. The end of the Napoleonic Wars had brought major changes within the armed forces, raising new questions concerning the Admiralty's role during peacetime and about what to do with all its mariners after demobilization. The essential function of the Hydrographic Office, the professional interests of surveyors and officers, and the technological and geographical transformations taking place within the Royal Navy all figured into the ultimate acceptance and enthusiastic use of Whewell's cotidal maps. With so much invested in their creation, the incentive to use them within the

Admiralty ran deep. At least in the case of Captain Hewett, lost at sea, this investment reached as deep as the ocean floor.

For Whewell and the scientists, the personal and professional interests proved equally strong. From the time Richard Spencer's letter reached Whewell's lodgings in Cambridge, Whewell was engaged in a self-reflective examination of the practice of science and the role of the scientist. He was motivated by his own interest in establishing tidology as a relevant field of research and by his insistence that the spatial approach and the graphical method were the best way to proceed. In his attempt to fashion his own niche as an active researcher, he was forced to deal with pressing questions that would form the basis of his *History* and his *Philosophy*, including the best means of analyzing large amounts of data and the relation between theory and observation. These in turn rested on questions concerning the role of associate laborers, hierarchy and collaboration, the amateur and the professional. Whewell's cotidal maps perfectly hid these relations, masking his diverse interests and intents. They nevertheless constituted the "thin, almost immaterial, sharp end of a huge instrument" that gave the final map its ultimate power.[104] If the authority of the map was not yet assured through this backdrop of people, institutions, equipment, and motivations, Whewell could add one last pillar: the definition of science itself.

Whewell's paper outlining the results of the 1835 experiment reenacted the process of creating scientific knowledge.[105] It supported the practice of induction according to the definition Whewell was trying to foster. At just this time, between 1833 and 1836, Whewell wrote his *History* and formulated his *Philosophy*. The first step in all the sciences, represented to Whewell through the most pressing problem in tidology, was what he termed "colligation," taking a mass of facts and organizing them into mathematical laws that could explain the phenomenon. Ultimately, bringing together facts (things) through the creative power of the intellect (thoughts) represented the most important step in discovery. This "fundamental antithesis of knowledge" required an act of the intellect that established a precise connection among diverse phenomena of nature.[106] Scientists accomplished this by constructing guiding ideas that united the facts into a meaningful and operable whole. "The knowledge of such connexions," declared Whewell, "accumulated and systematized, is Science."[107]

Whewell's paper exactly imitated this narrative, beginning with a similar interplay of theory and observation, mimicking his "fundamental antithesis of knowledge." It proceeded from disorder to order, from masses of facts to laws of the phenomenon. The visual graph acted as the theoretical guide directing which observations needed to be gathered—how, when, where, and for how long. Consistent with his philosophy of science, he imposed a rule on the data, in the form of isotidal lines, bringing order and regularity to the observations. At the center of all this was the scientist. "I solicited a repetition of the coast-guard tide observations in June 1835," Whewell began his reenactment, masking the process involved, "and also ventured to recommend that a request should be made to other maritime nations."[108] Whewell altered the story to emphasize the role of the theorist, a presentation that enhances the power of his argument and by extension the authority of his isomap. The scientist "colligated" the freshly gathered data through a visual graph to find the connection among diverse phenomena. This was science, accumulated and systematized.

Whewell's work on graphical representation was more representative than original. Though he advanced significantly on earlier methods introduced by William Playfair, Johann Lambert, and Alexander von Humboldt, other researchers were following a similar course.[109] John Herschel described the graphical method in comparable terms in 1833 in his first paper on nebular astronomy from observations acquired at the Cape of Good Hope.[110] As Humboldt's illustrations began to appear more systematically in the 1830s and 1840s through numerous pirated reproductions, the graphical method began to affect a wide range of sciences.[111] The case of thermodynamics is particularly instructive. Though Sadi Carnot had described the work of an idealized steam engine in 1824 in his *Reflections on the Motive Power of Heat*, it was only in 1834 that Émile Claperyon depicted it graphically in an indicator diagram (which, significantly, is what comes to mind when one thinks of the "Carnot cycle"). In geology, the cross section took on a similar explanatory burden, and in natural history, the transition from illustration to graph also happened at about this time.[112] For instance, though Darwin described his *On the Origin of Species* as "one long argument," the crux of that argument appeared in chapter 4, "Natural Selection," and an important means of presentation was a single diagram.[113] Like Whewell's cotidal maps, Darwin's visual was relational and had great explanatory power. Through this one visual, Darwin summarized descent, divergence, extinction, even competition and struggle among closely related species.[114]

The graphical method proved especially well suited to representing the results of a spatial approach to science. Humboldt and others at the turn of the eighteenth century had revolutionized it. Whewell demonstrated its power through its systematic use in his tidal studies and popularized it through his *Philosophy*. While perfect for Whewell's spatial approach to science, the graphical method also proved ideal for the British Admiralty. The synoptic view of the ocean conferred by the isotidal map matched the Admiralty's need for intelligence of the world's oceans and coastlines, enabling British naval vessels to extend their control of the estuaries, bays, and ports of Britain's expanding possessions. As Bruno Latour deftly noted, "there is nothing you can *dominate* as easily as a flat surface of a few square meters."[115]

## Conclusion

The seamless transition of Whewell's isomaps from the realm of science to the dominion of the Admiralty served as a model for other geophysical investigations of the ocean. In his opening address at the anniversary meeting of the Royal Society on 30 November 1837, the president began with just this point by discussing Whewell's global tidal research:

I gladly seize this opportunity of bearing testimony, occupying as I do the highest scientific station in this country, to the readiness which the Lords of the Treasury and the Admiralty have shown on this and on every other occasion to forward scientific inquiries, and particularly such as are connected with the advancement of *astronomy and navigation*.... I rejoice, Gentlemen, in such manifestations of the sympathy of the government of this great country for the progress of science, and I trust that its influence will be felt in the cordial union and cooperation of philosophers, in planning and in executing those *great systems of observations, whether simultaneous* or not, which are still requisite *to fill up some of those blank spaces which occupy so large a portion in the map of human knowledge.*[116]

The interplay between astronomy and navigation in this passage, between the map of human knowledge and the actual map used to bound the ocean, linked the aims of the scientists and the government. The president of the Royal Society rejoiced in the union between the Admiralty and scientific investigations, where the planning and execution of "great systems of observations" simultaneously filled the "blank spaces" in both human knowledge and the world's oceans.

Britain emerged from the Napoleonic Wars as the greatest maritime power in history. Its navy ruled much of the ocean. The Admiralty developed a growing interest in science and scientific pursuits as a result of advances in naval technology and the growing need to transport ships, cargo, and crew safely around the oceans and coastlines of the world.

The changing nature of the British navy, the realities of empire, and the desire for secure routes for Britain's export trade pushed the British government to set up worldwide coordinated observations in terrestrial magnetism, meteorology, and ocean tides and waves.[117] Soldiers, scientists, missionaries, and mariners participated in a new spatial approach to investigating nature's laws, one that covered vast geographical distances and relied on the accurate measurement of interrelated variables across space.

Whewell transformed the study of the tides from a temporal science into a spatial science by co-opting Humboldt's methods and harnessing the most geographically expansive institution in the world. He advanced a revolutionary approach to tidology that became a prominent type of science practiced in the Victorian era. As David Harvey and other geographers have noted, once the spatial dimension of the research changed, so did the resulting concepts and theories. Whewell and Beaufort forced a global framework onto the tides, even though local topography—the width of the channel, the contours of the estuary's floor, the depth of the river's entrance—seemed to produce wildly different tides on Europe's coastlines and bays. Changing the boundaries of the study was thus a strategic decision by the British elite, one that fit well with the resources of the British Admiralty, leading to a more useful way to conceptualize the ocean. Whewell's spatial approach set the stage for future involvement of the Admiralty in other large-scale geophysical initiatives—in meteorology, for instance, and terrestrial magnetism. Indeed, the British searched for the laws of the varying magnetic needle with such fervor that it resembled a religious calling, and the search ultimately became known as the "Magnetic Crusade."

Researchers like Whewell used these global projects to advance physical astronomy and to assert their own authority within the scientific community and their own vision of science in the halls of government. But the actual practice depended crucially on the collaboration of the British Admiralty, especially Francis Beaufort and his staff at the Hydrographic Office, including the multitude of interested surveyors and officers stationed throughout the globe. Beaufort was, after all, the hydrographer to the Admiralty; his job was to keep the ocean safe for military and commercial navigation. When Beaufort became hydrographer in 1829, however, apart from the soundings and surveys undertaken near the coastlines, the ocean was empty, a perilous void that mariners were forced to cross. Throughout the next several decades, scientists filled and bounded the ocean with graphical isolines of all types: tidal, barometric, thermal, and magnetic. This allowed them to extend their data points outward over the oceans where no actual observations could be made. This new conceptualization and the tangible products it created legitimized both the spatial turn in science and the role of the new scientist, reconstructing the ocean into one that was simplified and manageable. It was not the actual ocean; it was far better.

SIX

# Calculated Collaborations

[Tide] tables, which in every other province of physics are the result of the knowledge which our men of science have accumulated for us, are, in this department,
published by persons possessing and professing no theoretical views on the subject; and the methods on which they are calculated are not only not a portion
of our published knowledge, but are guarded as secrets, and handed down as
private property from one generation to another.

WILLIAM WHEWELL, *PHILOSOPHICAL TRANSACTIONS*, 9 JANUARY 1834

The Tide Tables were formerly quite empirical, and cheaply manufactured by
the Almanac Makers. At length Sir John Lubbock and Professor Whewell devoted
their great talent to the subject, and the Admiralty seconded their efforts by ap-
pointing two extra clerks from Somerset House to discuss for them the registers
of many thousand tides, and to carry out their views by very laborious computations. By these means definite rules were obtained, and the tide predictions now
published by the Admiralty approach the truth as nearly as possible.

SIR FRANCIS BEAUFORT TO THE LORDS OF THE ADMIRALTY, 27 JULY 1841

Taken together, the passages above reveal a prejudice that
saw tide tables, formerly cheap, empirical, and arcane, as
attaining an acceptable level of accuracy and hence scientific legitimacy only after the theorists in London and Cambridge "devoted their great talent to the subject." This
echoes earlier statements made by Peter Roget, the secretary of the Royal Society, who insisted to the Admiralty
that local tide tables were "calculated by unpublished methods upon unknown data, and without any progressive improvement in their accuracy."[1] These criticisms regarding
earlier tide tables rested on an indefatigable confidence in
the progress of science and the power of the state. The new
scientists not only ordered the ocean into a rational and

controllable grid, they made it safe for mariners far and wide by making accurate tide tables possible. Roget believed the government should construct these tables, work that Beaufort argued should be done within the Hydrographic Office. Such rationalizing and centralizing usually implied a devaluation of local knowledge.[2] A lack of theory meant a lack of accuracy; amateur tables were dangerous tables.

Yet it was local tide calculators, not elite theorists, who advanced the study of the tides in the late eighteenth and early nineteenth centuries, perfecting Bernoulli's equilibrium theory through their own extended observations. In fact, as Whewell always acknowledged, the tables in port cities such as Liverpool and Bristol were exceedingly, even exceptionally, accurate. Through their correspondence, in published papers, and ultimately through their funding strategies, Beaufort and Whewell tried not to curb but to harness the skills of these local tide table makers and professional calculators. Indeed, what puzzled Whewell in the opening quotation was the difference between the study of the tides and every other "province of physics." In other fields, scientists constructed all the needed tables; in tidology, local calculators living near the sea performed this essential task.

Whewell began his second paper on the tides, published in 1834, with just this conundrum in mind. He posited an "imaginary condition" in which "some great natural or moral convulsion" swept away all existing knowledge save a few general notions. "If, in this state of things," Whewell conjectured, "a few persons should, by their own sagacity and labour, or by the aid of some traditionary secret, attain to the power of predicting phenomena with tolerable correctness, we may imagine that they would use their peculiar skill for purposes of gain, and that they would not readily admit the world at large to the knowledge of the secret which gave them a superiority over the rest of their countrymen. Our knowledge of the tides, at the present time, exactly realizes this imaginary condition."[3]

Far from speaking with "any disrespect" toward these sagacious entrepreneurs, Whewell admitted they had advanced far beyond the theorists in Britain. In this he differed from Lubbock, who consistently abjured the contributions of local tide calculators in print, angered by their secrecy. Whewell felt more envy than anger. In a remarkable concession by one of the leading advocates for science, Whewell wrote alarmingly: "The circumstance most worthy of remark, is that on such a subject our men of science should be ignorant of, and unable to discover, that which persons of much less elevated pretensions know and apply; that the laws which are to be collected either by the observations of facts, or by the deductions of theory, should not be known to our philosophers by either method, and yet should be in the possession of other persons, to a considerable extent."[4]

For the natural philosophers in England who were defining a discipline, that secrets existed in such a pertinent sphere of the most advanced science was certainly troublesome. The legitimacy of science was at stake. But the only way Whewell denigrated local knowledge was in noting the calculators' "less elevated pretensions." He implied that the theorists' own pretensions were somehow more elevated, focusing on theory rather than practicality, on general rather than local knowledge. Unlike critics like Lubbock, Whewell acknowledged that a lack of theory did not necessarily equate to a lack of accuracy; amateur tables were all they had.

The previous chapter introduced the practice of tidology as a collaborative venture between British savants and the British Admiralty, reaffirming the fundamental contribution of the Cambridge Network (Whewell and Lubbock) and the scientific servicemen (Beaufort) to science in the Victorian era.[5] Yet there was a class of workers who labored alongside the scientific servicemen and scientific elite, a group that included dockyard officials, harbor masters, and the hundreds of seamen who collected the data. The results of the 1835 tide experiment relied heavily on expert calculators, particularly Joseph Foss Dessiou, who oversaw all the reductions for Whewell. Dessiou supervised several other calculators to organize the observations, discuss the data, draw the curves, and finish the maps. Whewell practiced tidology from the mid-1830s through the 1840s by intense collaboration with these "persons of much less elevated pretensions." He incorporated into his methodology the socially diverse spaces of the dockyard and the bustling port cities that Britain's industrialization fostered.

Throughout his research in tidal studies, Whewell worked closely with numerous tide calculators, a disparate group he called "subordinate labourers."[6] Some were members of the armed services and developed their mathematical and scientific acumen by participating in ordnance and trigonometrical surveys. Others were commercially inclined civilians who lived and worked near the sea. Their ingenuity with difficult calculations made these calculators indispensable for Whewell's tidology. They were "subordinate" only because Whewell and others wanted to push their own notions of the "scientist" as superior.[7] To this end, Whewell consistently created a distinction between his work as a theorist and the calculators' work in analyzing data. The work of these diligent men of numbers, however, demonstrates that no such sharp division existed in practice. Their active engagement with the theorists highlights their contributions in physical astronomy more generally, including John Herschel's work in meteorology, Edward Sabine's work in terrestrial magnetism, and Humboldt's work in all its variety.[8]

Whewell collaborated with numerous civilian and commercially inclined tide table makers living and working near the sea, nicely represented by Thomas Bywater, a Liverpool tide table calculator. Bywater was part of an expanding group of local practitioners who incorporated the theoretical work of Lubbock and Whewell into their local tide tables for personal gain. Whewell did not compensate Bywater, who would have tried to improve the accuracy of his tables with or without the impetus from theorists in London and Cambridge.

Local tide table makers participated in Whewell's research for the financial reward involved in keeping Britain's ports and estuaries safe for navigation. They were drawn to its commercial aspect. A select few of these calculators were paid for their work through Whewell, usually with grants from the British Association. Whewell's extensive work with Thomas Gamlen Bunt, a land surveyor and expert calculator from Bristol, demonstrates the various contributions professional calculators made to geophysical research. Like Bywater, Bunt was especially adept at sophisticated but tedious mathematical calculations. He attended the meetings of the Bristol Institution and became a member of the British Association for the Advancement of Science, but he had to struggle incessantly to enter the world of elite science within the halls of the Royal Society.

The last calculators to collaborate with Whewell worked within the Hydrographic Office in London and were paid for their work as part of their official duties for the Admiralty. Daniel Ross, who collaborated with him throughout the 1840s, was a member of the armed services and worked as one of several tide calculators under the direction of Francis Beaufort. He was hired to help with tide tables when the Admiralty first became involved in tidal studies, and in 1833 Beaufort placed him in charge of the *Admiralty Tide Tables*, first published that year. He then worked under Dessiou on the calculations for the 1835 multinational tide experiment. When Dessiou became too ill to continue his tidal reductions, Ross took over as the main tidal calculator in the Hydrographic Office. Though he was initially paid only by the Admiralty, his work quickly outgrew his official duties, and like Bunt, he also received compensation directly from Whewell.

The contributions of Thomas Bywater, Thomas Gamlen Bunt, and Daniel Ross show that the designation "calculator" is both fitting and misleading. The calculator in the mid-nineteenth century performed a

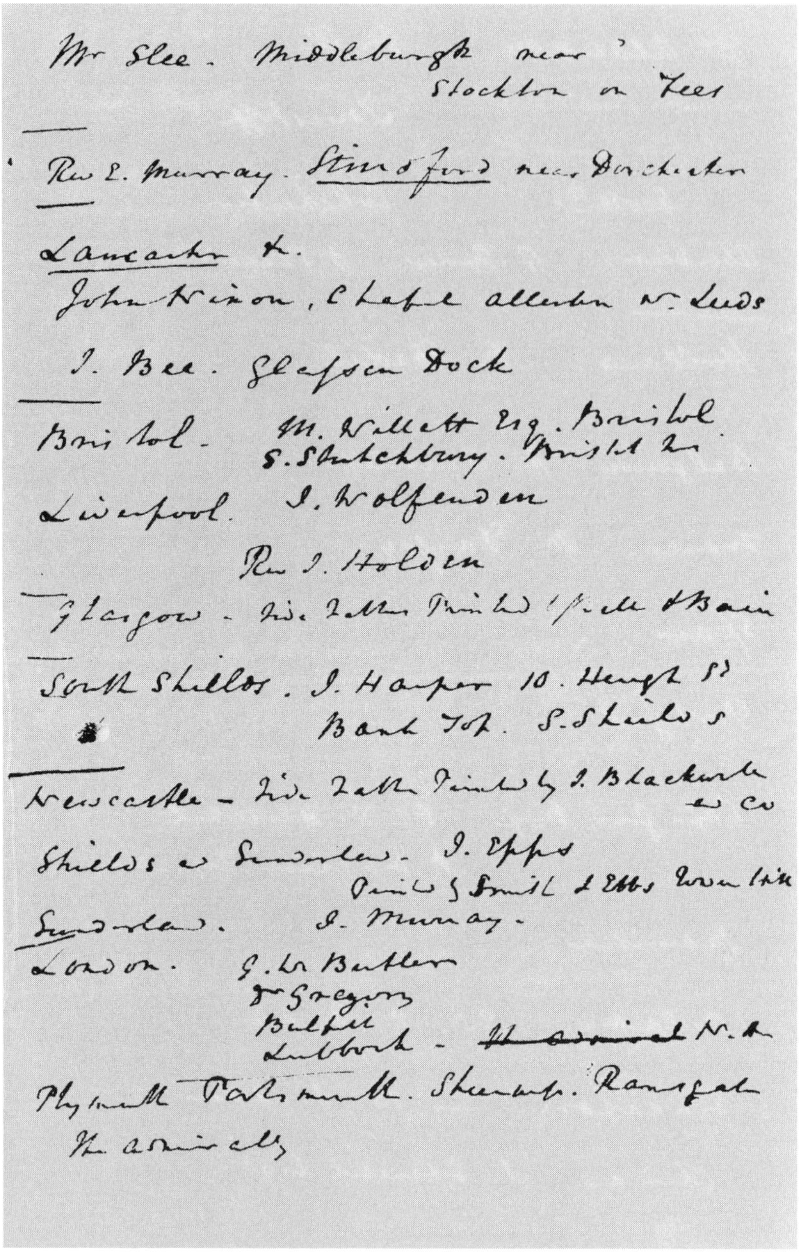
Mr Slee. Middleburgh near
Stockton on Tees

Rev E. Murray. Stinsford near Dorchester

Lancaster &c.

John Nixon, Chapel allerton nr Leeds

J. Bee. Glasson Dock

Bristol. M. Willett Esq. Bristol
S. Stutchbury. Bristol In

Liverpool. J. Wolfenden

Rev J. Holden

Glasgow. Tide Tables Printed by Bell & Bain

South Shields. J. Harper 10. Heugh St
Bank Top. S. Shields

Newcastle – Tide Tables Printed by J. Blackwell
and Co

Shields and Sunderland. J. Epps
Printed by Smith & Ebbs Tower Hill

Sunderland. J. Murray.

London. G. W. Butler
Dr Gregory
Bulpit
Lubbock – ~~the admiral~~ N. A.

Plymouth Portsmouth. Sheerness. Ramsgate

the admiralty

Figure 6.1. William Whewell's list of persons he hoped could help with his tidal research. The list includes several local tide calculators, including Mr. Willett in Bristol, George Holden in Liverpool, Mr. Bulpit in London, and the "tide tables printed" in Glasgow, Newcastle, and elsewhere. Reproduced from Whewell's own notes, Whewell Papers, Trinity College, Cambridge, England, R.6.20/35.

function similar to what electronic computers do today. But they were, of course, much more than computers in human form. They were in the field, collecting tide observations and incorporating them into the theory of the tides beyond what Whewell could ever have accomplished from his lodgings in Trinity College, Cambridge.[9] This was effectively true for Daniel Ross. Working out of the Hydrographic Office, he was the first person besides the actual observer to ponder the data in hand. Since he had access to everything coming into the Admiralty, he could make valuable connections on the spot and even suggest to Beaufort and Whewell what tidal data needed to be gathered and discussed. For Bunt and Bywater, this was literally true. They both lived in port cities and actually observed the tides in question. They actively contributed to Whewell's research program, gathered the necessary data, tested his theories, advanced his methods, and suggested new avenues of research.[10]

The work of these calculators went far beyond the practical application of science to produce a commercially viable product. Local tide table calculators did not rely solely on "traditionary secrets," nor did they contest the rationalizing and centralizing occurring in London and Cambridge. Rather, they worked in unison with the theorists as an integral part of the scientific process. The focus on different producers and consumers of science helps break down the simplistic dichotomies between theory and practice, the amateur and the professional.[11] It highlights the arbitrariness of the distinction between local and expert knowledge while continuing to advance the historiography of "subordinate" contributions to science. A culture of individualism has masked the participation of these collaborators, who were pushed out by the defining process going on in Cambridge and London.[12] The maritime spaces of port cities flowed together with the elite scientific spaces in Trinity College and the Hydrographic Office. Examining these participants' work as part of a larger scientific *process* uncovers their involvement and more fully demonstrates the significance of collaboration among different groups in Victorian physical astronomy.[13]

Severe competition between local tide table makers pushed them to incorporate the theorists' methods into their tables. They calculated their involvement to ensure the accuracy of their commercial product. The theorists, in turn, used the difficult numerical reductions of the local tide table makers to test their own theories and thus advance the theoretical understanding of the tides. The study of the tides relied on this mutually beneficial relationship. To analyze the essential contributions these tide calculators made to science, however, we can no longer focus only on London and Cambridge. Science was also practiced in cities such as

Liverpool and Bristol, where the science that developed matched the commercial and maritime interests of the port.

## Liverpool and Local Tide Calculators

Liverpool rose steadily as a commercial center throughout the eighteenth and early nineteenth centuries. Intense maritime activity defined the ever-expanding city, situated on England's northwest coast and dominated by the river Mersey. As Thomas Baines, author of the *History of the Commerce and Town of Liverpool*, boasted in 1852, "the commerce of Liverpool extends to every port of any importance in every quarter of the globe."[14] The port's direct access to the Atlantic made it the principal seat of commerce with both Ireland and the New World.[15] Whereas in Bristol those responsible for the harbor—the Society of Merchant Venturers and the Corporation of Bristol—were slow to act, the Liverpool Corporation invested large sums in developing and maintaining its port.[16] Its geographical position, combined with the massive rebuilding of its docks in the early 1800s, helped it overtake Bristol as the leading port on England's west coast.[17]

The dock expansion had produced effects like those caused by the bounding of the Thames, and the difficulty of navigating the approaches to Liverpool, already extreme, increased as vessels' hulls sank deeper. Graham H. Hills, staff-commander in the Royal Navy and marine surveyor in Liverpool, noted that "the harbours are for the most part tidal harbours, suited only to small craft, and that throughout these coasts there is a remarkable deficiency of harbours accessible at all times of tide."[18] Joseph Yates warned of similar problems in the approaches to the city. "The intricacy of the access to the River Mersey is well known," Yates acknowledged. "It arises from the accumulation, outside of its embouchure, of numerous beds of sand, which are frequently, and suddenly, changing their position and elevation, to the great horror and confusion of navigation."[19] The natural conditions of the river Mersey had not conformed to the rational vision of Liverpool's civil engineers and commercial elite. The changing currents and contours and extreme tidal variation represented the kind of confusion that the scientist and state sought to reduce to a grid.

Technological innovations in shipbuilding only added to this "horror and confusion." John Laird, a shipbuilder from Birkenhead, was an early and active proponent of the use of iron in ships' hulls.[20] And along with iron came steam. In May 1815 steam navigation first appeared on the Mersey, and by 1837 so many steamers were using the port that the Corporation of Liverpool set aside two docks for the exclusive use of steam vessels.[21] The much larger class of ship occasioned by iron and steam further limited coastal and river navigation and made the approaches to the docks all the more treacherous. Moreover, the number of vessels using the port grew steadily each year. By 1830, over one thousand vessels navigated the Mersey each month, two-thirds of them during high water hours.[22] Rough seas and an unforgiving tide created a notoriously difficult approach as vessels large and small vied to dock and unload their wares during the few hours of high tide.

The number of vessels, their growing size and weight, and their paramount importance to the trade of Liverpool made the city a "nautical vortex" that ebbed and flowed twice a day.[23] Whether wealthy residents or workers, most people in Liverpool were caught in this vortex, since the flow of the city's goods depended on a knowledge of its tides. Understanding, predicting, and ultimately harnessing that twice-daily tide fell to the expertise of local tide table makers living and working in the bustling working-class city. By the mid-1830s, they had managed to produce the most accurate tide tables in Great Britain. Even before the scientists in London and Cambridge resurrected the theoretical study of the tides, Liverpool had become a center of tidal analysis.[24]

From the theorists' perspective, the industriousness of these local calculators was a valuable resource. Tide tables, like all other tables in physical astronomy, acted as both an end and a means. While the professed *end* was the safety of navigators, these same tables were the only *means* by which natural philosophers could test their theories. Thus, when Whewell entered tidology, he was especially eager to obtain accurate tables from Liverpool to compare with his initial hypotheses. In particular, Whewell had in mind the tide tables of George Holden, the most accurate ones in the port. Holden was the person "of much less elevated pretensions" to whom Whewell referred in his 1834 publication, the calculator with the "practical sagacity" and "peculiar skill" Whewell wanted to harness. Holden based his tables on twenty-five years of observations faithfully kept by William Hutchinson, a privateer turned harbor master in late eighteenth-century Liverpool.[25] When Whewell and Lubbock first entered the field of tidology, they tried repeatedly to get their hands on these observations. After more than two years of wrangling, beginning with friendly correspondence and ending with strong-arm tactics, they finally succeeded. They used them to compute the main tidal constants for the port and to calculate the tables published in the *Admiralty Tide Tables*.[26]

Hutchinson's observations, however, were never the sole preserve of scholars in London and Cambridge; their most productive use came from tide table calculators living in the port city itself. When Lubbock and Whewell finally secured Hutchinson's observations, the first several pages of the manuscript had been ripped out. Hutchinson had offered the first five years' (three thousand) observations to the Reverend George Holden, perpetual curate of a chapel in Tatham Fells, and his brother, Richard Holden, a mathematics teacher in Liverpool. They applied Bernoulli's equilibrium theory to the observations to calculate their "Liverpool Tide Tables," first published in 1770, using methods they never divulged.[27] They passed down the methods to George's son, and by the mid-1830s they were in the charge of George's grandson, George Holden, a prolific writer and celebrated theologian.

Whewell tried repeatedly to discover the methods passed down through the Holden family tree. George Holden assured Whewell he would explain his methods in full, but he could not do so within the compass of a letter.[28] He offered to send Whewell all his methods in the future but stressed that they must be for Whewell's inspection only. The Corporation of Liverpool paid him £50 for the tables each year, as much as he made as curator of a free school in Maghull. If his methods were ever made public, Holden would lose half his income. Although Whewell promised secrecy, Holden thought better of his generosity and rescinded his offer. He told Whewell that the observations (Hutchinson's ripped-out pages) had regrettably been lost and again stressed that financial interest bound him to silence. "Nevertheless every man should rejoice at the progress of science; and if I can in any way contribute to it by shewing you my Rules etc, I shall be most happy to do so."[29] But Whewell would have to come to Liverpool and talk to Holden in person. Perhaps realizing he would never get the information he wanted, Whewell declined. He was sympathetic to the entrepreneurial interests of the tide table calculators, perhaps owing to his own relatively humble beginnings. Commercial interests had motivated Holden, and his tables were accurate because accuracy begat profit.

Holden's tide tables were relatively sophisticated. They had been passed down through three generations and represented the best of local knowledge. They included corrections for the moon's parallax and declination and reflected a practical understanding of Bernoulli's advances on Newton's equilibrium theory. They failed to correct for the diurnal inequality, however, a significant omission. The diurnal inequality is the difference between the two successive high tides each day and is relatively pronounced in Liverpool. As ships grew in size and weight and the number of vessels docking at Liverpool reached hundreds each day, an "immense number of vessels" were forced to dock during the evening high tide.[30] Because Holden's tables gave the heights of the tides only between 6:00 a.m. and 6:00 p.m. and the evening tides differed in both height and time, competition became inevitable.

Other calculators living in Liverpool used these limitations in Holden's tables to great advantage. One such local competitor, Thomas Bywater, used Whewell's and Lubbock's published papers to recalculate his "Liverpool Tide Tables," exploiting Holden's assumed lack of modern scientific techniques. Bywater had read the many papers on the tides Whewell published in the *Nautical Magazine* and obtained Lubbock's methods from a paper Lubbock had send to David Wylie, the secretary of the Liverpool Literary and Philosophical Society.[31] In the first quarter of the nineteenth century, the Liverpool Lit and Phil focused heavily on the "Lit" at the expense of the "Phil." Literary conversation accompanied by porter and oysters set a precedent that Bywater could not overcome.[32] He therefore took it upon himself to address the subject in a scientific manner, corresponding directly with both Lubbock and Whewell in November 1835. He applied several months of tide observations he had acquired from Jesse Hartley, a celebrated Liverpool dock engineer, to the theorists' work to produce tide tables that he hoped would compete with Holden's. Bywater acknowledged his debt to Lubbock and Whewell in his calculated tables as a "practical proof" of their method's "great utility," though his courtesy barely hid his desire for reciprocity.[33] He offered Hartley's observations, his own methods, and his calculated tables to the theorists in hope of an equally beneficial return.

Since Bywater had "deviated from the plan" adopted by Whewell and Lubbock, he wrote to Whewell as his first set of tables were going to press, offering to stop the printing for "any improvements you may suggest."[34] He also wanted "the kind of Table Mr. Dessiou has formed" for the diurnal inequality on England's west coast. Whewell spent some time and energy on Bywater's tables, writing back with several suggestions. When it came to Dessiou's latest work on the diurnal inequality, however, Whewell had to admit that the results were not as satisfactory as he had hoped. In his fifth series of researches on the tides, "On the Solar Inequality, and on the Diurnal Inequality of the Tides at Liverpool," published in the *Philosophical Transactions* in 1836, Whewell could only hazard a guess at the inequality's correct mathematical form. Bywater compared his own observations with Dessiou's calculated tables and could write to Whewell that "the line of diurnal variation (which you have particularly alluded) was so strongly marked through the whole of these observations and referred so clearly to the varying position of the moon that it left no doubt in my mind respecting the principle on which this Diurnal difference depended."[35] Bywater then sent Whewell his own tide table that included the diurnal inequality in its predictions of the heights and times of high water.

Bywater, with an openness not always present in the competitive tide table industry, also emphasized that one of Whewell's problems might be with the epoch he had directed Dessiou to use.[36] While grappling with this difficulty, Whewell wrote to Lubbock referring to the diurnal inequality at Liverpool. "It appears to me quite clear," Whewell acknowledged, "that the theoretical and the observed curve would approach very near by putting the epoch of the former about 3 hours forward."[37] Whewell then added that "this is very nearly what I collected from Bywater's results." He made it sound as if he had uncovered the correct epoch himself by rooting through Bywater's work, but in reality Bywater had told him the correction was needed. Whewell returned to the subject of the diurnal inequality in his seventh series, "On the Diurnal Inequality of the Height of the Tide Especially at Plymouth and at Singapore," published in the *Philosophical Transactions* in 1837. Whewell ended his paper by examining all the reports of the diurnal inequality he could accumulate, including those at Liverpool from Thomas Bywater and those at Bristol from Thomas Gamlen Bunt. From all these observations, he was finally able to abstract a law for the diurnal variation, his most prized conclusion in his twenty-year research project on the tides.[38]

Thomas Bywater died unexpectedly in 1837, before Whewell's results on the diurnal inequality went to press. Bywater's methods, however, lived on in the "Liverpool Commercial Almanac and Tide Tables," calculated by Alexander Brown. Brown also started corresponding with both Whewell and Lubbock, stressing that he had "stated in that part of the Almanac referring to the tides that I have employed your tables in the computation."[39] In return, the theorists sent Brown all their published theoretical papers and suggested improvements to his tables.[40] Brown then sent his finished predictions to the theorists, along with his methods of calculation, so they could test the accuracy of their hypotheses. He was being shrewd, intent on continuing Bywater's advantageous relationship with the theorists, spurred by the same interests. Britain's global ascendancy depended on its ability to maneuver safely through its ports and estuaries, and supporting such maneuverability could be lucrative. Brown was occupying a commercial niche.

For their part, Thomas Bywater and Alexander Brown received a ready-made apparatus for constructing their tables, complete with the latest

Difference in the Heights of the Morning and Evening Tides at Liverpool — R.6.20.45

| | January P.M H.W | January A.M H.W | February P.M H.W | February A.M H.W | March P.M H.W | March A.M H.W | April P.M H.W | April A.M H.W | May P.M H.W | May A.M H.W | June P.M H.W | June A.M H.W |
|---|---|---|---|---|---|---|---|---|---|---|---|---|
| 30 | − .41 | + .61 | − .52 | + .86 | − .75 | − .03 | − .18 | + .28 | − .33 | + .01 | + .20 | − .33 |
| 1 30 | − .74 | + .50 | − .10 | + .83 | − .82 | + .69 | − .89 | + .24 | − .11 | .50 | + .16 | .45 |
| 2 30 | − .06 | .93 | − .73 | + .56 | − .38 | .32 | − .70 | .08 | + .05 | .37 | .04 | .68 |
| 3 30 | − .22 | + .78 | − .79 | + .52 | − .41 | + .21 | − .21 | .05 | + .09 | .33 | + .35 | .54 |
| 4 50 | − .56 | + .65 | − .32 | + .16 | − .39 | + .12 | − .07 | .29 | + .19 | .33 | + .38 | .59 |
| 5 30 | − .60 | + .82 | − .63 | + .40 | − .43 | + .08 | − .17 | .27 | + .21 | .45 | + .16 | .34 |
| 6 30 | − .51 | + .28 | − .12 | + .42 | − .11 | .32 | + .01 | .24 | + .32 | .26 | + .26 | .49 |
| 7 30 | − .62 | .48 | + .06 | + .07 | − .16 | − .15 | + .08 | .11 | + .28 | .08 | + .29 | .14 |
| 8 50 | − .17 | + .17 | + .33 | − .24 | 0 | − .13 | + .02 | + .03 | + .20 | .06 | + .02 | + .06 |
| 9 30 | − .07 | − .03 | + .41 | − .17 | + .07 | + .28 | + .17 | + .05 | + .10 | + .15 | − .10 | + .20 |
| 10 50 | + .03 | − .08 | + .23 | − .06 | + .27 | − .54 | + .06 | − .07 | + .16 | + .17 | − .12 | + .24 |
| 11 50 | + .30 | − .07 | + .22 | − .02 | + .07 | − .09 | − .19 | − .18 | − .06 | + .35 | − .14 | + .33 |

| | January A.M H.W | January P.M H.W | February A.M H.W | February P.M H.W | March A.M H.W | March P.M H.W | April A.M H.W | April P.M H.W | May A.M H.W | May P.M H.W | June A.M H.W | June P.M H.W |
|---|---|---|---|---|---|---|---|---|---|---|---|---|
| 30 | + .03 | − .33 | + .28 | − .31 | + .23 | − .18 | − .08 | .21 | + .09 | .34 | − .67 | + .25 |
| 1 30 | − .09 | .41 | + .71 | + .19 | + .44 | + .01 | + .22 | + .20 | + .06 | .46 | − .42 | .45 |
| 2 30 | + .25 | .40 | + .62 | .28 | + .28 | − .03 | + .24 | + .21 | − .16 | .46 | − .59 | + .60 |
| 3 30 | + .39 | .18 | + .07 | − .67 | + .44 | 0 | + .10 | + .25 | − .28 | + .59 | − .65 | + .36 |
| 4 30 | + .36 | .52 | + .73 | − .24 | + .15 | + .09 | − .12 | + .49 | − .19 | + .46 | − .39 | + .90 |
| 5 30 | + .48 | .50 | + .21 | − .02 | − .06 | + .09 | − .18 | + .44 | − .51 | + .54 | − .63 | + .62 |
| 6 30 | + .26 | + .31 | − .11 | + .32 | + .03 | + .65 | − .14 | + .41 | − .78 | + .68 | − .04 | + .76 |
| 7 50 | + .16 | + .37 | − .35 | + .02 | − .17 | + .57 | − .38 | + .27 | − .80 | + .49 | − .17 | + .56 |
| 8 50 | + .06 | − .03 | − .06 | + .33 | − .27 | .22 | − .50 | + .27 | − .48 | + .18 | − .27 | .14 |
| 9 30 | − .11 | + .13 | − .49 | + .40 | − .20 | + .17 | − .46 | + .26 | − .52 | + .23 | − .16 | + .03 |
| 10 30 | − .46 | + .58 | .15 | + .15 | − .64 | + .08 | − .20 | + .15 | − .13 | + .01 | − .08 | .09 |
| 11 30 | − .47 | + .69 | .48 | − .09 | − .31 | + .23 | − .53 | + .15 | − .16 | .13 | + .03 | − .31 |

Figure 6.2. A table showing the diurnal inequality, the difference in the heights of the morning and evening tides, at the port of Liverpool for the first six months of 1836. Whewell obtained the differences from Thomas Bywater's published tide tables for the port. Reproduced from Whewell Papers, enclosed in R.6.20/23.

theoretical work by two of the main authorities on the subject, one the vice president of the Royal Society of London, the other a professor from Trinity College, Cambridge. They unabashedly printed the theorists' names and academic positions in their tide tables, boasting that these theoretical advances yielded accuracy beyond all competing tables, including Holden's. Their tables included, for instance, the high and low water for both night and day tides as well as the diurnal inequality.[41] Note that these local calculators would have accomplished this with or without guidance from the theorists. As early as 1835, while Whewell was just beginning to work on the diurnal inequality, Bywater included this in his tide tables for the port. He *then* contacted Whewell, tested

Whewell's theoretical conclusions through his own predictions, and offered advice on how to iron out the discrepancies.

This was all free work for the theorists. In tidology, constructing tables was the only way to test theory, and if Brown and Bywater had not already done these extremely time-consuming numerical calculations, Lubbock and Whewell would have had to hire someone to do them. And as noted earlier, paying calculators was a constant problem, so they welcomed the free work done on the periphery. The theorists' work, moreover, was being used to heighten the accuracy of local tide tables, and they were acknowledged in print. Whewell and Lubbock used this recognition to obtain further funding from the British Association and ultimately the British Admiralty. Far from disparaging the local tide calculators, Whewell always encouraged their efforts, noting in print their "usefulness" to theoretical developments in tidology.[42]

Their usefulness had direct and immediate consequences for the city of Liverpool by helping to tame the city's notorious approach to its port. By the second quarter of the nineteenth century, Liverpool had become a port second only to London. The type of science it fostered followed suit. A lively tidology was developed by people not considered "scientists" by today's standards, in places more closely linked to the commercial and maritime interests of the city than the gatherings of its Philosophical and Literary Society. As Thomas Stuart Traill suggested in the opening address to the British Association meeting held in Liverpool in 1837, it also brought notice and recognition to the port city.[43] The collaboration between the scientists and calculators in Liverpool also furthered the overall success of the research program back in Cambridge and London, blurring the lines between local and expert knowledge. Fueled by the expansion of global capitalism, local practitioners initiated correspondence with the theorists to incorporate the latest scientific methods into their tables—traditional knowledge at work. The theorists, in turn, tested their latest theories with locally produced tables—expert knowledge at its best.

### Thomas Gamlen Bunt and the Bristol Tides

A heavy reliance on local tide table calculators is apparent throughout Whewell's tidal research. As his initial difficulties with George Holden suggest, however, not all the local calculators were as forthcoming as Bywater and Brown. Whewell could also rely on the good graces of the Admiralty, which allowed both Joseph Dessiou and Daniel Ross to work under his guidance. But like Bywater and Brown, the tide calculators at the Admiralty had specific aims—constructing accurate tables to guard the safety of British vessels at home and abroad—that often conflicted with Whewell's "higher pretensions."

To control the calculators, Whewell needed to control the purse strings, and he turned to the British Association to contribute the funds. By the mid-1830s, Lubbock was running out of steam. As his duties at his London bank became more pressing, he slowly dropped out of tidal investigations; he returned to the subject only briefly in 1839 to gather his former materials and write a textbook on the tides.[44] Lubbock's waning interest left Whewell as the main tidal theorist in Britain, and thus the main beneficiary of calculations and funds coming from both the British Association and the Hydrographic Office. The British Association meeting in 1836 was held in Bristol, and Whewell told Henry Phillips, the secretary of the Association, that he wanted to give a paper on his tide research in general and Bristol in particular. "I also want to apply for some money to spend upon the Bristol tides," Whewell added. "There is a most zealous and intelligent calculator, called Bunt, who lives in Small Street Court, Bristol, who has been working at them for some time, and whom I wish to keep at work."[45]

Whewell's contrasting relationships with Lubbock and Bunt shed considerable light on the process of investigation in physical astronomy. The correspondence between Whewell and Lubbock reveals mutual professional respect. They commented on each other's methods, and if they disagreed, they tried to smooth out their differences by following their own lines of research. Whewell initially had followed Lubbock's methods, but as he became more sure-footed in the field, he distanced himself from his former student. By the mid-1830s, they had their own distinct research programs and stayed out of each other's way. "If I am allowed to proceed my own way I know what I am about" was Lubbock's caustic response when Vernon Harcourt attempted to grant the funds from the British Association to a committee rather than to Lubbock alone.[46] The correspondence between Whewell and Bunt, however, conveys a strikingly different feel.

Over sixty letters from Bunt to Whewell exist, and besides offering a rare glimpse into the relationship between a man of science and a local calculator, they also reveal the personal affinity between the two. More than anyone else including Lubbock, Bunt served as Whewell's collaborator in tidology. Detailed discussions of the mathematics involved, the methods used, the errors, even the joys, disappointments, and successes of scientific research, fill their lengthy correspondence. As with most collaborations, their intense study bound them together, giving each a stake in the other's achievements. Bunt's letters were filled with flattery and deference, while Whewell's reiterated time and again that he could not have proceeded without Bunt's efforts and, more surprisingly, his guidance.[47]

A meticulous mathematical intellect, Bunt was an unassuming man of "retiring habits" who "shunned notoriety and cultivated but a limited circle of acquaintance."[48] He was also of strict, even severe, religious faith, refusing to take tide readings on Sundays.[49] He opened a timber and land surveying business in Bristol in 1827, executing surveys for private and government contractors, including some of the leveling operations for the British Association in the late 1830s. Though Bunt's principal vocation was land surveying, he spent much of his leisure in scientific and mechanical pursuits. In 1832 he published locally an account of a "Planetarium," a mechanical invention that demonstrated the positions and nodes of the planetary orbits and predicted the transits of Mercury and Venus. Whewell would later write flatteringly of Bunt's invention, showing it to both John Herschel and George Biddell Airy.[50] Its initial notice in a lecture by the Reverend Lant Carpenter, a founding member of the Bristol Literary and Philosophical Society, first introduced Bunt to the Bristol scientific scene.

The Bristol Literary and Philosophical Society was attached to the Bristol Institution, a highly exclusive body whose members had to buy £25 shares.[51] The Society's aim was not to advance science but to solidify the power of the social elite—a social and political agenda. Its members came from the ruling classes. As shares dwindled in the 1830s, however, membership became slightly more democratized, and instrument makers, booksellers, and others gained limited access.[52] Bunt was nominated in November 1832 and unanimously elected an "associate" in February 1833, about the same time Whewell was preparing his first paper on the tides for publication. "Associates" did not have to pay the annual subscription of two guineas but still had full access and free admission to all parts of the Bristol Institution except on special occasions. Bunt assisted in the installation of scientific instruments for the Institution and was eventually introduced to the study of the tides by the curator of the museum, Samuel Stutchbury.

The range of the tide in the Bristol Channel is among the greatest in the world, reaching a height of thirty-seven feet at the dockyard in Bristol, six miles up the river Avon. Millions of tons of water rushed in and out of the Avon every six hours. Liverpool had overtaken Bristol as the leading port on Britain's west coast, partly owing to the difficulties

Figure 6.3. A photograph of a steamer, the *Gypsy,* stranded by the outgoing tide and wrecked at Horseshoe Point, a particularly dangerous part of the river Avon. Reproduced from a photograph on display at the Bristol Industrial Museum.

of navigating these severe tides. Figure 6.3, a photograph of a steamer split in half and lying aground in the middle of the river Avon, illustrates these dangers. The steamer's pilot had misjudged the tides. Only a few hours earlier, the steamer's hull, buoyed by the massive inflow of the tide up the river, would have barely sunk below the place where the onlookers are standing on the now-exposed mudflats on each side of the crippled vessel. A miscalculation of only a few feet in the height of the tide could lead to disaster. Predicting the ever-changing tides in the approach to the Bristol docks therefore became increasingly important and lucrative, especially as the vessels entering the river got heavier and drew more water. The Bristol Institution took an early interest in tidal studies because most of its members made their living through the port. In the second quarter of the nineteenth century, Bristol's leaders dedicated themselves to reversing the city's decline, intent on keeping the port safe and open to river traffic. This meant remedying their previous neglect and giving special attention to the tides.

Thomas Gamlen Bunt's initial interests matched those of the Bristol Institution. He was fascinated with scientific topics, especially as they pertained to the sea, but he was first motivated by the commercial side of the investigation. On the death of Mr. Willett, "the publisher of the Tide Tables, <u>professedly</u> calculated for the port of Bristol," Bunt aimed to fill this commercial void.[53] Bunt's emphasis on "professedly" suggests that Willett's tables were not calculated at all; they were probably loosely based on Bernoulli's equilibrium theory combined with a year or two of local observations. It also suggests that Bunt wanted to use modern scientific methods of calculation. To this end, Stutchbury wrote to Whewell introducing Bunt as "a very worthy man, who has at our solicitation undertaken to calculate a set of tables for our Port, which is now a particular desideratum; the author of those we have thereto had, being lately deceased leaves us without any. The nearest port for which tables have been calculated being Liverpool. . . . Should you sir approve of the mode which he has adopted, it will be extremely gratifying to us to be able to state so."[54]

Stutchbury's letter underscores the backward state of tidology as it existed in Bristol in 1835. The port had no reliable tide tables, making the entrance to its docks exceedingly dangerous. To add insult to possible injury, it was forced to rely on tide tables for Liverpool, in many respects a rival port on England's west coast. The Bristol Institution wanted to place Whewell's name on the publication of its local tables as a testimony of their accuracy, and Bunt wanted to correspond with the leading tidologist in Britain.

Bunt had already used Whewell's previous publications in the *Philosophical Transactions* to complete a tide table for the following year (1836), and he sent Whewell a few graphs to demonstrate his command of the graphical method of data representation.[55] Whewell was duly impressed and responded in glowing terms, which Bunt then included not only in his published tables, but also in the Bristol newspapers, advertising his "new and improved" tide tables.[56] Using Whewell's research as his theoretical foundation, Bunt undertook a complete overhaul of the Bristol predictions, incorporating the parallax and declination of the sun and moon as well as the diurnal inequality. He was on the cusp of the latest scientific advances in tidology and used this argument as the main strategy to sell his tables in Bristol. He became widely known among those who made their living through the port as the author of "Bunt's Tide Tables."[57]

As he worked, Bunt made significant advances on Whewell's methods. On several occasions a "difficulty" occurred to him, and in the most

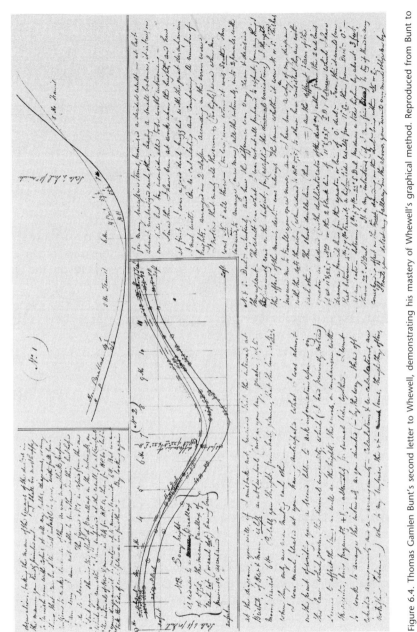

Figure 6.4. Thomas Gamlen Bunt's second letter to Whewell, demonstrating his mastery of Whewell's graphical method. Reproduced from Bunt to Whewell, 29 October 1835, R.6.20/87.

apologetic terms, he suggested alternative routes for further investigations.[58] Lubbock and Whewell, for instance, had found the moon's parallax inequality by determining the mean of all the curves for observations covering an entire year. Once Bunt turned to this inequality, "some improvements in the process having suggested themselves as I proceeded," he was led to discuss the curves *at each hour of transit.*[59] These differed considerably from one another and from the mean of the whole year, and Bunt warned Whewell that these differences materially affected the results. Bunt calculated new tables that he enclosed for Whewell's inspection, along with an analysis of how they improved the predictions in relation to Willett's earlier tables.

The success of Bunt's tide tables stopped with their accuracy; he made very little money from the venture, though it cost him a considerable amount of time. He had complained to Whewell earlier that "these oft repeated arrangements and re-arrangements, calculations and re-calculations are no trifling labour."[60] With so much time spent on the calculations, and with such little monetary return, Bunt warned Whewell that he might be forced to "return to the country"—to go back to his surveying.[61] He had already moved to an examination of the parallax and declination curves for the solar inequalities for the Bristol tides, however, and again he found errors in Whewell's analysis.[62] His results matched observational data better than Whewell's, but again he feared he would not be able to proceed. "The sacrifice of time," he protested," which I have hitherto disregarded, is beginning to be seriously felt by me." He aimed to produce accurate tables for the port, however, and he was willing to press on, eager to hear Whewell's thoughts on his improved methods.[63]

Whewell replied with several queries about Bunt's exact approach, though he seemed most concerned with Bunt's hints that he might not be able to proceed. Bunt's language quickly reverted to deference. "I should be sorry if a momentary expression in a recent letter of mine should have seemed like the language of complaint," he wrote, stressing that Whewell's own interest constituted his "chief stimulus for proceeding."[64] The tension between the calculator and his patron highlights the different and often contradictory motivations behind their work. Bunt wanted to earn money, whereas Whewell was concerned with scientific advance. Emboldened by his earlier successes, however, Bunt ended his letter with further corrections to Whewell's work. "Though I am quite afraid of wearying you, on the one hand, or of dogmatising, on the other, I cannot forbear one remark more on the declinations," insisting on a different epoch to be used for the declination corrections for both the moon and the sun.

Although Bunt's calculations so far had resulted in only minor advances on Whewell's work, Whewell had found a competent workman who shared his enthusiasm. He did not want Bunt to return to the country, but the funds from the British Association were dwindling. This prompted Whewell to apply to the Society of Merchant Venturers in Bristol, with a personal testimony "to the great service in reference to the Theory of the Tides" that Bunt's meticulous exertion had produced.[65]

Whewell proposed that the Merchant Venturers offer Bunt £100 a year. Bunt received only £25, which, moreover, "must be considered as a Donation and not as in anywise implying any engagement for an annual subscription."[66] Whewell was not deterred. He restated his support of Bunt in a memorial sent to the mayor, alderman, and councillors of the city and county of Bristol "as to the desirableness of a regular pecuniary provision" for Bunt.[67] Signed jointly by Whewell and Stutchbury, the memorial stressed the need for accurate tidal data to match the "increasing spirit, and the opening prospects" of the port. After reviewing Bunt's work on the tides, it ended in the strongest of terms: "The Port of Bristol is indebted to him for the correction—indeed almost revelation— of its Tide Tables."[68] Whewell went out of his way to solicit funds for Bunt's tidal computations, emphasizing the significance of Bunt's work to his own investigations, the success of Liverpool, and the progress of science in general.

Whewell's lobbying the city of Bristol met with limited success, and he again was forced to turn to the deep pockets of the British Association. He had used the entire grant (£150) for 1836 solely on Bunt, and in his report to the Association that year, he focused more on Bunt's work than on his own, exhibiting a model of Bunt's self-registering tide gauge that was then under construction. At the next meeting, held in Liverpool the next year, Whewell again applied for funds, receiving an additional £75 for Bunt's investigations. This time, however, he suggested that Bunt use the money to discuss the observations from the self-registering tide gauge that he had finally completed.

In 1834 Bunt had begun working on his tide gauge, an improved version of Palmer's gauge erected at the London Docks in 1832. Bunt erected his gauge on the property of the Society of Merchant Venturers at the Hotwells, a mile below Bristol, in the summer of 1837. According to Whewell, Bunt's gauge was "incomparably the most accurate and complete of any that has yet been employed," and he had a complete

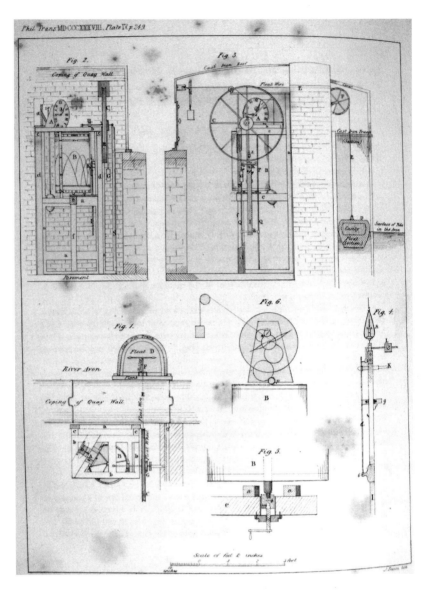

Figure 6.5. Thomas Gamlen Bunt's self-registering tide gauge. Notice the representation of the rise and fall of the tide, labeled B in figure 2. Reproduced from William Whewell, "Description of a New Tide-Gauge, Constructed by Mr. T. G. Bunt, and Erected on the Eastern Bank of the River Avon, in Front of the Hotwell House, Bristol, 1837," *Philosophical Transactions* 128 (1838): 249-51.

Figure 6.6. Model of Thomas Gamlen Bunt's self-registering tide gauge of 1837. It is on display at the Bristol Industrial Museum.

description published in the *Philosophical Transactions* in 1838.[69] These machines added significantly to tidal theory by recording the tides throughout their entire rise and fall, during the day and at night and for extended periods.[70] Moreover, they did not simply measure the height of the tide; they also measured the time and the wind direction. Bunt's gauge of 1837 even measured the barometric pressure and the moon's transit across the meridian.[71]

Bunt's contributions to tidology reached far beyond acquiring accurate tidal observations; he was the first researcher to construct a gauge and then discuss the observations immediately after it registered data. Moreover, by reading the tides directly from his tide gauge, Bunt already had his results in the form of curves. He was thus able to bypass tables altogether and reveal quantitative relationships in the data without the interference of a human observer. He was therefore able to add directly to Whewell's methods of graphical representation.

When describing his graphical method, Whewell always acknowledged the problems caused by the confluence of forces acting on the tides. The many inequalities often masked the correct mathematical laws. He counteracted this difficulty by combining his method of curves, described in the previous chapter, with what he termed the "method of residues." When confronted with a large set of observations, Whewell first had them converted from tables into curves. He then constructed a curve entirely from theory, representing the main force acting on the tides (the semimenstrual inequality produced by the declination of the moon). Next he compared the observational and theoretical curves. Since more than one force acted on the data, the semimenstrual curve never entirely accounted for the observational curve. A "residue" was left. Whewell then placed only this residue in the form of a curve and repeated the process using the second most important inequality (the parallax of the moon), on and on until the entire curve was fully represented.

In his *Philosophy*, Whewell explained that astronomers followed a similar process (though without the graphical method) to find the inequalities in the motion of the moon, a task that took many centuries, beginning with Ptolemy and continuing until Whewell's own time. "In the examination of the tides, on the other hand, this method has been applied systematically and at once."[72] The method of residues significantly reduced the time needed to bring the tides to rule. It placed special emphasis, however, on the "steady hand" of the scientists to represent the initial observational curve correctly.[73]

It was in this step that Bunt's curves from his tide gauge proved so fruitful. He began with the actual curve from his tide gauge, thus bypassing the transformation from table to graph. A steady hand was no longer required. Bunt thereby corrected for human error in making observations. He then superimposed "residual" graphs obtained by using Whewell's theoretical work until the residues exactly matched the curve from the tide gauge. Bunt realized the "superior accuracy" of his curves, which allowed for more precision in determining residual quantities.[74] While in the middle of their data analysis, he wrote to Whewell: "One reason why I was desirous that you should see the curves from the machine observations, while the discussion was in progress, was this;— because their great regularity enables the eye to trace the effects of the lunar parallax and declination, in the *first residue*, much more easily than could be done formerly; and may therefore give some hints for the

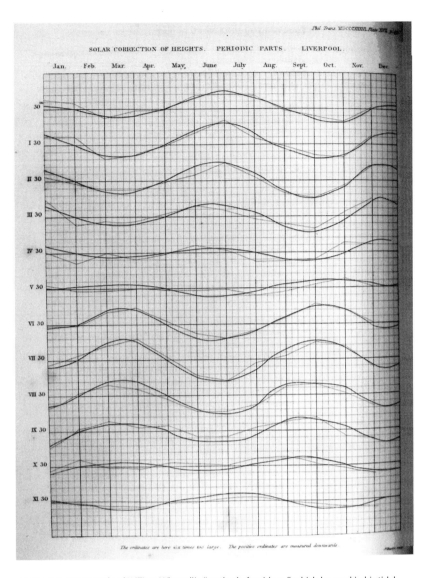

Figure 6.7. An example of William Whewell's "method of residues," which he used in his tidal studies throughout the 1830s and described in full in his *Philosophy of the Inductive Sciences* in 1840. William Whewell, "Researches on the Tides," 5th ser., "On the Solar Inequality, and on the Diurnal Inequality of the Tides at Liverpool," *Philosophical Transactions* 126 (1836): 137.

improvement of the *Epoch*."[75] He went on to explain part of the process of finding the different residual inequalities: "By *numerical calculations*, combined with the use of those curves drawn on transparent paper, and *sliding* over one another, which I remember to have shown you, I found all the times for 1838. . . . These slides exceedingly facilitate the calculation, which I could not have got through, had I worked from numerical tables of double entry . . . according to the usual method."[76] The "usual method" was Whewell's, a method Bunt showed to be obsolete.

The advantages Bunt obtained, then, both from the accuracy of his self-registering tide gauge and from reducing those observations immediately after they had been registered, allowed him to move beyond Whewell's own methods. Whewell acknowledged Bunt's contributions in print on several occasions, especially in the *British Association Reports*, which often contained more direct quotations from Bunt than Whewell's own words. Bunt's methods were so important that Whewell quoted him in full for several pages in his "Researches on the Tides," ninth series, "On the Determination of the Laws of the Tides from Short Series of Observations," read before the Royal Society on 14 June 1838 and published in the prestigious *Philosophical Transactions* that same year.[77]

Though Whewell introduced Bunt's approach as a method "practised at my suggestion," the method was not Whewell's. But neither was it entirely Bunt's. The two had collaborated for over two years to come up with the final means of discussing the data. Whewell gave Bunt specific directions, but Whewell was not himself in direct contact with the observations. Since they lived in different regions of England and rarely saw each other, Bunt often had to take Whewell's directions as *suggestions*. Bunt improvised according to the accuracy of the data, which were sometimes rather good, at other times erroneous. He often attempted several methods of discussion at once, and if he found himself unsure how to proceed, he experimented. The extent of Whewell's other obligations between 1836 and 1838 was extraordinary, and Bunt had a surveying business to run. If he had waited for Whewell's suggestions every time he ran into trouble, nothing would have gotten done. In a sense, Bunt's confidence to proceed on his own defined what it meant to be a good calculator.

Being a good calculator, however, took time. "I am almost ashamed to tell you the time I have consumed on this subject," Bunt wrote to Whewell early in 1837, "lest you should consider me an idle, or at least, a sluggish workman."[78] He rarely charged Whewell for all his labors, especially if they led to a dead end. "A good deal more time than what I have set down has been consumed on occasional investigations that have occurred to me, that are not worth recording," Bunt admitted, inwardly bemoaning the effort. He gave Whewell "full authority" over his charges, suggesting that if Whewell wanted further investigations that would overrun the sum granted by the British Association, Bunt would gladly work an extra week or two without pay. "I generally rise about 5 a.m., as it is; if I want a little more time, I must try the anterior period of 4," a pun on the epochs used in tidal computations, humorous perhaps only to tidologists.

Distance and deference, correspondence and collaborative interests all affected Whewell's work as well, though in a different manner.[79] Whewell rarely dealt directly with the data, and what the Victorians called "discussing" the data often proved extremely complicated. It required actual discussion, conversation, ideas to be talked about rather than demonstrated. Questions arose that demanded immediate answers. Which epoch, why those years, and why only the southern ports? Why the means rather than the actual heights, and why only to find the parallax corrections and not the declinations? Can the discussion be limited to daytime tides, and which residual quantity should be calculated first? Bunt explained in excruciating detail every aspect of the procedure leading to the multicolored graphs stenciled on wax paper that he sent to Whewell. But correspondence had its limits, and since the necessary dialogue was missing, he often failed to convey his approach clearly.

In doing the work for his "ninth series," Whewell did not understand Bunt's methods in their entirety, and he wrote to say so. "I will try to answer the questions it contains in a few days," Bunt responded, "and likewise endeavour to give you some account of the process by which the curves of declination and parallax correction were obtained, and the semimenstrual curve."[80] Often Whewell could do no better than quote Bunt's own description; thus, his quoting of Bunt for three pages in his "ninth series." It is telling that George Biddell Airy, the astronomer royal, when describing the method of residues in his article "Tides and Waves" in the *Encyclopaedia Metropolitana*, also quoted not Whewell, but Whewell quoting Bunt.

Whewell and Airy both profited from Bunt's seeming ease with numbers throughout the 1840s, and Whewell continued to work assiduously as Bunt's patron by acquiring grants from the British Association. Whewell was "very desirous of continuing to profit by [Bunt's] labours" and applied for an additional £50 at the 1841 meeting of the Association.[81] By then, however, Whewell was counting his successes, and he asked Bunt what he thought were the most important areas for further investigation. Bunt replied that they should compare the heights of the tides with atmospheric pressure, calculations he had already begun.[82] He had already found the barometric effect, and by using the method of residues, he was able to give its mathematical proportion.[83] He also determined that the contemporaneous barometric effect should be used, not, as with other inequalities, an epoch preceding the day of high water.

Bunt's work in tidology proved highly successful. He perfected the instruments, recorded and analyzed the data, advanced Whewell's methods of analysis, and suggested new areas of research. Most important for Bunt, it provided steady and profitable employment. Between 1836 and 1844, Whewell secured hundreds of pounds sterling for Bunt's tidal computations and leveling operations. Although this sum included Bunt's own expenses for hiring assistants to perform some of the more mundane calculations, this was a very good living. He also received some compensation from the Bristol Institution, the Society of Merchant Venturers, and the Corporation of Bristol, largely through Whewell's efforts on his behalf.

Bunt's work in tidology also introduced him to the British scientific community. He gave numerous papers at British Association meetings and met some of the top scientists in Britain, including Herschel and Airy. Through these meetings, Bunt attained a limited but significant status within the learned community. Twenty-five years after Bunt's and Whewell's collaboration, he published his own paper in the *Philosophical Transactions* on the diurnal inequality of the tides in the port of Bristol.[84] That same year, the British Association set up a committee to improve the harmonic analysis of tides under the leadership of Sir William Thomson, and Bunt served as a member. It is noteworthy, however, that these final achievements—his published paper and his committee work—both occurred only in 1867, the year after Whewell's death.

Bunt's status as an "associate"—not an actual fellow—of the Bristol Institution followed from his lower-middle-class origins, which made it difficult for him to actively participate in Bristol's notoriously aloof learned culture. Since Bunt did not pay dues to the Institution, he did not financially support Bristol's scientific culture. Rather, the members of the Bristol Institution supported him. Stutchbury first introduced him to Whewell, and both he and Whewell attempted to raise financial support for Bunt through local patronage, particularly the Society of Merchant Venturers and the Corporation of Bristol. Bunt collaborated closely with Whewell, but always as a "subordinate labourer." He was paid for his time-consuming calculations, but because he was paid, his work ultimately belonged to Whewell. Whewell unabashedly published Bunt's methods as his own, compensating him with added work rather than added prestige. In the process, Bunt advanced Whewell's science in important ways, representative of the way local calculators contributed to geophysical research in general. Although the Bristol elite had trouble crossing socioeconomic borders, the practice of tidology did not. Research on the tides in the early Victorian era could not be accomplished solely in Cambridge or London. Theorists like Whewell needed calculators like Bunt, and Bunt, in turn, used the interests of the scientists and the Admiralty in the theoretical study of the tides for his own economic gain. He filled a commercial void in Bristol by producing the port's tide tables, using his skill with difficult calculations to raise his own socioeconomic status.

### Daniel Ross and the Hydrographic Office

Perhaps the most telling testimony to Bunt's energies came when Whewell and Lubbock were discussing plans for persuading the government to employ a full-time superintendent for extending and improving the British tide tables. "Of all the persons whom I know," wrote Whewell, "I should think Mr. Bunt the most likely to be useful. He is a very fair mathematician and a man of great resource and great information, besides being very fond of the subject of the tides. *He has improved almost all the methods I have taught him.* I am quite confident the subject, in his hands, would make surprising progress."[85] The subject did make surprising progress in Bunt's hands, but not because he worked as a full-time superintendent for the Admiralty. That job was conferred on Daniel Ross.

Like Bunt, Ross was trained as a surveyor and employed by Whewell as an expert calculator. He differed from Bunt, however, in that he worked at the Hydrographic Office, was paid by the government, and had access to all the Admiralty publications. And his position within the Admiralty had its advantages. Whewell did not know the extent of the data arriving in Beaufort's office, the personalities of the observers, or the accuracy of their observations. Ross knew these things as a matter of course. He was therefore in a better position than Whewell to determine what should and should not be discussed.[86]

The Admiralty assigned Ross to help with tide calculations in 1831, and in 1833 Beaufort placed him in charge of the *Admiralty Tide Tables*, first published that year. He began discussing data specifically for Whewell only in 1835, including a large amount of work in 1837 on the

Plymouth tides. Since this work tended to overlap with the Admiralty's interests, he was not paid extra for these services. He followed Whewell's graphical method to lay down the tides in curves, find the requisite inequalities, and compute predictions. Ross's work for Whewell, however, began to move far beyond anything that could be considered specific to the Admiralty. More to the point, it fell in line with Bunt's work for the British Association, for which Bunt was paid from grants designed for such purposes.

In October 1839, Ross noticed some peculiar anomalies in a set of observations for the port of Leith that had been registered for more than thirteen years. This time Whewell suggested that Ross keep track of his time and perform the operations outside his official duties.[87] Ross worked on the Leith tides out of his home on and off for twelve months before and after work, and at the British Association meeting in 1840, Whewell applied for £50 to support Ross's calculations.[88] At the following meeting of the Association in 1841, while acting as president Whewell reported their results.[89] Although Whewell presented the report, Ross had done all the work. "The present Report," Whewell began, "will refer to a new mode of presenting the corrections of the height of high water for lunar parallax and declination. . . . Mr. Ross suggested to me the advantage of such a table, and has constructed it from the Leith observations."[90]

By using a double-entry table, Ross could show that the correction for declination varied as the square of the declination, while that for parallax varied linearly. He could further demonstrate that the same held true for the observations at London, Liverpool, Plymouth, and Bristol. "The agreement of these results," boasted Whewell, "cannot but be considered as decisive evidence of the correctness of the tables which we have obtained, as to their form and general law."[91] Ross's diligent and innovative work endeared him to Whewell, who sent him a constant supply of work along with remuneration from the British Association.

Ross's work demonstrated the kind of practical innovation that the Admiralty calculators brought to early Victorian data analysis. The correspondence between Whewell and Ross was not always steady, and Ross often forged his own path. He discussed data he thought was important the day it arrived in the Hydrographic Office. One evening while working on the East Wallaby tides out of his home, Ross noticed a diurnal inequality in both height and time of low and high water, the first case ever reported. Noting this peculiarity, he sent Whewell the curves the next morning, suggesting that the Australian tides would be interesting to investigate. He also informed Whewell that Captain William Beechey and Captain Robert Fitzroy "must have a great quantity of Tides" in that region.[92] Ross knew where the surveyors had been, and who had collected the most accurate data. Whewell took up the tides of Australia with fervor, presenting a paper to the Royal Society four months later, in December, and publishing it as his Bakerian Lecture in the *Philosophical Transactions* for 1848. His paper included the reduction of the Australian data suggested and performed by Ross as well as a separate section on the diurnal inequality in the South Pacific.[93]

Like Bunt's work on the effects of atmospheric pressure, Ross's suggestion that Whewell take up the Australian tides is just one of many examples where the tide calculator determined the research agenda. With an abundance of unreduced data before him, Ross was the first to see the patterns. This further stimulated novel methods of analysis and led to additions to theory. In July 1847 Ross noticed a peculiar similarity between the curves for the mean spring and mean neap high water at Brest, and with the mean spring and neap low water.[94] They both differed by five feet, four inches. Being responsible for creating the tide tables published by the Hydrographic Office had one great benefit. There were always tide observations lying around. He immediately set down similar curves for numerous ports in the English Channel and found that the same constant (five feet, four inches) appeared in the curves for Boulogne, Holyhead, and Calais. He then produced predictions for these ports based on those already made for Brest and compared them with observations in his office. They matched beautifully. Ross had hit on a rule that let him predict the tides for any place where the difference between spring heights and neap heights was the same. For instance, one could use the predictions at Portsmouth, where the tides were known with precision, to calculate the predictions for Cork, Waterford, Inverness, and numerous places along the French coast. "In fact," Ross reported to Whewell, "if I had sufficient observations to get correct curves I might make it almost universal."[95]

Whewell was skeptical. He wanted to know *exactly* how far Ross's rule extended, so Ross showed him the numerous tables he had already produced. Ross was sure of the work, confident in his results, and wanted to include the predictions in the *Admiralty Tide Tables*. "It has occupied me a very considerable time and I should be unwilling after so much trouble and labour to find it thrown on one side."[96] It seems that even a calculator's pretensions could be elevated. After consulting the data and drawing curves for himself, Whewell approved Ross's discovery. Accordingly, Ross greatly expanded the *Admiralty Tide Tables* for 1850.

He was able to give the heights of high water for nearly one hundred additional places, including forty on the coast of France, forty more in England and Scotland, and over twenty in Ireland.[97]

Whewell's final publication on the tides in the *Philosophical Transactions*, "On the Results of Continued Tide Observations at Several Places on the British Coasts," was not his own work.[98] The paper was a detailed explication of Ross's rule. The title page of Whewell's article is reproduced in figure 6.8. Whewell usually began his published papers by referring readers to the theoretical additions his work made to the science of tidology. Notice, however, that Whewell introduced this paper by noting Ross's "great labour and perseverance," whose results Whewell gladly published because of their "use to mariners." The paper was an addition to the practice of navigation, not to the theory of the tides. Its aim was to produce accurate tables, good for the maritime interests of an aspiring nation. Whewell was making a distinction between his own theoretical work and the practical work of calculators.

No such sharp distinction existed in practice, however. Like the results of the 1835 tide experiment described in the previous chapter, Whewell's further research on the tides flowed seamlessly between the science of tidology and the practical construction of tide tables. Scientists made the ocean safe for navigation by constructing such tables, work that relied heavily on the industry and skill of tide table calculators, whether stationed at the Hydrographic Office or in the maritime ports on England's coasts. As he did with Bunt, Whewell paid Ross for his calculations, and they were therefore Whewell's results, published as Whewell's "fourteenth series of researches," not as Ross's first. But it was only through the close collaboration with calculators that Whewell obtained results novel enough to publish. Thomas Gamlen Bunt's specialty had been data analysis. Ross's specialty, as this publication suggests, was data organization. Whewell's specialty, in his later years, was keeping the two men steadily employed.

## History, Tables, and the Advancement of Science

Daniel Ross served as the first superintendent of the *Admiralty Tide Tables*, a new position created only in the mid-1830s. But he also occupied a professional position with an illustrious history. Calculating tables for the good of practical navigation had been the responsibility of the British government at least since the founding of the Royal Greenwich Observatory in the late seventeenth century. Government commissioned

## XI. Researches on the Tides.—Fourteenth Series.

## On the Results of continued Tide Observations at several places on the British Coasts.

### By the Rev. W. Whewell, D.D., F.R.S.

Received October 24, 1849,—Read January 31, 1850.

TIDE observations made at several different parts of the British and the neighbouring shores, and in some instances continued for a considerable period, have been
discussed by Mr. D. Ross of the Hydrographer's Office, with great labour and perseverance; and as the results which his labours afford may be of use to mariners, I
offer to the Royal Society a brief statement of these results.

The discussions at present referred to relate to the height of high water, and the
variations which this height undergoes in proceeding from springs to neaps and from
neaps to springs. It is found, by examining the observations at 120 places and throwing the heights into curves, that the curve is very nearly of the same form at all these
places. Hence the semimensual series of heights at any place affords a rule for the
series of heights at all other places where the difference of spring height and neap
height is the same. For instance, Portsmouth, where the difference of spring height
and neap height is 2 feet 8 inches, is a rule for Cork,
Waterford, Inverness, Bantry, Boucout on the
French coast, and other places.

And the Tables of the height of high water at
one of these places suffice for all the others, a constant being of course added or subtracted according
to the position of the zero-point from which the
heights at each place are measured.

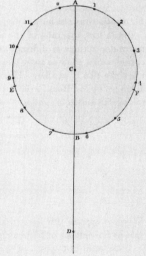

The series of heights of high water for a semi-
lunation also agrees very exactly, as to the form of
the curve, with the equilibrium theory. The following construction gives this curve.

With centre C and radius CA (half the difference
of the height at spring and neaps), describe a circle;
and in AC produced take CD to CA as 12 to 5.
Divide the circumference of the circle into twelve
hours, representing the twelve hours of moon's
transit; and join D with each of these divisions.
The lines thus drawn to the hours will give the
heights of high water for each hour of the moon's

Figure 6.8. The title page of Whewell's final publication on the tides in the *Philosophical
Transactions*. The article is a detailed account of Daniel Ross's investigations, including a diagram
Ross invented to expand the *Admiralty Tide Tables* for the year 1850. Reproduced from William
Whewell, "Researches on the Tides," 14th ser., "On the Results of Continued Tide Observations
at Several Places on the British Coasts," *Philosophical Transactions* 140 (1850): 227.

astronomical tables to determine the positions of the sun and moon, the transits of Venus, and any other tables that could help mariners find their way in the open ocean. These tables were *predictive*, published in the *Nautical Almanac* several years ahead of time for the use of mariners on long voyages. They were necessarily based on theories of the motion of the celestial bodies and could also be used to test those theories. Though Ross formed predictive tables for the tides, not astronomical positions, his work served the same purpose. Ross's professional post therefore had over a century of history, and as Whewell noted in his *History* and *Philosophy*, that was an important precedent indeed.

Constructing tables to test theory occupied a central place in Whewell's science, which goes far to explain why the process was so important to his history and philosophy of scientific discovery. Whewell wrote his *History* and outlined his *Philosophy* at the height of his tidal studies; he had his own research at the forefront of his mind. Taken together, these works offered a "heuristic" designed to guide a science through its several stages of development. Whewell traced each science through its prelude, inductive epoch, and sequel, using as his model astronomy, the only complete science. Within astronomy, the inductive epoch of Kepler served as Whewell's favorite example. Though Kepler did not yet understand the true force causing the motion of the planets, through a "boldness and license in guessing" he determined the correct motion of the planet Mars; then he had to verify the initial "guess" and extend it first to the orbits of Mercury and then to the other planets. But according to Whewell, the *"real verification* of the new doctrine concerning the orbits and motions of the heavenly bodies was, of course, to be found in the *construction of tables* of those motions, and in the continued *comparison of such tables with observation."*[99] This verification of calculated tables through direct observation represented an essential step that made the science "inductive," notwithstanding the highly hypothetical way the theory was first developed.

Only by creating tables could astronomers verify their initial hypotheses, and after publishing his *Epitome Astronomiae Copernicanae* in 1622, Kepler turned to constructing his *Rudolphine Tables*, published in 1627. This step allowed astronomers after Kepler to both verify and improve the knowledge of planetary motion by demonstrating inequalities and perturbations, then finding the laws to account for them.[100] The construction of tables, especially those by Jeremiah Horrocks published in 1673, led to the eventual acceptance of Kepler's theory, which astronomers extended and applied to the motions of all other planets.

Astronomers, according to Whewell, had followed this same step-by-step procedure to advance all other questions in observational astronomy.

The sequel to Kepler's inductive epoch formed the prelude to "the last and most splendid period in the progress of astronomy," the formulation of the cause of such motion.[101] Whewell's history of the inductive epoch of Newton likewise described Newton's inventive faculty in formulating his powerful hypothesis of universal gravitation, which Newton superimposed on Kepler's laws to account for the true cause of planetary motion. And again, the sequel to the epoch of Newton, its verification and completion, served as a heuristic for scientific advance. The first test of the Newtonian theory, not surprisingly, came in constructing tables of the moon derived from theory and comparing these tables with direct observation.[102]

Whewell viewed tidology as a subject of historical analysis in two distinct ways. It represented the final stage in the sequel to Newton's inductive epoch, the last great pillar of physical astronomy left to be erected. Thus in his *History* Whewell introduced the tides in the last section on the sequel to the inductive epoch of Newton: "We come, finally, to that result, in which most remains to be done for the verification of the general law of attraction; the subject of the Tides."[103] But tidology had its own history, with its own prelude, inductive epoch, and sequel; it required its own process of induction. Whewell outlined the attempts the Royal Society and the French Academy had made to collect observations to test Newton's theory of the tides in the first half of the eighteenth century. He admitted, however, that Newton's own theory "had not been at that time sufficiently developed" for these initial steps to prove fruitful. This deficiency was allayed to a large extent by the prize essays of Leonard Euler, Daniel Bernoulli, and Colin Maclaurin in 1740. Bernoulli, especially, provided "the means of bringing this subject to the same test to which all the other consequences of gravitation had been subjected:—namely, the calculation of tables, and the continued and orderly comparison of these with observations."[104] Verification through constructing tables had proved essential in all past epochs in astronomy, and the tides were no exception.

This reading of Whewell's *History* and *Philosophy* is consistent with past commentaries, though it stresses different aspects that become apparent only through an analysis of Whewell's personal scientific investigations. History was instructive, and it taught that only by taking each science step by step through its several stages of development could scientists ensure true progress. Whewell's own investigations in tidology influenced his views on the history of induction, a history that focused heavily on preludes and sequels and the seminal importance of testing theory with tables. Two points need to be stressed here concerning the relation between Whewell's tidology and his *History* and *Philosophy*. The first is that Whewell's work in tidology depended on collaboration with professional calculators to produce accurate tables to test theory. Whewell could not accomplish the meticulous and labor-intensive investigations on his own. He required local tide table calculators. The second is the importance of these tables in Whewell's assessment of the advancement of science. Without such tables, scientists would languish, unable to test their theories or advance on previous research. Calculators, whether stationed in London or in the provinces, were essential to the overall success of physical astronomy.

## Conclusion

When Lubbock entered tidology in 1830, the Admiralty had no tide calculators and did not produce tide tables. Captains of Her Majesty's vessels were expected to fend for themselves, which meant acquiring local tide tables for all the ports they intended to visit. Lubbock's and Whewell's initial efforts prompted Beaufort to assign members of his staff exclusively to the accumulation and reduction of tidal data. He transferred Joseph Foss Dessiou to tidal studies in 1831 and hired Daniel Ross that same year solely to work on tidal analysis and the creation of tables. The theorists offered valuable guidance to the tide calculators, supplying them with the mathematical constants needed for the *Admiralty Tide Tables*. Ross used Whewell's theoretical work to determine the establishments for ports across Great Britain, and he created and updated the tide tables when new observations came into the Admiralty from surveys at home and abroad. Beaufort then sent William Stratford the corrections for the *Nautical Almanac* and Charles Knight the corrections for the *British Almanac*. By 1850, tide tables out of the Hydrographic Office covered over one hundred ports, including most of Great Britain and numerous ports in Europe and overseas.

This centralizing did not go unchallenged. In the *Nautical Magazine* for 1839, K. B. Martin bemoaned the variance between practice and theory so "strikingly exemplified" in the study of the tides.[105] More than once he had observed a "distressed mariner with his crippled ship" attempting to navigate the "hazardous" coasts of Britain, where the hourly variation of the tide combined with shifting sandbars often had "evil consequences." Martin, a harbor master at the Royal Harbor at Rams-gate, believed he knew better than most that "in this respect, *local tide tables* constructed on the spot are always found to be most correct, and it would be well, if they were more generally referred to by the compilers of almanacks."[106] Thus far, according to Martin, the theorists in London and Cambridge had only managed to make things worse: "This is a seri-ous affair for a heavy ship. It is very consoling truly to the commander, when he lands to be told by a mathematician, or astronomer, that the tide tables are correct by Apogee and Perigee, by phases of the moon, and increasing or decreasing influences of the prime luminary; and if the tides *are not so*, they *ought to be so*."[107]

Though Martin was reacting against the seemingly inevitable loss of local knowledge, he unwittingly contributed to the prejudice against lo-cal tide table makers. Such a bias posited the scientists, in the midst of defining their discipline, as transforming the "traditionary secrets" of lo-cal tide table makers into "scientific knowledge" for the good of physical astronomy. It claimed that local knowledge at the periphery was neces-sarily giving way to science in the center. Such a conceit misrepresented the complex exchange of information among the individuals involved and the dynamic, innovative, and collaborative nature of "traditional" knowledge.

In some respects Martin's criticisms were justified. The commercially inclined tide table calculators, without any initial guidance from sci-entists, created tables that were by far the most accurate in the coun-try. Their local knowledge served them well; they lived, worked, and breathed the tides and had used their peculiar skills to move far beyond the natural philosophers in predictive accuracy. As Martin suggested, al-manac makers would have done well to refer to them. But here is where his criticisms also proved misguided. Almanac makers *did* refer to local tables. By 1839 the *Admiralty Tide Tables*, the *Nautical Almanac*, and the *British Almanac* included theoretical predictions tested by and compared with local tide tables. Since the other major almanacs often simply pi-rated these results, Martin's struggle was moot.

Moreover, local tide tables used the theorists' work as well, blurring the distinction between local and expert knowledge. Local tide table makers incorporated the latest methods from the theorists into their tables as part of their "traditional" practice. They had to. They were in-volved in one of the most difficult enterprises in science, attempting to match predictions with observations in an area of study that included numerous and complex variables. Constructing tide tables was a compet-itive and lucrative business, and calculators relied on the latest scientific methods, along with the names of eminent philosophers, to gain an edge. When Lubbock and Whewell took up the theoretical study, the local tide table makers ardently fell in line. As quickly as Lubbock and Whewell published their latest results, tide table makers from Liverpool and Bristol incorporated them into their predictions. They were the first, for instance, to include the diurnal inequality in their tables and extend those tables to evening predictions. This affected not only Whewell's tidology but also the way he explained scientific advance in his *History* and *Philosophy*.

Through an intense study of the history of astronomy, Whewell realized that his graphical method sped up data analysis from the centuries needed to determine the true motion of the moon to the years needed in tidology. Throughout the eighteenth century England and France provided the funds, people, and equipment to produce the tables required to verify the moon's motion, a process centralized in the national observatories.[108] In tidology, the collaboration between local tide table makers and the theorists shortened this significantly. The verification of theorists' results was no longer the preserve of national governments and their observatories; it was now within the purview of tide table makers near the sea. This surprising difference between tidology and other branches of astronomy, according to Whewell, was the "circumstance most worthy of remark" when he entered the field, and he remarked on it at every turn, often directly quoting tide calculators in print. Whewell solved the conundrum by incorporating their contributions into his methodology. He secured outlets for their publications and searched out funding at the local and national levels for their time-consuming discussions.

The work of Thomas Bywater, Thomas Gamlen Bunt, and Daniel Ross demonstrates the kind of practical innovation that calculators brought to early Victorian data analysis. They used their unique position as the first to come in contact with new observations to advance novel methods of analysis, propose additions to theory, and in the case of both Bunt and Ross, suggest new avenues of research. Their collaboration was calculated to ensure the most accurate tide tables possible, and in the process, they became actively involved in the theoretical advancement of science. These calculators moved smoothly between predictions for their tide tables and contributions to tidal theory, between numbers in their account books and figures in Whewell's tidal equations. Their efforts expertly demonstrate their lack of any distinction between theory and practice or between the amateur and the professional, even though

the scientists in Cambridge and London, especially Whewell, were trying to make such distinctions. That the scientists ultimately succeeded
should not obscure the contributions these subordinate laborers made to
Victorian science. *How* the scientists demarcated their role from that of
the associate and subordinate laborers is the subject of the next chapter.

# SEVEN

# Creating Space for the "Scientist"

Many the wonders but nothing walks stranger than man.
He crosses the sea in the winter's storm,
Making his path through the roaring waves.
SOPHOCLES, *TRACHINIAE* (CA. 430 BCE)

In the fall of 1835, as William Whewell collaborated with several computers to bring order to the observational data from the multinational tide experiment, the scientific elite in Britain were in the midst of a highly charged debate concerning the role of calculators and observers in the discovery process. Five years earlier, in his controversial text on the decline of science in England, Charles Babbage had discoursed at some length on the collaborative nature of scientific discovery. He was struggling with the construction of his first difference engine, a machine specifically built to rid the sciences of the more mundane tasks of calculation, and he moved easily in his discussion from the manufacturing of machines to the production of thoughts. Both, he mused, relied on a division of labor. "The progress of knowledge convinced the world that the system of the division of labour and of cooperation was as applicable to science, as it has been found available for the improvement of manufactures."[1] Yet Babbage advocated a hierarchy within the division of labor that subordinated the calculator and observer—those who were taught their skills and whose task could ultimately be reduced to a machine. "To discover new principles," Babbage wrote, "and to detect the undiscovered laws by which nature operates, is another and a higher task, and requires intellectual qualifications of a very different order: the labour of the one is like that of the computer of an almanac; the inquiries of the other resemble more the researches of the accomplished analyst, who has invented the formulae by which those computations are performed."[2] Scientific discovery relied on a broad base, including computers, observers, and artisans (inventors of mechanical instruments). According to Babbage, however, the "higher task" of uncovering nature's laws required an uncommon sagacity found only in a select few, a point to which a number of the British scientific elite took offense.

The debate came to a climax the year of Whewell's and Beaufort's great tide experiment. Francis Baily, one of the founding members of the Royal Astronomical Society of London, published *An Account of the Rev. John Flamsteed, the First Astronomer Royal*, a spirited defense of the role of expert calculators and practical observers in the process of scientific discovery. According to Baily, those interested in privileging theory over observation, individual genius over sound methodology, had misrepresented Flamsteed's life and work. The influential Scottish natural philosopher David Brewster, for instance, in his *Life of Isaac Newton* (1831), had considered Flamsteed an obstructionist, interfering with Newton's attempt to perfect his lunar theory. Baily's defense of Flamsteed captured the sentiments of those intent on safeguarding the role of the indefatigable observer in the scientific discovery process. For Baily, science was a highly empirical affair, resting squarely on those who unceremoniously plodded through the meticulous but often tedious procedures that, according to him, made up the bulk of modern science.[3]

The Newton-Flamsteed debate that flared in the mid-1830s was argued on several interrelated levels: the place of observation and calculation in the discovery process, the enduring question of the ownership of intellectual property, and the proper methodology of scientific discovery.[4] The view of science as a division of intellectual labor accorded with the outlook of the British Association for the Advancement of Science, whose original founders opted to portray the practice of science as open to all. Even the most humble researchers stationed at home or abroad could further scientific knowledge. By using its growing coffers to fund the reduction of data, the British Association affirmed the importance of calculation and the need for accurate observations.

The question of the proper role of calculators and observers symbolized the tension between two decidedly contradictory visions of the discovery process, and it centered on the merits of popular involvement in the sciences. As the cultural fascination with science grew, the issue of who could actively participate was open to debate. For members of the British Association and the Royal Astronomical Society, science was an empirical process where a controlled division of intellectual labor increased productivity. Theory and hypotheses were anathema; a secure methodology for scientific advance rested on the accumulation and reduction of observational data. Yet this egalitarian view was at odds with the attempt by Babbage, Herschel, Whewell, and others of the scientific elite to maintain an intellectual high ground for the full-time devotee. They were attempting to demarcate legitimate from illegitimate science, explicitly trying to define what it was to be a "scientist."[5] The term did not include women, popular writers, or those on the periphery; it no longer included the aristocratic amateur or the practical inventor. The title rested on men with theoretical insights and mathematical training— the education of those sitting for the mathematical tripos at Cambridge.

At the center of the debate stood William Whewell, one of the premier spokesmen for science in the early Victorian era, and one of those largely responsible for establishing the hierarchical nature of the scientific enterprise. As a Peelite conservative and an Anglican minister, Whewell has been viewed as the premier elitist, the one who barred the door of science to amateurs and artisans.[6] He was also, however, the premier tidologist in Britain, as familiar as anyone with the vital contributions of computers, artisans, and accurate observers in addressing questions in physical astronomy, and his own research informed all his discussions concerning the mental hierarchy of the discovery process. As geographically oriented research become prominent in Britain, Whewell expanded his spatial and disciplinary shaping of tidology to an equally ambitious shaping of the terrain of the "scientist," a term he coined at the outset of his tidal investigations and spent his entire career attempting to define.

Science for Whewell was from the beginning a collaborative and hierarchical affair, and the elite theorists deserved recognition above the calculators and artisans who were paid for their work. He purposefully delineated the study of the tides as "the last great bastion of physical astronomy," the peculiar preserve of theorists, accessible only to himself and others (like Airy, Babbage, and Herschel) with an eye toward theory. He thereby contributed to a definition of the scientist that effectively excluded artisans and calculators. Yet the obvious contribution these associate laborers made to theoretical developments in tidology informed his negotiation of the intellectual and social role of the scientist. He attempted to harness their skills, and rather than dissociate them from the scientific process, he incorporated their work into his published proclamations on the creation of scientific knowledge.

Figure 7.1. William Whewell, master of Trinity College, Cambridge, as portrayed in the *Illustrated London News* in 1843. He gained prominence throughout the 1830s and 1840s as a spokesman for the sciences. *Illustrated London News* 3 (28 October 1843): 284.

Whewell had a close and evolving relationship with scientific instruments; he was active in designing and maintaining an anemometer to measure wind velocity and direction, and he benefited immensely from the invention of the self-registering tide gauge. He also relied heavily on expert computers: Thomas Bunt, for instance, proved crucial for his tidology and was arguably Whewell's main scientific collaborator. The way Whewell delineated the practice of science in his *History* and his *Philosophy*, however, did not include mechanical instruments or calculation. He became a champion of science but remained silent on technology.[7] He acknowledged his debt to expert computers but incorporated their work into what he termed "colligation," which consciously privileged theory at the expense of calculation. Whewell's definition of the scientist rested on a larger historical vision that emphasized a slow, arduous advance toward necessary truths. Rather than making a single inductive inference or performing an isolated act of colligation, the scientist nurtured a specialized field of research through its many stages of development. To become a scientist required focusing on a large-scale and historically informed train of research.[8]

Whewell's own research emerged squarely within the economics and politics of imperialism, guided by the Admiralty's growing requirements for intelligence. He practiced an expansive and expensive type of science that relied heavily on government participation and support. His initial work in tidology allowed him to speak on methodology and practice as an active researcher and also eased his entry into Admiralty circles. As he extended his research to the world's oceans, he helped the Admiralty maneuver safely between imperial possessions. As one of the premier evaluators of science, Whewell likewise placed himself at the forefront in discussing the social and intellectual role of the modern scientist. In graphing the world's oceans, he also helped navigate the social process of knowledge creation.

Throughout his research on the world's tides, Whewell delineated space in two ways. He helped transform the spatial scope of science while simultaneously expanding the terrain of the scientist. In Britain's transition from the first industrial nation to the mightiest imperial power, the scientist emerged as the authority on questions relating to the nautical spaces between Britain's vast possessions. Indeed, for a maritime nation, nowhere was man's power over nature more apparent than in his ability to domesticate the oceans. By the second half of the nineteenth century, scientists were the ones who helped man cross the "sea in the winter's storm" and guide his path "through the roaring waves."

## Scientists and "Subordinate Laborers"

When members of the British Association traveled to Cambridge for its third meeting in the fall of 1833, Whewell was just embarking on his tidal research, a budding man of science who had risen to a professorship at Cambridge. Lubbock and Dessiou were at the height of their controversy with the proprietors of the *Nautical Almanac* over paying calculators, and both Whewell and Lubbock were trying to find funds for reducing the Liverpool observations uncovered during the Association's first meeting. Whewell was also planning his first tide experiment around the coasts of Great Britain, a massive operation that required accurate observers, precise timepieces and measuring devices, and a corps of human computers. He well knew the significance of observers, instruments, and calculators and was attuned to the division of labor needed in a spatial science like the study of tides. Significantly, it was also during this meeting that Whewell first argued for a special term to denote an active researcher, whom he opted to call a "scientist."[9]

As one of the vice presidents of the Association, Whewell gave the keynote address to open the Cambridge meeting. His address was decidedly not a disputation on the accessibility of science or the need to share intellectual labor; he did not posture in front of the provincial observers in attendance. Rather, he used the address to highlight the primacy of theory in the discovery process, attacking those who insisted on a division of intellectual labor. Though he would affirm the contributions of "subordinate labourers" in advancing science, he unambiguously placed them below those of the theorists. Whewell declared, "We must all start from our actual positions, and we cannot accelerate our advance by any method of giving to each man his mile of the march. Yet something we may do: we may take care that those who come ready and willing for the road, shall start from the proper point and in the proper direction."[10] The "actual positions" differed for the subordinate laborer and the theorist, and the "proper direction" was determined by the theorist.

Whewell, then, did not pander to the rank and file or bow to the power of the masses. He focused on how to advance those sciences that were at the very beginning of their march toward certain knowledge. He began with astronomy, "not only the queen of sciences, but, in a stricter sense of the term, the only perfect science," and delineated a hierarchy for the sciences, concentrating on their positions within the ascent toward necessary truths. Those sciences in their formative stages—tidology, meteorology, and other research "to be promoted by combined labour such as that which it is a main object of this Association to stimulate and organize"—demanded above all patience and judiciousness. "Instead of developing theories, we have to establish them; instead of determining our data and rules with the last accuracy, we have to obtain first approximations to them."[11] This could not be accomplished either by jumping forward with theory or by merely gathering data but required an intermingling of both: "A combination of theory with facts, of general views with experimental industry, is requisite, even in subordinate contributors to science."[12] This antithesis between theory and facts, exclaimed Whewell, "has probably in its turn contributed to delude and perplex; to make men's observations and speculations useless and fruitless." According to Whewell, it was only through the "connexion and relation of facts" that the observer could determine which circumstances ought to be noted and recorded. This phrase—"connexion and relation of facts"—would soon hold a prominent place in both Whewell's *History* and his *Philosophy*. In this address he was ascribing the process not only to heads of the subcommittees and leaders of the Association, but also to the subordinate contributors to British science. "Every labourer in the field of science, however humble, must direct his labours by some theoretical views, original or adopted." Acquiring observations and reducing them did not make a scientist; theory did.

Whewell would soon open both his *History* and his *Philosophy* with this "fundamental antithesis between facts and theory," the beginning point for advances in every field of science. And it was exactly here that the British Association, with its ample resources, could make the biggest difference in the cultivation of knowledge. "There are no subjects," Whewell preached, "in which we may look more hopefully to an advance in sound theoretical views, than those in which the demands of practice make men willing to experiment on an expensive scale, with keenness and perseverance; and reward every addition of our knowledge with an addition to our power. And even they—for undoubtedly there are many such—who require no such bribe as an inducement to their own exertions, may still be glad that such fund should exist, as a means of engaging and recompensing subordinate labourers."[13]

Whewell is here speaking of his and Lubbock's research on the tides. The simultaneous Coast Guard observations around the entire coast of Great Britain, which Whewell was then contemplating, certainly constituted an "experiment on an expensive scale," one that required "engaging and recompensing" "subordinate labourers." The main function of the British Association, then, was not to promote an egalitarian view of science, where each was given his mile of the march. Its role, according to Whewell, was to place the full-time devotee in a better position to combine theory and observation by making funds available for hiring subordinate laborers. Whewell advocated a science where men such as John Herschel and George Biddell Airy were undeniably in a higher position than the unnamed calculators and provincial observers listening to his keynote address.

The 1833 British Association meeting in Cambridge therefore held many firsts for Whewell. There he first coined the term "scientist"; he first voiced his "fundamental antithesis of knowledge," which would form the foundation of his philosophy of science; and he first spoke directly to, and about, "subordinate labourers." All three must be viewed together when determining not only Whewell's evaluation of the process of scientific discovery, but also the social fabric of early Victorian science. The significance of Whewell's own research to the formation of his scientific methodology cannot be overestimated. He was at the height of his tidal studies in 1834-36, the same years when he wrote his *History* and outlined his *Philosophy*. His first ratiocinations concerning the discovery process were accomplished with this research at the forefront of his mind. It proved fundamental to his philosophy, his history, his views of discovery, everything that is usually associated with Whewell's work as a metascientist.

Immediately after the British Association meeting, Whewell began to map out his *History* based on "exemplary sciences," fortifying the hierarchy of the inductive sciences that he had outlined at the meeting.[14] Much of the narrative had in fact been written. Whewell either had personally been involved in scientific research (geodesy, crystallography, mineralogy, meteorology, and for several months, tidology) or had written textbooks (*Mechanics* and *Dynamics*). In all these endeavors, he studied the history of those sciences, noting specifically how they fit within the progression of scientific advance. As he prepared for his tide experiment around the coast of Great Britain in early 1834, he turned to his *History* in earnest.[15] Whewell explained to his confidant Richard Jones late in the year that he was especially focusing on the "history of mechanics and astronomy," since they were the most "important and instructive" subjects if one was concerned with a philosophy of discovery. Throughout his career, Whewell unabashedly used history to advance both his philosophy and his scientific work. He acknowledged, for instance, that his *History* was begun as a precursor to and material for his projected *Philosophy*. His history of the inductive sciences not only would survey the past accomplishments of certain disciplines, but would also "bring before the reader the present form and extent, the future hopes and prospects of science, as well as its past progress."[16] His *History* was meant to demonstrate how to "improve and increase" scientific knowledge; past efforts were the foundation for future discovery, and by closely following the lessons from the best examples of past science, a researcher could advance those fields at the beginning of their ascent toward inductive truth.

Induction, as Whewell defined it, entailed the "successive steps of generalization" where facts once accumulated were "colligated" by the discoverer's imagination; it entailed both accumulating facts and ordering them through hypotheses. Every science had its "inductive epochs" where facts were ordered and generalized through the active powers of the mind. To Whewell, bringing together facts and thoughts, the act of "colligation" leading to an inductive epoch, was best exemplified by astronomy. He never wavered on this point. Astronomy had reached a state of maturity that other sciences could look toward and try to emulate.

Whewell was influenced by the philosophy of Immanuel Kant, who gave a prominent role to the human imagination in unifying sense experience. Like Kant, Whewell emphasized the creative role of the scientist and the need for unifying principles that went beyond the empirical evidence. As he explained to Richard Jones, "to arrive at knowledge of science we must have besides impressions of sense, certain mental bonds of connexion, ideal relations, combinatory modes of conception, sciential conditions."[17] The scientist's mind unified observation into a coherent pattern, leading to laws and ultimately to theories that were necessarily true.

The hero of Whewell's narrative, surprisingly, was not Newton but Kepler, partly because of the self-reflective way Kepler narrated his errors, offering Whewell insight into the mental process of colligation.[18] The "colligation of facts" entailed much more than organizing and systematizing elementary data. Whewell explained that in cases like Kepler's, "the particular facts are not merely brought together, but there is a New Element added to the combination by the very act of thought by which they are combined."[19] Thus, every case of true scientific progress required an act of invention by the scientist; there was always some conception superimposed on the facts. Whewell described colligation as akin to "happy guesses," usually masked by both the scientific researcher and the passage of time. Kepler, however, left a detailed record of every happy guess he made, arduously following (and reporting) no fewer than nineteen hypotheses before hitting on the true doctrine of the ellipse. Whewell concluded his discussion of colligation by emphasizing that "boldness and license in guessing" was not only warranted but indispensable to advancing knowledge. "Real discoveries are thus mixed with baseless assumptions; profound sagacity is combined with fanciful conjecture; not rarely, or in peculiar instances, but commonly, and in most cases; probably in all, if we could read the thoughts of discoverers as we read the books of Kepler."[20] But thankfully, there were guides who could help researchers with their happy guesses and fanciful conjectures, and again Kepler provided the best example.

Kepler took center stage in Whewell's analysis because Kepler's tortuous route to his planetary laws allowed Whewell to highlight the distinction between phenomenological laws and causal laws and thus the difficult and time-consuming procedure of inductive generalization. The first plateau in every science required discerning what took place, not why it occurred. "Kepler discovered that the planets describe ellipses, before Newton explained why they select this particular curve."[21] Whewell noted that the discovery of phenomenological laws often constituted the whole of a science for a very long time, and it was only through "great talents and great efforts" that advances were eventually made. Though Kepler was "by far the most conspicuous instance of success in such researches," other routes to causal laws could be found; the trick was to bring a new element into the search. Whewell, of course, was at this exact time searching for the phenomenological laws of the tides, the same act of "colligation" that Kepler had so successfully accomplished for planetary motion. Whewell quite consciously viewed himself as the Kepler of tides.

Whewell stressed in the opening pages of his *Philosophy* that he was not offering an "art of discovery," for no certain path to progress existed, but rather a historically informed explication of the method by which "truths, now universally recognised, have really been discovered."[22] By studying each science's progress, the scientific researcher would have the means of deciding where it was in its development and what was needed to advance it. This explains why Whewell held astronomy in such high esteem. It was a finished product that could be emulated, a model by which other sciences could determine their condition and mark their path. Whewell's process of scientific discovery required knowing the history of each subject up to its present condition, when his philosophy of scientific methodology could prove useful in its advance. The scientist was part of a historically informed process that furthered a particular part of science, depending on where it was in its inductive path. A true "scientist" knew where his science was in this progressive march and knew where his own research needed to begin and end. A good scientist, in other words, was an equally good historian.

By the time Whewell published his *History* in 1837 and his *Philosophy* in 1840, he had collaborated closely with Thomas Gamlen Bunt and other expert computers. Whewell's research in tidology rested on a combination of advances in observation, theory, *and* calculation. He always acknowledged Bunt's contributions in personal correspondence and in papers presented to the British Association and the Royal Society, stating on one occasion that Bunt had "advanced all the methods" he had taught him. By centering both his *History* and his *Philosophy* on the fundamental antithesis between theory and observation, however, Whewell successfully subordinated the processes Bunt performed under the theoretical act of colligation. Indeed, in his *Philosophy* he separated out the work of calculators under "special methods of induction applicable to quantity."[23] Whewell introduced these methods—the method of curves, the method of means, the method of least squares, and the method of residues—as part of a discussion of the colligation of facts, which represented the most important task of the scientist, thereby transforming calculation and the role of the computer into questions of theory. In his *History* and *Philosophy*, Whewell does not mention collaboration or the division of intellectual labor. He consciously removed the social aspect

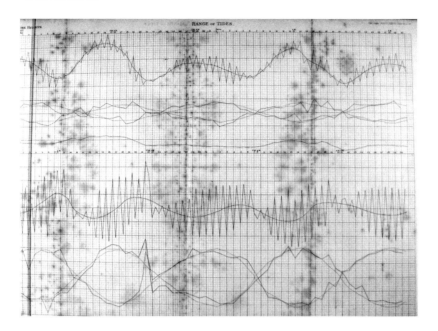

Figure 7.2. Whewell used the graphical method throughout his tidal research, and that work was largely accomplished by computers. William Whewell, "Researches on the Tides," 7th ser., "On the Diurnal Inequality of the Height of the Tide, Especially at Plymouth and at Singapore; and on the Mean Level of the Sea," *Philosophical Transactions* 127 (1837): 75–87.

of the scientific enterprise in favor of the intellectual process of combining theory and observation, part of his larger effort of separating the "scientist," who ruled over the historical act of colligation, from "subordinate laborers," who failed to see the overarching historical process of investigation. Science progressed only through the active minds of scientists, because they understood the historical progression of each field. Perhaps the most obvious rhetorical difference between papers presented in the Victorian era and those presented today is in the lengthy historical introductions to the research, a strategy meant to exemplify just this arduous path toward inductive generalizations. That Whewell and others took to numbering their entries in scientific journals—sixth series, seventh series—made the process explicit.

## Scientists and Artisans

William Whewell was one of the leading figures in the Victorian era's battle over the contested space for science and the new scientist. While his own research depended crucially on cooperation with Bunt and other subordinate laborers, he distanced his work from theirs by stressing the historical progress of science and its reliance on colligation, the most important aspect of advancing theory. His research in tidology had also placed him in close contact with artisans (the makers of machines), and he initially drew a similar distinction between the artisan and the scientist. At the same British Association meeting where he outlined the role of subordinate laborers, he acknowledged the importance of art (technology) but pronounced it subordinate to science. "Art has ever been the mother of Science," Whewell averred, "the comely and busy mother of a daughter of a far loftier and serener beauty."[24] In mining and mineral products, in glass and steel production, and in similar technologies, the practical processes always preceded the theoretical understanding. But for Whewell it was exactly the theoretical understanding that made it science, and he said nothing about the role of technology in advancing scientific inquiry. In this Whewell was following other members of the scientific elite in the audience of the British Association meeting. When the Association was initially formed, its leading members paid little heed to the mechanical arts, and while instrument makers and inventors could contribute to its meetings, they were increasingly subordinate within the mental hierarchy of knowledge creation.[25]

Whewell's own experiences in tidology throughout the 1830s started to change his conception of the formative role of "scientific instruments," a term that also emerged only in the 1830s.[26] Whewell was especially indebted to self-registering tide gauges, and though self-registering instruments had been used for some time, they became an essential part of the fact-gathering initiatives in the geophysical sciences only in the nineteenth century.[27] The self-registering tide gauge, moreover, was far from a passive agent, merely supplying the data needed for tidal theory. Rather, its proliferation guided the research of natural philosophers and helped determine what type of science was possible. When Lubbock and Whewell entered tidology, for instance, they had access to only high water data, so they limited their investigation to a hydrostatic approach, advancing the Newton-Bernoulli equilibrium theory.[28] Self-registering gauges, however, traced all the irregularities of the tide throughout the entire rise and fall, the exact type of data required for a dynamic theory of the tides. Only after data were culled from these instruments could the astronomer royal, George Biddell Airy, improve on Laplace's hydrodynamic and essentially correct theory of the tides.

The self-registering tide gauge was invented in the first years of the 1830s by dockyard engineers—J. Mitchell at Sheerness and Henry Palmer at the London Docks—and perfected by Whewell's expert calculator, Thomas Gamlen Bunt. The technicians built the machines to meet the practical needs of their job, but they also wanted to contribute to the theoretical understanding of the tides. Palmer, for instance, was interested foremost in the *rate* of the rise and fall of the tide—a question of theory—and a description of his gauge made it into the *Philosophical Transactions of the Royal Society of London*, the premier publication in science at that time. The artisans' own use of the machines advanced the methods of data analysis and contributed to important aspects of tidal theory, including the efficacy and acceptance of the graphical method of representing data.

Whewell, of course, realized all this, and it put him in a difficult position. "As the cartographer of the various scientific disciplines," Jack Morrell stressed, "Whewell defined boundaries and contours."[29] One of the boundaries Whewell set throughout his published research and other writing was a strict demarcation between science and technology. Whewell, however, was also an inventor of machines, at home with winches and pulleys and practical contrivances of all kinds.[30] In the same years that Bunt was perfecting the self-registering tide gauge, Whewell was working on a self-registering instrument of his own to improve the accuracy of tidal measurements: an anemometer to register the velocity and direction of the wind. Like Bunt, Whewell began designing his instrument in 1834 and exhibited it to the same British Association meetings in 1835 and 1836 where Bunt presented his tide gauge. The two instruments, moreover, worked on the same design: the wind or tide was converted by means of wheels and pulleys to a pencil attached to a slowly revolving cylinder that traced a line (graph) representing the changes over time.

John Newman, a flourishing mathematical instrument maker in London, built Whewell's first device, and William Simms, one of the last instrument makers to become a fellow of the Royal Society, fashioned later models.[31] Machines copied from Whewell's design were used by James David Forbes and J. Rankine at Edinburgh, and Thomas A. Southwood and William Snow Harris at Plymouth. Whewell read a description of his anemometer to the Cambridge Philosophical Society in May 1837, along with preliminary results from observations made at the Cambridge Observatory by Professor James Challis, and those Whewell supervised at the Cambridge Philosophical Society carried out by Mr. Crouch, the Society's housekeeper.[32] As Whewell related, the British weather got the best of him; the pencil tracing the readings made "not a single line, but a broad path of irregular forms," and he could draw no conclusions from the initial data. He continued to wrestle with the design of his machine and advanced it enough to apply his graphical methods of data analysis, the same methods he was perfecting in his tidal research. "It may be observed," he wrote, "that the graphical method offers at once the mean direction of the wind, and the resolution of the winds into their cardinal parts." He ended by emphasizing the essential role of the recording instrument in all meteorological research: "If the Anemometers of the kind now described were fixed in various parts of the world, . . . this portion of meteorology, and probably other portions which are connected with this, would soon make great progress."

Great progress, however, would have to wait, owing to the difficulty of keeping the instrument in working order. After working on it for two more years, Whewell could report only that his instrument was "frequently thrown out of repair in the places where it was erected."[33] Meanwhile, both Southwood and Rankine were having similar problems. Southwood, for instance, noted that "there should be two of every part in order to prevent a break in the observations."[34] Whewell concluded his meteorological studies with a sullen note on the difficulties of using and maintaining such devices: "In concluding I may observe that the labour of registering, reducing, and constructing the course of the wind in this manner is far from inconsiderable; and that the constant alterations which the instrument requires to perfect it and keep it in order, can hardly be given, except by some person who calculates the results soon after they are made, so as to discern the defects when they occur."[35]

Here as elsewhere, Whewell appreciated the intricacies of instruments and the difficulty of keeping them functioning. They were useless, however, without a theory-driven observer ready to reduce the observations once recorded.

Shortly after Whewell had his self-registering instrument built and distributed to several places on the British coast, George Biddell Airy asked him about the measurements to be used not in meteorology or the tides, but in terrestrial magnetism. Whewell's reply is instructive:

One of my pieces of wisdom that I wish to impart to you is a strong recommendation to you touching your magnetic observations, namely, that you will adopt some contrivance by which you can have self-registering observations of the daily changes of diurnal variation. . . . I am sure that with regard to all these half-formed sciences, such as terrestrial magnetism is, we shall never come to any knowledge worth having, except we have the whole course of the facts. Self-registering machines are the only

# DIRECTIONS FOR OBSERVING

# WHEWELL'S ANEMOMETER.

1. PLACE the instrument in a situation well exposed on all sides, and fix it so that when the wind is South, the pencil is on the line S on the barrel. This may be done by clamping the weathercock part of the instrument with the pencil on the line S, then turning the box till the tail points due north, and then fixing it in that position.

2. Read off the instrument every day at a constant hour.

The pencil in descending will make a broad path, in consequence of the wavering of the wind. The darkest part of this path must be taken; and from this, the *direction* of the wind determined, by reference to the points of the compass marked at the bottom of the cylinder; and, as the wind changes, the directions of the successive strips of wind must be noted.

3. To read off the *amount* of the wind in each of these successive strips; —slide the lower index so that the point is upon the top of the first strip; then slide the upper index to touch the lower; then slide the lower index to the bottom of the first strip, or the top of the second strip of wind; then read off on the graduated rod, the interval (in tenths of inches,) through which the lower index has moved; then again slide down the upper index to touch the lower, slide down the lower index to the bottom of the second strip, and read off the interval;—and so on. Write down these intervals under the corresponding directions of the strips of wind, observed as above.

4. When the pencil has reached the bottom of the barrel, it must be moved to the top, by unscrewing the clamping screw of the nut, removing it to the top of the barrel, and clamping it.

At the same time the barrel must be cleaned, by rubbing it with a soaped cloth enclosing a smooth wooden rubber.

5. The following is suggested as a simple way of marking the points of the compass; for example, from the North to the East the points may be

N. N*e*. NNE. NE*n*. NE. NE*e*. ENE. E*n*. E.

and so on for the other quadrants.

The only ambiguities which can arise by this method, are N*e*, N*w*, S*e*, S*w*; which must be distinguished from NE, NW, SE, SW.

N*e* is N *by* E; and so of the rest.

6. The following form may be used for tabulating the readings off.

| 1836. October. | ANEMOMETER. | | | | |
|---|---|---|---|---|---|
| 19 | S*e*. | S*w*. | SW. | W*n*. | NW. |
|    | 4 | 20 | 3 | 3 | 2 |
| 20 | WNW. | | | | |
|    | 4 | | | | |
| 21 | SE*s*. | | | | |
|    | 3 | | | | |

Figure 7.3. Instruction sheet: "Directions for Observing with Whewell's Anemometer." Whewell constructed his anemometer partly to help advance the accurate collection of tidal data. Whewell Papers R.6.20/3.

inventions worth having for those changes, of which the laws are as yet all in a mass of confusion. And all that I have seen of magnetic observations leads me to believe that with a self-registering apparatus, you may obtain very important results.[36]

Whewell had spent the previous five years ensconced in the "half-formed science" of tidology, where the laws were "all in a mass of confusion," and as his labor on his own anemometer suggests, he knew well the importance of self-registering machines.

When Airy turned his interest to the tides in the winter of 1840, he sought to reconcile the observations of the tides with his own unique theory of waves in canals, a hydrodynamic approach. For this he needed observations throughout the entire rise and fall, observations he could obtain only by accepting Whewell's "piece of wisdom" to use self-registering gauges. He wrote to Captain Smyth of the Royal Navy, stationed in the Bristol Channel, asking "whether amongst the various stations of the Harbour you have a self-registering tide gauge. If not, I pray you to set one up."[37] No such instrument existed at his station, and Smyth suggested that Airy contact the publisher of the Bristol tide tables, Thomas Bunt, who was living in the city.[38] Airy wrote to Bunt in late December, asking for full details of his instrument, including the cost of erecting such a gauge, the name of a workman in London who could build one, and a copy of the sheets it produced.[39] Bunt replied that the machine was built locally in Bristol, and he was unaware of anyone in London who had constructed such a device, but he offered Airy the original sheets extracted from his own gauge.[40]

Airy was elated at the accuracy of the data and at the range the self-registering gauge could measure. He asked Whewell in early January for a complete list of all self-registering tide gauges throughout the globe, remarking what a "great pity" it was that "there is not an infinity of tide gauges and gauges symmetrically distributed over the coasts and seas."[41] Airy had come to realize that the type of machine registering the data conditioned what theories a researcher could choose to investigate the tides. "I must remark that the subject of Tides is a very wide one," Airy wrote to a friend late in 1842, "and that different persons will fix upon different parts of it according [to] their tastes, *their means of observation*, or their means of calculation. . . . But will you permit me to ask whether you have ever thought of erecting a self-registering tide gauge? This would give all that I have mentioned, completely, without any trouble but that of keeping the machinery in order."[42] With data culled from these gauges, Airy was able to formulate a hydrodynamic theory of the tides, resurrecting Laplace's essentially correct hydrodynamic theory that had lain dormant for half a century. Subsequent theorists did not return to the Newton-Bernoulli theory.

In his treatment of the tides in the *Encyclopaedia Metropolitana* of 1845, Airy spent an inordinate amount of time on the importance of self-registering machines. The article ran to 145 pages, and Hugh James Rose, the editor of the *Encyclopaedia*, became increasingly anxious.[43] Owing to "the extreme inconvenience to which the extent of your Treatise would subject the work," Rose politely asked Airy if he could cut the section on tide gauges.[44] Airy would have no part of it: "Very few people are aware of the existence of such a thing, and it is one of the most important new instruments with Tides...I propose to you therefore, leave the Article as it is, and cut down my payment well. I should not like to be without any, because I have spent £40 or £50 in journeys to look at the Tides, but above that do as you please."[45]

Airy, always concerned with payment for his work at the Royal Observatory, was willing to take a cut in pay to avoid cutting tide gauges. He even added to the article, including several plates of Bunt's tide gauge and examples of its registry. Self-registering tide gauges proved to be far from passive instruments used to collect data. Rather, they changed the character of investigations in early physical astronomy, allowing for different research objectives and approaches.

Through his own research and throughout his correspondence with researchers like Airy, Whewell likewise acknowledged the ultimate significance of self-registering instruments in creating scientific knowledge. As I will demonstrate below, his views on the relationship between science and technology appeared much more sophisticated by the time he spoke to the Great Exhibition in 1851. Yet throughout his major tidal researches, his reviews and essays, and especially his *History* and *Philosophy*, Whewell severely underplayed the role of technology and the artisan. In book 12 of his *Philosophy*, he commented at length on the methods of exact and systematic observation "by which facts are collected as form the materials of precise scientific propositions."[46] For Whewell, these "materials" were always, "from various causes, inaccurate." What was required was sophisticated techniques of analyzing the data once they were gathered. Observational data acquired meaning only through a procedure Whewell termed "manipulation."

The process of applying practically methods of experiment and observation is termed Manipulation; and the value of observations depends much upon the proficiency of the observer in this art. This skill appears, as we have said, not only in devising means and modes of measuring results, but also in inventing and executing arrangements by which elements are subjected to such conditions as the investigation requires: in finding and using some material combination by which nature shall be asked the question which we have in our minds.[47]

The "question which we have in our minds" was always primary for Whewell; theory determined the apparatus used, not the other way around. This, of course, fits with Whewell's entire conception of the act of scientific discovery. As an anglicized Kantian, Whewell emphasized the creative role of the mind and the need for bold unifying conjectures that brought meaning to the empirical observations. Observations—and the machines that measured them—were mere fuel for the theoretical flame.

As with the calculators, so with the artisans; Whewell successfully transformed their innovative techniques and significant contributions into questions of theory. The role of the calculator was subsumed under "colligation" and that of the artisan under "manipulation," both guided by the active mind of the researcher. In this manner Whewell distanced his own work from that of the instrument maker, placing theoretical discussion of the data far above fact gathering. But there was no such distinction in practice, as can be seen in the work of Airy, Palmer, Whewell, Bunt, and others. For Whewell, scientists did not collect data, they informed it though the manipulation of techniques and sound theoretical guidance. Or, to put it slightly differently, the scientist could be found at Cambridge or Greenwich, not Bristol or the London docks.

While theorists like Whewell and Airy recognized the importance of self-registering gauges, the makers of those machines were decidedly not part of their conception of the scientist. Whewell, for one, argued for a stratification within the scientific enterprise, with theorists on top and engineers and inventors of machines tucked somewhere on the tiers below. Whereas mathematical instrument makers had achieved a relatively elevated status within the scientific community toward the end of the eighteenth century, in the early nineteenth they entered a decline. The organizers of science increasingly excluded them, and by midcentury the era of the elite instrument maker receiving accolades and fellowships from the Royal Society had come to a close.[48]

## Scientists and the State

Whewell created a hierarchy within the practice of science to demarcate who could legitimately make knowledge claims about the natural world.

While artisans and subordinate laborer made essential contributions to science, the mental operations involved were different from those of the scientist. Scientists followed a train of research through its many stages of scientific development and knew exactly where in the history of each discipline their own research fit. Researchers thereby established a research space that they then followed from beginning to end, leaving the subject, in Whewell's words relating to his own tidal investigations, "in a condition very different from that in which I found it."[49] Scientists had an overall view of the relation between observation and theory and were thus in a better position to combine the two. Whewell often compared the process to weaving. "The separate filaments must be drawn into a connected thread, and the threads woven into an ample web, before it can form the drapery of science."[50] Instrument makers helped gather the threads; calculators helped order them; but only the scientist was the expert weaver.

In like manner, Whewell mapped out a hierarchy within the sciences themselves. Astronomy was the only complete science, followed by those sciences dependent on mechanics and dynamics for their formulation. This included the sciences of physical astronomy—terrestrial magnetism, tidology, and meteorology among others—the same ones that made up Section A of the British Association for the Advancement of Science. Increasingly, the individuals who practiced physical astronomy—Whewell, Herschel, Airy, and others—became the leaders of the British Association; they served on its committees, presided over its meetings, gave its annual addresses, and of course benefited the most from its monetary grants.[51]

These same sciences, not coincidentally, comprised the subjects national governments needed most for navigating the world's oceans. Indeed, it is significant that the hierarchy of the sciences outlined by Whewell and funded by the British Association matched the list compiled for inclusion in the *Admiralty Manual of Scientific Enquiry,* published in 1849. The *Manual* began with astronomy, followed by those topics most crucial for navigation, such as magnetism, hydrography, and the tides. Only then appeared the sciences of natural history, beginning with geography and geology, then zoology and botany, and ending with the human sciences. Beaufort's close connections to the British scientific elite let him marshal experts to prepare the volume, including John Herschel as general editor. Airy on astronomy, Edward Sabine on terrestrial magnetism, Whewell on the tides, and Herschel on meteorology are only a few who contributed.

That the *Manual* was published by and for the Admiralty is not surprising; it symbolized above all the close connection between the scientist and the state, and the Admiralty's willingness to commit resources—both people and money—to global scientific initiatives. Beginning in the last decades of the eighteenth century, natural philosophers attempted to position themselves as the purveyors of useful knowledge, and they saw the ocean as a convenient space for establishing credibility.[52] As the politics of imperialism intensified, the Admiralty's need for scientific and technical expertise grew more profound.[53] Changes in maritime operations were increasingly technological (iron and steam) and the needed areas of expertise more scientific (magnetism, tides, meteorology). The imperialistic enterprise heightened the political and economic status of science, and the Admiralty repeatedly turned to science and scientists to aid in the imperial business of overseas expansion.[54] The result was an increasingly strong bond between the Admiralty and the scientific community. In the first half of the nineteenth century, they worked out the intricacies of a still inchoate relationship. By the last decades of the century, the Admiralty embraced science without reserve.

Through his twenty-year research project on the tides, Whewell helped transform science into an international undertaking that relied heavily on government participation and support. The product of his researches, the isotidal map, passed easily between scientists and mariners, helping to establish scientists as the arbiters of knowledge in the nautical spaces between Britain's expanding possessions. Thus Whewell's work extended science not only geographically over the world's oceans, but also intellectually within the Admiralty, leading to increased support in the new, empirically based projects in physical astronomy. The Admiralty was caught up in a broad cultural fascination with science and a growing respect for the scientific mission. The modern scientist, in turn, was shaped by the military's growing requirements for intelligence and control of the world's oceans. Empire and the modern scientist matured together.

All maritime nations contributed to the growing relationship between scientists and the state. As the earth's oceans became culturally, economically, and politically relevant, by the end of the century the scientist was responsible for defining and managing the physical properties of the sea. The ocean's tides attracted the attention of several distinguished scientists who based their methods of analysis directly on Whewell's spatial approach, using simultaneous observations over large areas of the globe. Airy, as Britain's astronomer royal and the premier scientist working for the state, used his position to answer questions ranging from ship's magnetism to horology, and he devoted considerable time to the nation's tides. Beyond Britain, Alexander Dallas Bache published on aspects of the sea ranging from magnetism to hydrography, and he laid the foundation for American tidology. The work of Airy in Great Britain and Bache in the United States demonstrates that as the growth of trade and the need to protect that trade expanded, governments had to grapple with highly scientific problems, and they found scientific specialists eager to assist. Their work also demonstrates the fundamental role of artisans and computers in their collaborative work and demonstrates how the early Victorians delineated the social and epistemic space of the scientist.

Airy and Whewell initiated a lifelong friendship at Cambridge, where both received their degrees, became fellows, and accepted professorships. Airy graduated senior wrangler in 1823 and was elected a fellow of Trinity College in 1824. He received the Lucasian professorship two years later and the Plumian professorship and directorship of the Cambridge Observatory in 1828. He was adept at both observational and physical astronomy, but his true strength lay in his organizational abilities.[55] He left Cambridge in 1835 to succeed John Pond as astronomer royal, where his concern for the practical application of astronomy led him to work closely with the Admiralty, especially Francis Beaufort and the Hydrographic Office.[56] Late in his life, he admitted that the greatest service he had rendered was to meet the practical requirements of the state: "the distribution of accurate time, the improvement of marine timekeepers, the observations and communications which tend to the advantage of Geography and Navigation, and the study, in a practical sense, of the ramification of Magnetism."[57] Airy's interest in subjects eminently practical, combined with his mathematical acumen and emphasis on order, made the tides a perfect subject of research. He published three papers on the tides in the *Philosophical Transactions* between 1842 and 1845 and a 155-page opus in the *Encyclopaedia Metropolitana* of 1845 outlining a hydrodynamic theory based on the motion of waves in canals.[58] He wrote in his autobiography that he "had always something on hand about Tides,"[59] an interest he sustained until very late in his life. In 1865 he wrote an essay in which he returned to a topic first pursued by Edmond Halley, dating the invasion of Britain by Julius Caesar through the wonderful regularity of the ocean's ebb and flow.[60]

From his initial interest in tidal theory, Airy dissented from Lubbock's and Whewell's focus on the Bernoulli equilibrium theory and based his theoretical approach on the work of Pierre-Simon Laplace. Yet Airy profited greatly from Whewell's investigations, especially his use of simultaneous measurements and his advocacy of self-registering gauges. After a first paper on the tides in the river Thames, Airy adapted his research to help Colonel Thomas Colby, director of the Trigonometrical Survey

of Ireland, answer practical questions about the level of the sea. To aid
Colby's surveying work in Ireland, Airy chose to replicate for Ireland
what Whewell had done for all of Great Britain. He established simultaneous observations of the tides at twenty-two stations equidistant from
each other around the entire coast. Colby placed noncommissioned officers and privates of the Corps of Royal Sappers and Miners (Royal Engineers) at Airy's service to take observations of the tides, day and night, for
two months, and Airy had the officers return skeleton forms similar to
Whewell's directly to the Royal Observatory at Greenwich. "The whole
number of observations exceeds two hundred thousand," Airy boasted;
*"the circumstances of place, simultaneity, extent of plan, and uniformity of
plan, appear to give them extraordinary value."*[61] Once the observations had
been made, Airy used numerous calculators employed at the Royal Observatory to reduce them. Relying on the graphical method, especially
the method of residuals, Airy determined the law of the diurnal tide
across the coast of Ireland, helped establish the mean level of the sea
as the standard datum point used in leveling operations, and confirmed
the relationship between the height of the tide and barometric pressure,
a law he attributed (incorrectly) to Whewell and Bunt.[62]

Airy's use of simultaneous observation, his methods of data analysis,
his reliance on self-registering instruments, and even his terminology
were profoundly influenced by Whewell's initial investigations in the
tides. Yet Airy's greatest debt to Whewell was simply that he took up
the subject in the first place. Whewell had made the tides a hot research
topic, of significant practical importance to the Admiralty. Airy was a
man of order, and as an astronomer and excellent mathematician, he
was drawn to the masses of numbers and the chance to bring order and
consistency to this untamed field in physical astronomy. He also might
have been provoked into addressing the subject by Lubbock's and
Whewell's insistence on comparing their results with the Bernoulli equilibrium theory, which Airy obviously detested. And as astronomer royal,
with a strong sense of duty and a long history of helping the Admiralty,
Airy not only was doing good science, he was doing a good job. Airy
was one of the foremost proponents of the Admiralty's debt to science,
and he argued throughout his tenure as astronomer royal that the government should increase its funding of science. Arguably, Airy took up
the tides because the subject fit his personality and was a field where he
could use his station as the premier astronomer in England to help the
Admiralty control the world's coastlines and oceans.

Airy and Whewell (along with Herschel and Babbage) formed part of
what Susan Faye Cannon has termed the "Cambridge Network," a small

but extremely influential cadre of physical scientists who presided over Section A of the British Association.[63] Their lofty aim was nothing short of controlling the British scientific establishment, including its patronage and government involvement. They attempted to promote the importance of science while raising the status of the scientist. A decade later, a similar group formed in the United States. Known as the Lazzaroni, derived from a term for Italian beggars, the group attempted to organize science in the United States. As the historian Robert Bruce has noted, their aims were in every way similar to those of the Cambridge Network: "setting scientific standards to separate 'charlatans' from 'real working men'; raising the status of scientists; increasing support for science; winning recruits to it; enlisting and guiding government in the cause; and organizing scientists themselves, under an elite leadership, to further these ends."[64] Their central figure, the one they referred to simply as "the Chief," was Alexander Dallas Bache, the great-grandson of Benjamin Franklin.

The period 1840 to 1860 was the time when American science advanced from dependence to interdependence, the same period when the youthful nation developed a growing interest in the science of the sea.[65] The success of the East Coast cod fisheries and whaling industries and the rapidly expanding merchant marine meant that the United States garnered much of its wealth from the ocean. Like Whewell, Alexander Dallas Bache established himself as the spokesman for science during this period, and he symbolized the growing connection in the United States between the advance of science, the creation of an intellectual role for the scientist, and the need for the U.S. Navy to gain knowledge and control of the ocean.

Though Whewell and Bache did not meet until Bache's tour of Europe in 1837, their correspondence began in 1834 after Francis Beaufort asked the United States to take part in the multinational venture of 1835. Bache was a young officer of the Topographical Engineers, and like Beaufort, he was working his way up the military ranks through his interest in science. As head of the tidal department of the topographical engineers, he responded enthusiastically to Beaufort's request. He reprinted a tract on the tides that Whewell had sent him in the *Journal of the Franklin Institute* and distributed copies to people in the military and hydrographical engineers who were engaged in surveying.[66] In Beaufort's and Whewell's international endeavor, the United States collected more tide observations than any other country after Great Britain and Ireland, including twenty-eight stations covering two thousand miles from Maine to the mouth of the Mississippi.

Before the 1835 multinational venture, the United States paid little attention to the tides, since such work had "never been suggested or entertained."[67] When Bache became superintendent of the Coast Survey in 1843, he made resurrecting tidal studies in the United States a government priority. His first concern was to collect observational data for the safety of American shipping, but he also was dedicated to appropriating the subject from Britain and upgrading America's scientific reputation.[68] He directed Charles Henry Davis to collect and examine the extant tidal records in the U.S. Coast Survey and report on the feasibility of determining "an approximation to the course of the general tidal wave" on the coasts of the United States.[69] Davis wanted to forge a fruitful connection between the scientific establishment and the U.S. Navy. He helped establish both the Nautical Almanac Office, serving as its first superintendent, and the National Academy of Sciences.[70] When Bache asked him to collect and examine the tidal records, he was still a promising naval officer, charged with overseeing the tidal department of the U.S. Coast Survey.[71] While serving with Bache, Davis published several important papers concerned with the geological effects of the tides on the American coastline,[72] but he found little to answer Bache's queries.

Bache was a man of immense political, intellectual, and social clout, and he took it upon himself to "create a system" and "to arrange the forms of reductions."[73] He separated the coast of the United States into nine sections, establishing two types of tidal stations in each. The first were "more permanent tidal stations, where long series of observations were made for determining the tidal constants, & from full discussions of the phenomena such as had been given in [Whewell's] papers & those of Mr. Lubbock."[74] At the second set of stations, he initiated simultaneous but short-term observations, "not only for furnishing the mean or corrected establishment of observations, but for investigating the motion of the tide wave in general or in particular cases."[75] The system Bache created mimicked the research project that Whewell had followed throughout the 1830s and 1840s. Bache systematized Whewell's earlier efforts and made them permanent. No such complete system existed in Britain. The length of the British coast was paltry compared with the American coastline, and Bache transformed the British model to fit the American context.

Not only were Bache's initial questions and methods of analysis directly related to Whewell's own research, but he also based his systematic network of tide stations on Whewell's combination of simultaneous and long-term observations. Bache had observers fill in skeleton forms published by the Coast Survey that were then sent back and discussed "according to the methods of Dr. Whewell."[76] He then hired external calculators, such as L. W. Meech and Benjamin Pierce, to perform the more difficult reductions.[77] These calculators reduced the data based largely on Whewell's "ninth series," where Whewell explained the procedure for finding the effects of lunar parallax and declination from short-term observations. This included combining the graphical method of curves with the method of residuals.[78]

As head of the Coast Survey, Bache fostered a geographically oriented style of science that became prominent in the United States by midcentury.[79] He actively involved the Coast Survey in international cooperative endeavors, including the Magnetic Crusade in the 1840s and multinational meteorological research in the 1850s, spearheaded by his rival, Matthew Fontaine Maury. The study of the tides, however, was Bache's favorite. Tidology ranks above terrestrial magnetism, surveying, and triangulation in the number of published reports of the Coast Survey up to 1860.[80] Bache expanded his own investigations of the tides to several journals, including the *Coast Survey Annual Reports* and the *American Journal of Science and Arts*. Between 1850 and 1857, Bache published thirteen papers on the tides, most focusing specifically on Whewell's cotidal lines.[81] On his determination of the cotidal lines for the Atlantic coast, Bache commented: "My attention has been called, also, by the request of a valued friend, the Master of Trinity College, Cambridge, to some attempt of this sort, and his labors in connection with the subject on our own coast have entitled his request to the most respectful consideration."[82]

As Bache's words suggest, he esteemed Whewell's efforts highly, but he also used Whewell's work "on our coasts" to underscore the need for tidal research in the United States. Bache published the first tide tables from the Coast Survey Office beginning in 1855 and continued to publish them throughout his tenure as superintendent.[83] Bache's work on the tides advanced both scientific theory and its practical application, a trend found throughout his work at the Coast Survey.

When Alexander Dallas Bache took over the U.S. Coast Survey in 1843, Whewell had completed his most energetic tidal research. The British model—meaning Whewell's model—prevailed. By 1856, the year of Whewell's last publication on the tides, Bache had established over six hundred tidal stations in the United States, Mexico, and Alaska, covering the Pacific, Atlantic, and Gulf coasts.[84] Bache used as a guide not Herschel's edition of the *Admiralty Manual of Scientific Enquiry*, but the *Smithsonian Directions of Meteorological Observations*. American science, and American tidology, had come of age. Most important, the individuals involved, particularly Bache and Davis, were the ones most concerned in situating the new scientist fully within the scope of government patronage and support. Bache constructed, systematized, and then expanded a system first introduced into tidology by Whewell, his "valued friend," while Whewell viewed Bache as an equal who with "great energy and success" was advancing tidology in a part of the world that was of particular interest to Whewell.[85] Both tried to organize science in their own countries as a hierarchical and collaborative project run by elite theorists with special ties to their governments.[86]

Airy, Bache, and Whewell all had similar approaches to the study of the tides. Whewell's multinational tide experiment underscores not only his own participation, but also the efforts of Francis Beaufort of the Hydrographic Office and the servicemen of the Preventive Coast Guard. Likewise, a detailed analysis of Airy's tidal work reveals his achievements and the efforts of Thomas Colby of the Ordnance Survey and the noncommissioned officers of the Royal Sappers and Miners. Both led to a closer connection between the Admiralty and the elite theorists in Britain. These same lessons hold true for Bache. The resources at his disposal as head of the U.S. Coast Survey, combined with his position as the chief of the Lazzaroni, helped fortify geographical science as a significant type of science practiced in the United States.

The term "scientist" did not readily catch on it Britain; most thought it "un-English," partly because it was used so readily in the United States.[87] Two Americans, the astronomer Benjamin Gould and the philologist Fitzedward Hall, both independently coined the term again in the 1850s. Whether accepted or not, words are guided by the context in which they are invented; they appear when they do for a reason. In Britain in the early 1830s and in the United States in the early 1850s, science was increasingly becoming institutionalized and engrained in the public sphere. Its devotees needed a name to legitimize their practice, to strengthen their position for government funding, and to organize participation in what was increasingly becoming an international endeavor. "While science is without organization," Bache reported to the American Association for the Advancement of Science, "it is without power."[88] The Cambridge Network and the Lazzaroni, in different countries and separated by two decades, both sought to organize science under an elite company of scientists, and by linking it closely with government, they ended up conferring remarkable power on their own nations.

## "Scientists" and Spectacle

After Whewell coined the term "scientist" at the British Association meeting in 1833, he used it in print only three times: first in a review of Mary Somerville's *On the Connexion of Physical Sciences* in 1834, then in his own *Philosophy of the Inductive Sciences* in 1840, and finally in his inaugural lecture on the significance of the Great Exhibition in 1851. In the intervening two decades, Whewell negotiated a space for the scientist in two ways. One was social and explicit, demarcating the scientist from the artisan and the subordinate laborer in an overt attempt to manage large-scale participation in science. In this he was joined by other elite scientists—most notably his Cambridge coterie—in an attempt to heighten the status of the scientist. The other was intellectual and implicit, marking scientists as significant to the imperial process. In this he was joined by a larger group of scholars throughout Great Britain and the world who were intent on advancing the cause of science by closely associating it with the progress of nations.

Whewell's management of the social and intellectual roles of the scientist came together in the fall of 1851 at the Great Exhibition on the Progress of Art and Science. Hailed as the first world's fair, the Great Exhibition showcased Europe's premier examples of technological advance. By midcentury, Britain was well on its way to becoming the greatest imperial nation in the modern world, and the exhibition was a chance to show off its own superiority. It presented powerful images of Britain's and other nations' industrial prowess, offering an unparalleled synoptic view of human—and especially British—progress.[89] The method of display matched Whewell's own synoptic brand of science. "By annihilating the space which separates different nations," Whewell proclaimed, "we produce a spectacle in which is also annihilated the time which separates one stage of a nation's progress from another."[90] In his cotidal maps of the globe, Whewell used time as a means of ordering and hence flattening space. Here Whewell reversed the process, asserting that the spatial arrangement of the exhibition could be used to flatten time. Space and time always figured prominently in Whewell's mind.

The Great Exhibition took place at the magnificent Crystal Palace in Hyde Park, London, where hundreds of thousands of people from different social backgrounds and nationalities walked awe-struck through the multitude of halls filled with scientific instruments and industrial machinery. After the international exhibition had ended, its organizers invited commentators to lecture on the morals to be drawn from it. William Whewell, by then the master of Trinity College, Cambridge,

Figure 7.4. An outside view of the north transept of the Crystal Palace, the site of the Great Exhibition in the summer and fall of 1851. Whewell used the iron and glass structure as a paradigm for demarcating science from industry, the scientist from the artisan, and Britain from the rest of the world. Reproduced from *The Great Exhibition of the World's Industry, Held in London in 1851, Described and Illustrated by Beautiful Steel Engravings, from Daguerreotypes by Beard, Mayall, etc. etc. etc.* (London: John Tallis, 1852). Henry Madden Library, California State University, Fresno.

and at the height of his career, delivered the inaugural address on 26 November 1851. As Richard Yeo exclaimed, "Whewell gave one of his most extraordinary performances."[91]

The way Whewell managed the relationship between science, technology, and empire had larger implications for the hierarchy of scientific work in general and the definition of the new scientist in particular. As an architect of the "scientist," Whewell used the gigantic iron and glass Crystal Palace to once again distinguish science from industry, the scientist from the artisan, and Britain from the rest of the world. Because his own research depended so crucially on artisans, he was especially concerned with the appropriate relationship between them and the emerging scientists. As an advocate of the scientist, moreover, he was interested in consolidating the artisans' cooperation without conferring heightened status on them, a complex negotiation present in many of the sciences specializing during this period.[92] He had to sustain

Figure 7.5. Inside the transept of the Great Exhibition looking north. Whewell used his position as an architect of the "scientist" to help Victorians navigate the seemingly endless halls of the Crystal Palace, where science, technology, and empire were all on display. Reproduced from John Tallis, *Tallis's History and Description of the Crystal Palace, and the Exhibition of the World's Industry in 1851; Illustrated by Beautiful Steel Engravings from Original Drawing and Daguerreotypes by Beard, Mayall, etc.* (London: J. Tallis, 1852). Henry Madden Library, California State University, Fresno.

the collaboration of the instrument makers while creating a leading and formative role for the centralized expert. Thus Whewell tried to convince each side that cooperation was in its best interests, without antagonizing either.

The Crystal Palace exhibited the wonders of modern technology, a subject on which Whewell had remained all but silent throughout his career. By the early 1850s he was recognized within the nascent scientific community as a spokesman, and his views as the champion of science were well known. He therefore distanced himself as a mere "spectator," using the word three times in the opening paragraph alone. As an "unconnected spectator of the great spectacle," Whewell could wander through the crowds and instruments in the Crystal Palace taking a disinterested view toward synthesis. He combined this image of an unconnected spectator with that of an explorer who traveled the globe to view all of human invention.[93] "And now let us, in the license of epical imagination, suppose such an Ulysses—much seeing, much wandering, much enduring—to come to some island of Calypso, some well inhabited city, . . . and there to find that the image of the world and its arts, which he had vainly tried to build up in his mind, exhibited before his bodily eye in a vast crystal frame." Through this heroic Ulyssean traveler—much seeing, much wandering—Whewell synthesized the fruits of industry through an authorial persona very different from the one he claimed in his scientific work. From this vantage point, Whewell tried to locate the "moral" of the grand spectacle, beginning with the relationship between science and technology.

The Victorian scientific elite often misrepresented this relationship, insisting that science was largely responsible for technological advance and thus economic progress. This view largely formed the popular perception as well. Though Whewell's earlier published works displayed an almost total disregard for technology, in 1851 his views appeared far more nuanced. Whewell continued his discussion with a rhetorical device that forced scientists and artisans to view their relationship in an entirely novel way.[94] Through an elaborate analogy relating technology to poetry and science to criticism, Whewell outlined his role: "Perhaps such remarks as I have to make may rather be likened to the criticism which comes after the drama. For, as you know, Criticism does come after Poetry; the age of Criticism after the age of Poetry; . . . language is picturesque and affecting, first; it is philosophical and critical afterwards:—it is first concrete, then abstract:—it acts first, it analyses afterwards."

The analogy compelled his audience to cross conceptual boundaries and to think about science and technology in a new light. By introducing his task with this elaborate analogy, Whewell transferred the values of the well-established hierarchy of poetry and criticism onto the hierarchy he was attempting to create between technology and science.[95] The first is picturesque and affecting, the second, philosophical and critical. He then defined not technology but science:

But to discover the laws of operative power in material productions, whether formed by man or brought into being by Nature herself, is the work of a science, and is indeed what we more especially term Science. . . . We have, instead of the Criticism which naturally comes after the general circulation of Poetry, the Science which naturally comes after a great exhibition of Art: two cases of succession connected by a very close and profound analogy. That this view of the natural and general succession of science to art, as of criticism to poetry, is not merely fanciful and analogical, we may easily convince ourselves by looking at the progress of art and of science in past times. For we see that, in general, art has preceded science. Men have executed great, and curious, and beautiful works before they had a scientific insight into the principles on which the success of their labours was founded.

He then repeated almost word for word the analysis of the relationship between science and technology that he had used in his British Association address almost two decades earlier. "Art was the mother of Science: the vigorous and comely mother of a daughter of far loftier and serener beauty." In direct contrast to the popular view, and the view offered by many of the scientific elite, technology here becomes the "mother of science," the progenitor of progress. Art, in fact, attains a position more important than science, since the latter is dependent on the former. And yet science retains its position as "a daughter of far loftier and serener beauty." Whewell thus maintained the heightened status of the scientist without diminishing the role of the artisan. In their own ways, both were important to the overall progress of knowledge.

But much had also changed in two decades. For instance, there was the meteoric rise by midcentury of the chemical industries, which owed their entire existence to the science of chemistry and the theoretical ingenuity of Claude Berthollet, Antoine Lavoisier, and others. Indeed, the massive industrial complexes popping up in Britain's cities dwarfed even the Crystal Palace. "Here," Whewell averred, "Art is the daughter of Science." Whewell turned his earlier comparison on its head—but with a twist. Whereas he had concluded in 1833 that science was "the fully developed blossom, of which art is the wonderfully involved bud," now in 1851 he concluded that "the tree of Art blossomed from the root of Science." What appears to be a rhetorical mirror image is not. In both cases, science brought analysis and insight; it still attained the position of criticism following the play. In like manner, and the main goal of Whewell's "Address," the scientist's job was to articulate the significance of the myriad of instruments and industry on display, to bring meaning to the spectacle. Indeed, it was only from such an exalted position that one could draw morals from the Great Exhibition.

After elaborating on the relationship between science and technology, Whewell turned to the benefits of the collaborative hierarchy by explaining the "morals" themselves. The first moral concerned the difference between the technologies advanced by highly "civilized" societies and the works advanced by "savage hands." For Whewell the difference was social rather than intellectual. "That in those countries the arts are mainly exercised to gratify the tastes of the few; with us, to supply the wants of the many." Thus, concluded Whewell, "we have, indeed, reached a point beyond theirs in the social progress of nations." Here then was a mighty moral, one that benefited all involved—the social progress of a great and powerful nation.

The second moral followed from the first and had to do, strangely enough, with the "classification" system of exhibits, whereby different types of instruments and machines were placed in different classes. Whewell was well versed in the classification of the sciences, an important aspect of both his *History* and his *Philosophy*, where he had attempted to judge the sciences based on their "fundamental ideas" and their places within the inductive march toward truth. Here he transplanted these same values onto technology. One "value and advantage" of such a classification was that "the manufacturer, the man of science, the artisan, the merchant, would have a settled common language, in which they could speak of the objects about which they were concerned." For Whewell, this would "facilitate and promote their working together." Bringing together "the insights of science to the instinct of art" was in every way similar to bringing together "thoughts and things," the definition of his fundamental antithesis of knowledge articulated earlier in his career. In both cases the synthetic mind (in this case, through the correct system of classification) "makes general propositions possible," the ultimate goal of discovery.

Collaboration was also where Whewell chose to end his lecture. It was this joining of forces of "artists and scientists (if I may use the word)" that "generated a community of view, a mutual respect, and a general sympathy." Whewell noted that such a collaboration was similar to a university setting. "Without underrating the effect of lectures and tasks, of professors and teachers," Whewell pronounced that the most precious result of a university setting was the "effect produced upon those who resort thither by their intercourse with, and influence upon, each other." Whewell then transformed the university into a metaphor: "To a University of which the Colleges are all the great workshops and workyards, the schools and societies of arts, manufactures, and commerce, of mining and building, of inventing and executing in every land—Colleges in which great chemists, great mechanics, great naturalists, great inventors, are already working, in a professional manner, to aid and develop all that capital, skill, and enterprise can do." Whereas the university was normally associated only with the scientist, here Whewell's university includes the mechanist and inventor working side by side with the naturalist and chemist. Whewell eloquently concluded with this forceful idea in mind: "We were students together at the Great University in 1851."

The triumphant ending of Whewell's 1851 address represented the optimism the British felt in their industrial might. But Whewell's speech was optimistic in another sense as well. The specialization and popularization of science during the preceding half century had transformed its social organization. Those who could actively participate in science had increasingly come under scrutiny. Whewell and others fostered a growing distinction of the scientist from the artisan. But as an organizer of global research that relied heavily on the inventors and makers of instruments, Whewell also realized the importance of keeping close ties between the two, and thus he employed a rhetoric specifically designed to promote collaboration. He argued that such a collaboration would lead to national and international unity and keep open the gates of progress beneficial for both groups. The artisans were not mere laborers but were an important part of the process of creating knowledge. The scientists likewise kept their status, which included ties to the artisan. This definition enabled the scientists to collaborate with artisans while convincing artisans that working with, and accepting the definition of, the new scientists would be in their own best interests.

The Great Exhibition symbolized the triumph of British technological advance. Whewell transformed the setting into a triumph of classification and order. He thus helped both the philosopher and the artisan navigate the spectacle of invention and commerce that was everywhere taking hold. James Glaisher, one of the judges of the exhibition, welcomed Whewell's focus on collaboration. "In proportion as is necessary to the interest of science that theory, observation, and experiment should march hand in hand, so is it equally essential that theoretical and practical men of science should come into contact with men to whom must be entrusted the construction of instruments necessary to the completion of their views."[96] Yet under Britain's seeming technological superiority lurked a tinge of dismay. The British won seventeen of the thirty-one council medals, but only four in Class X, encompassing "Philosophical Instruments and Processes Depending on Their Use."[97] Two of those four went to a single instrument maker, John Newman, one for his air pump and the other for his self-registering tide gauge. For some this lack of technological prowess in scientific instruments related directly to the growing distinction between scientists and artisans. A year after the Great Exhibition, the renowned natural philosopher Lyon Playfair warned, "In this country we have eminent 'practical' men and eminent 'scientific' men but they are not united and generally work in paths wholly distinct.... From this absence of connection there is often a want of mutual esteem and a misapprehension of their relative importance to each other."[98] This distinction, so prominent in today's hierarchy of knowledge, was a direct result of the early Victorians' attempts to delineate an active researcher, one who could help navigate the increasingly technological sophistication of the modern world.

## Conclusion

In the winter of 1836, Whewell was in the midst of preparing his *History* for publication, writing his *Philosophy*, and conducting his most energetic tidal studies. He referred mainly to his tidology when he applied for the Lowndean professorship of astronomy, which had become vacant that fall. "I have written to Lord Minto," Whewell explained to Beaufort, "respecting to him that I had given great time and pains to this subject [of the tides], that I believed the results were considered as of use, that I expected to go on with the same kind of work for some time, and that, besides any other claims I might have to the Lowndean professorship, I ventured to apply for it, in order that this appropriation of my time and thoughts might be a professional employment, recommended and awarded by a public office."[99]

Whewell hoped that Beaufort and Lord Minto would recommend his tidal researches to Lord Melbourne, one of the electors for the professorship. "I certainly should very much like the professorship," he wrote his confidant, Richard Jones, in the midst of mobilizing support, "very much on public grounds for Tidology."[100] The only "professional employment" for natural philosophers at that time was through just such a "public office," but this would change in the decades to follow. Science became a powerful force in the Victorian era, and its practitioners had already begun to mark out their position in the Western world.

The Lowndean professorship of astronomy went to George Peacock rather than to William Whewell. Though it was a severe disappointment, Whewell cast his net more broadly, publishing reviews and essays covering almost every field in the physical sciences, and thus helped define what it was to do science and to be a scientist. In his self-defined role as a definer of science, he created a hierarchy with the physical sciences on top and the other sciences on the tiers below. In his role as a definer of the scientist, he advocated a similar hierarchy with theorists on top and the artisan, subordinate laborer, and practical observer somewhere underneath. In both roles, he implicitly situated the activity fully within the sphere of government support. These hierarchies, and the link with government, are a central part of the sciences today.

Whewell relied heavily on subordinate computers, Admiralty fact gatherers, and mechanical inventors. The work of Thomas Gamlen Bunt in Bristol and Daniel Ross in London was integral to Whewell's tidal analysis. They did not passively sit on the data but used their unique position to advance novel methods of analysis and propose additions to theory. Yet while Whewell considered Alexander Bache, the great-grandson of Benjamin Franklin, his equal, he could dismiss his hired help as mere collectors or calculators.[101] Artisans, likewise, were increasingly taking on subordinate roles within the scientific community. They were not interested in working for or under scientists, but the Victorian scientific elite had deliberately limited their status. The scientists required their participation but saw them as mere subordinate laborers.

Just as words themselves are a product of their time, so are negotiations to define the social and intellectual meaning of words. The way Whewell defined "scientist" suggests that we need to look beyond tripartite models of class conflict.[102] In his rhetoric and practice, Whewell defined the scientist based on hierarchy and subordination relating to appropriate methodology. Though class rhetoric became more politicized in the first half of the nineteenth century, a collaborative, hierarchical view of society remained pervasive. What Lord John Russell called "a fair and gradual subordination of ranks" and William Gladstone hailed as "a state of graduated subordination" was British society as most contemporaries saw it, even in the highly charged 1830s and 1840s. This is certainly how Whewell chose to structure the collaborative nature of science and his definition of the modern scientist.

While Francis Baily tried to raise the status of the practical observer and expert calculator with his biography of Flamsteed published in the mid-1830s, Whewell countered with a move toward theory, a "higher" form of construction in the building of science. Science may be made up of distinct observations, but scientists supplied the "relations and connexions" holding them together. They could do this largely through an understanding of the history of each science, which demarcated where it was in its ascension toward truth, and of what path it needed to follow. "Every labourer in the field of science," Whewell insisted, "however humble, must direct his labours by some theoretical views, original or adopted." The problem—or for Whewell the solution—was that subordinate laborers did not know the history; they specialized in observational data and the techniques of their reductions. They saw only tides and waves, whereas the scientist had a view of the entire ocean. The metaphor is apt. As trade and the need to protect that trade expanded throughout the nineteenth century, governments had to grapple with problems of a highly scientific nature, and they were forced to rely more heavily on scientific specialists. Maritime nations' economies rested on the ability to cross the sea in winter's storms and find a path through the roaring waves. Whewell used this requirement to help the early Victorians navigate the social process of knowledge creation and to demarcate, in his new lexicon, a space for the modern "scientist."

# Conclusion:
# The Tides of Empire

He who commands the sea, commands the trade routes of the world.

He who commands the trade routes, commands the trade.

He who commands the trade, commands the riches of the world, and hence the world itself.

SIR WALTER RALEIGH, *HISTORIE OF THE WORLDE* (1616)

Sir Walter Raleigh's declaration on the importance of commanding the sea came at a time when control of the ocean was just beginning to weigh on the minds of the English elite. England had only a fledgling navy, mostly commercial vessels doubling as fighting ships, and at best an inchoate strategy for becoming an imperial power through long-distance trade. Raleigh wrote these words to Elizabeth I, the queen of England, who—not coincidentally—issued one of the first proclamations on the freedom of the seas. Rejecting the Spanish king's insistence that English vessels stay clear of Spanish waters, Elizabeth declared that "the use of the sea and air is common to all; neither can any title to the ocean belong to any people or private man, for as much as neither nature nor regard of the public use permitteth any possession thereof."[1] Not surprisingly, war with Spain soon followed. It was fought, like most major wars before the twentieth century, largely *on* the sea for possession *of* the sea. Commanding the sea to make it free and safe has had a long and turbulent history. By the mid-nineteenth century, at the height of free trade, "ruling the waves" meant ruling a vast empire.

272

Raleigh's words held particular significance because they reflected an inherent switch in mentality. Elizabeth, reigning over an unruly Parliament, had her sights fixed on ameliorating England's domestic problems. Raleigh, however, suggested that England turn outward toward the oceans. This demanded a change in the nation's geopolitical gaze, one that Britain— and all other nations with imperial designs—would be required to make from the sixteenth century to the present day. Scientists and the state shifted their attention from metropolitan and colonial positions on land to the largely unexplored spaces of the sea, a physical and intellectual space pregnant with commercial and imperial significance. In the second quarter of the nineteenth century, scientists helped control the ocean by creating an ocean space that paralleled the physical track and commercial interests of imperial powers. Modern conceptions of the ocean grew out of the close alliance between the advance of science and the creation of empire.[2]

Controlling the ocean conferred remarkable power on industrializing nations, particularly Britain, whose island geography fostered a maritime orientation that grew steadily throughout the late eighteenth and early nineteenth centuries. Britain's merchant marine doubled in size between 1786 and 1815.[3] In the half-century that followed, Britain established itself as carrier, banker, and insurer to the Western world.[4] The unprecedented growth of shipping that accompanied industrialization led to an oceanic "transport revolution" as significant as the preceding land-based revolution necessary for industrialization. Where land travel involved constructing roads and bridges, building canals, and dredging inland rivers, ocean transport depended on knowledge of jet streams, currents, and wind patterns. An intimate knowledge of the ocean's tides and currents, its storms and magnetic variations conferred control that translated directly into the ability to dominate lands and cultures connected by its rim.

Up to the nineteenth century, ships, cargo, and crew primarily traveled to distant destinations following known sea-lanes, and the prospect of being lost at sea or shipwrecked on an uncharted island filled the popular imagination. The sea was a vast, mysterious, and unrelenting space that had yet to become a place of scientific study. To view the ocean in this way, natural philosophers required access to empirical data, beginning with the coastline and extending far out to sea. Fascination with the ocean was manifested in the avid reading of the new maritime novels, the rise of yachting, the increasing popularity of ocean travel, and the crazes for marine natural history and home aquariums.[5] As the earth's oceans grew more economically and politically relevant, the spaces between landmasses likewise became both interesting and knowable. Beginning with their home waters, scientists extended their methods, viewing the marine frontier between continents as one connected whole. This extension allowed them to speak decisively about areas of the open sea where measurements were impossible or severely limited. In this way, they projected the laws obtained near the shore, laws that were coming to be known with great exactness, onto the unknown oceanscape. Tides, terrestrial magnetism, and meteorology all became popular research topics within the scientific community, reaching a peak by midcentury. Governing imperial possessions and competing for lucrative markets in far-off lands meant acquiring complete command of the vertical and horizontal spaces in between.

## Horizontal and Vertical Consciousness

Advancing on the technological and scientific accomplishments of the late eighteenth century, from the perfecting of the chronometer to the theory of the moon's motion, voyages of discovery expanded to most of the coastlines throughout the world, including the North polar regions. The nineteenth century was the great age of imperialism, and European ships traveled the oceans to establish trade routes and find raw materials for their nations' industrializing economies. The French, following the tradition set by Louis-Antoine de Bougainville and Jean-François de La Pérouse, continued their active interest in voyages of scientific discovery, and by midcentury Germany, the United States, and several of the Scandinavian countries were making similar voyages. But Britain had emerged from the Napoleonic Wars as the world's dominant maritime power, extending its reach over much of the globe. The nation's expansionist mentality required an equally all-encompassing scientific conception of the ocean that would allow them to traverse the open sea unimpeded.

After the successful naval battles of 1814, Britain controlled and patrolled larger swaths of the ocean than any previous sea power. In the second quarter of the century, the British used their military might to ensure an open system of trade and commerce, a stable European balance resting on the principle of "freedom of the seas."[6] After having had an unstoppable war machine based in Europe at the end of the eighteenth century, Britain cut back its navy to a less substantial, and thus less provocative, policing force with ships stationed around the world.[7] The Royal Navy became the protector of maritime commerce—in Paul Kennedy's words, "the defender of British interest in regions where organized government appeared to be lacking, a 'policeman' to a certain extent, but also a surveyor and a guide."[8] The oceans were to be both free and safe, a policy that helped usher in almost a hundred years of European peace, now known as Pax Britannica.

Strategic bases throughout the coastlines of the world were of immense value to the British naval and merchant marine. Britain had acquired such advantageously positioned outposts as Malta, Ceylon, and the Cape of Good Hope after the Napoleonic Wars, and it continued to strengthen its imperial reach by claiming islands in the Atlantic, Pacific, and Indian oceans. The case of the Falkland Islands is representative of early British imperialism. From this one set of islands, British warships could protect both the South Atlantic and Pacific trade routes.[9]

These bases allowed commerce to grow. Once trade expanded, the Admiralty sent expeditions to survey the coastline, ports, shoals, and bays of their distant territories. Britain's strength, as both an island and an imperial nation, demanded maneuverability throughout the world's oceans. The Admiralty thus needed to provide its mariners with correct soundings, tide tables, and sailing directions. The end of the Napoleonic Wars freed up many navy vessels to undertake the great surveys of the second quarter of the nineteenth century.[10] Advances in trade called forth the Admiralty's policing force; that force necessitated surveys; and those surveys demanded an understanding of the science of the ocean.

While mariners applied new technologies—the chronometer, iron, and steam—scientists adapted the spatial approach of Alexander von Humboldt to study the ocean's physical properties.[11] Humboldt set the tone for both the horizontal and vertical axes of nineteenth-century scientific exploration. Besides measuring interconnected natural phenomena, he insisted on gathering large amounts of observational data over wide areas. "Amidst the apparent disorder which seems to result from the influence of a multitude of local causes," Humboldt suggested, "the unchanging laws of nature become evident as soon as one surveys an extensive territory."[12] While traveling naturalists followed Humboldt's creed inland, scientists also extended his emphasis on data collection and mapping to the largely unexplored frontier of the ocean. The oceans were in many respects better adapted to Humboldt's spatial approach than were landmasses, where mountains and national borders hampered the study of meteorology and magnetism across vast geographic areas. The ocean, though at times forbidding and dangerous, was also smooth, constant, and open. The compilation and correlation of measurements, with due attention to their distribution across vast spaces, became the major characteristic of nineteenth-century ocean science.

Britain's maritime focus led to an expansive geographical (horizontal) enterprise. Yet this was also a time of ordnance surveys, of triangulations, and of leveling operations in general, such as for canals and railroads. A pressing and formidable vertical endeavor therefore existed alongside Britain's much discussed horizontal expansion. Two brief examples will demonstrate the difficulties. First, when Thomas Colby became head of the Ordnance Survey in the 1820s, he was asked to calculate the height of about three hundred hills across England and Wales.[13] But height above what? Second, in 1830 the Lords of the Admiralty directed John Augustus Lloyd to make accurate measurements of the height of the river Thames from Sheerness to London Bridge, partly in preparation for the removal of the old London Bridge and the erection of the new bridge.[14] For both operations, surveyors needed a zero point to measure from. The level of the sea was assumed to be the zero point, but the "level of the sea" proved a highly contentious concept. Groups with disparate interests used entirely different zero points. Admiralty charts were set to low water to point out the most dangerous shoals and sandbars, while harbormasters and pilots cared only for high water levels, since that was the information they needed to get vessels safely in and out of port. Scientists, in turn, argued that neither high water nor low water marks were sufficient as a standard for surveying, since they differed from hour to hour, day to day, month to month, and year to year. The British had a devil of a time flattening the sea.

The examples of Thomas Colby and John Augustus Lloyd demonstrate that a simple undertaking such as measuring a hill or building a bridge was bedeviled with vertical difficulties. Sea level turned out to be a slippery concept, especially in England where, owing to its massive tides, the level of the sea changed constantly. Indeed, the margin of error associated with determining the vertical datum point proved much greater than the margins of error used for the land leveling and surveying techniques then used by the Admiralty, a point previously overlooked. The Admiralty realized that the problem could be solved only through an adequate study of the science of the sea, specifically a theory of the tides. Only with sufficient tidal data, reduced and analyzed, could the British government build canals and railroads, erect bridges, and survey its possessions.

Both Lubbock and Whewell warned that the vertical datum point for surveying operations was a problem only tidology could solve. As Lubbock suggested: "The practice which at present obtains of referring the heights of building and mountains to *the level of the sea* or to *high or low water mark*, seems objectionable. The heights of *spring* or *neap* tides, although not subject to so much uncertainty, are also quantities too vague to be used with propriety as standards of reference."[15]

He could point to many examples. The most notorious case was the Isthmus of Panama, where the Atlantic and Pacific oceans were separated by roughly thirty miles. When Lloyd first surveyed the Isthmus in the early 1830s, he discovered that the tides on each side differed significantly.[16] On the shores of the Atlantic the tides rose less than a foot, while on the Pacific side they rose sixteen feet or more. The dream of connecting the Atlantic with the Pacific, first by a railway and ultimately by a canal, required a common measuring point. Using either low water or using high water gave wildly incommensurate values.[17]

When Whewell entered the debate, he used examples closer to England to bring the point home. In his "seventh series" published in 1837, Whewell investigated the diurnal inequality at several places on the globe, focusing especially on Plymouth and Singapore. The curves Bunt had drawn in the course of that investigation let Whewell offer his views concerning the proper level to be used by surveyors and civil engineers. Whewell began by warning that "'the level of the sea at low water,' a phrase sometimes used by surveyors, is altogether erroneous, and may lead to material error."[18] At St. David's Head, for instance, the farthest western point of Pembrokeshire (and now part of the Pembrokeshire Coast National Park), the tide could reach a range of thirty feet. Directly across the Irish Channel at Wicklow, however, the range was a mere three feet. "If the sea were level at low water," Whewell scoffed, "the difference of the mean heights on the two sides of the Channel (which is only about fifty miles) would be fourteen feet. Such an average elevation of one side of a narrow sea above the other is quite inconsistent with the laws of fluids." According to Whewell, this was of practical importance to both the surveyor and the mariner, and only the mean tide level would make sense in such a case. Although both the high and low tides varied considerably from month to month in Plymouth and Singapore, the mean tide level, according to Whewell, remained "far more nearly constant."

Whewell returned to the question in his "tenth series," published two years later.[19] By this time, determining sea level had become a major project carried out and funded through the British Association for the Advancement of Science. The Association had hired professional surveyors, of which Bunt was one, to carry out a level line between the north shore of Somerset and the south shore of Devon, stretching across the entire island. "This line has also been referred to the sea at its extremities," Whewell noted in his presentation to the Royal Society, "and the observations show that the height of mean water coincides, at least very nearly, at different places, as well as at the same place at different times."

Whewell added an analysis of six years of tide observations made at Plymouth by William Walker, the harbormaster at the royal naval dockyard, and discussed by Daniel Ross of the royal Hydrographic Office, to show that the mean tide level remained constant at Plymouth from year to year. Whewell believed this could serve as a zero point for all surveying and topographical work undertaken in Britain and its possessions.

A royal dockyard harbormaster and a Hydrographic Office computer had collaborated with a Cambridge University professor to finally flatten the sea, and the practical implications should not be missed. That Whewell first reported his results in a paper comparing Plymouth and Singapore suggested that the two places, separated by thousands of miles, had at least one thing in common: a mean tide level. Once even a short series of measurements were made at each port—a prospect outlined by Whewell in his "ninth series"—one could determine the mean tide. From the mean tide, one could then measure the height of buildings, hills, railroads, and bridges based on a common zero point.

George Biddell Airy took up a similar study in the early 1840s to help the director of the Trigonometrical Survey of Ireland, Colonel Thomas Colby, with his survey of the heights of hills and slopes of rivers.[20] Airy suggested that Colby institute a series of tide measurements "in order to refer the levels to the mean height of the sea, or to its height at some definite phase of the tide." Airy established simultaneous observations of the tides at twenty-two stations, and Colby posted members of the Corps of Royal Sappers and Miners to take observations. After this collaboration among the director of the Survey of Ireland, the astronomer royal, and the Royal Sappers and Miners, the mean sea level became the preferred zero point for leveling operations throughout the British Empire. Once the Ordnance Survey of Great Britain began, the mean sea level at Liverpool was selected as the datum plane for all the levels used throughout the country.[21] These same methods were then transferred across the ocean in the last quarter of the century when the mean level of the sea became the standard datum plane in the Survey of India.

The power conferred on the state by a standardized system of weights and measures linking the center with the periphery made imperial coordination possible on a global scale.[22] The first half of the nineteenth century was a great age of standardization, when the dimensions of space and time were fixed by imperial powers attempting to promote long-distance communication. Everything was measured from the center, a point that was as much political as practical. In 1833 the astronomer royal, John Pond, began the practice of dropping a red ball from the tower at Greenwich Observatory every day at 1:00 p.m. so mariners could

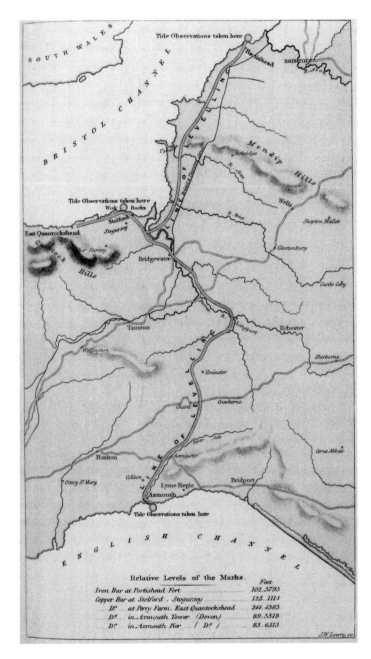

Figure C.1. The level line surveyed by Thomas Gamlen Bunt from the English Channel to the Bristol Channel, as part of the project within the British Association to find the relative level of land and sea. *British Association Report* 1838 (1839): 1–11.

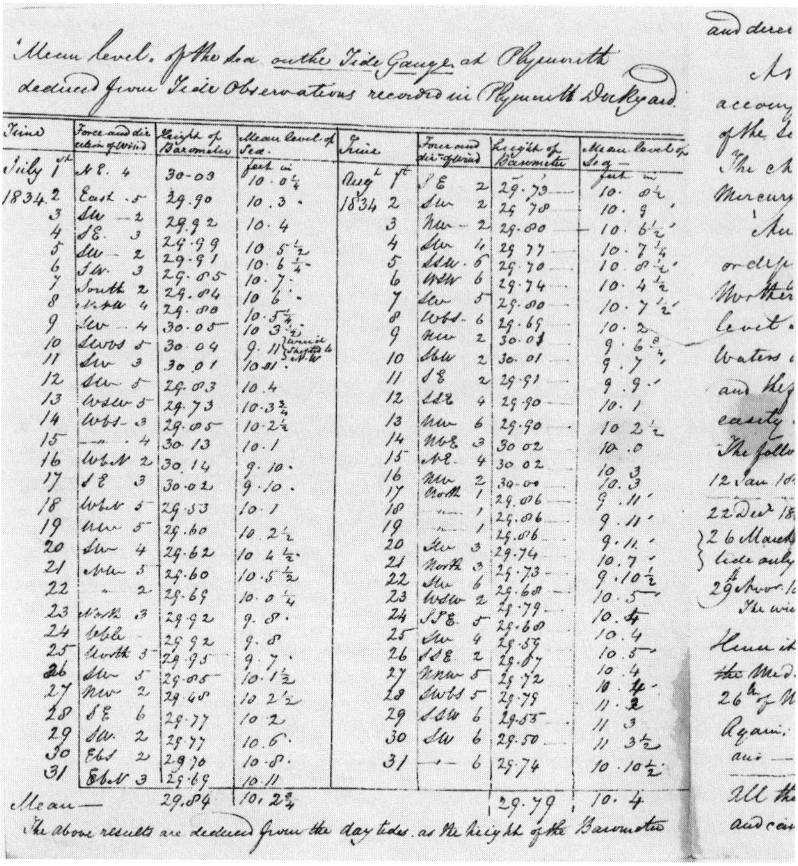

Mean Level of the Sea on the Tide Gauge at Plymouth deduced from Tide Observations recorded in Plymouth Dockyard.

| Time | Force and direction of Wind | Height of Barometer | Mean Level of Sea (feet in) | Time | Force and direction of Wind | Height of Barometer | Mean Level of Sea (feet in) |
|---|---|---|---|---|---|---|---|
| July 1st 1834 | NE 4 | 30·03 | 10·0¼ | Augt 1st 1834 | SE 2 | 29·73 | 10·0½ |
| 2 | East 5 | 29·90 | 10·3 | 2 | SW 2 | 29·78 | 10·9 |
| 3 | SW 2 | 29·92 | 10·4 | 3 | NW 2 | 29·80 | 10·6½ |
| 4 | SE 3 | 29·99 | 10·5½ | 4 | SW 4 | 29·77 | 10·7¼ |
| 5 | SW 2 | 29·91 | 10·6¼ | 5 | SSW 6 | 29·70 | 10·8½ |
| 6 | SW 3 | 29·85 | 10·7 | 6 | WSW 6 | 29·74 | 10·4½ |
| 7 | South 2 | 29·84 | 10 6 | 7 | SW 5 | 29·80 | 10·7½ |
| 8 | NNW 4 | 29·80 | 10·5¾ | 8 | WbS 6 | 29·69 | 10·2 |
| 9 | SW 4 | 30·05 | 10 3½ | 9 | NW 2 | 30·03 | 9·6¾ |
| 10 | SWbS 5 | 30·04 | 9·11 | 10 | SbW 2 | 30·01 | 9·7 |
| 11 | SW 3 | 30·01 | 10·01 | 11 | SE 2 | 29·91 | 9·9 |
| 12 | SW 5 | 29·83 | 10·4 | 12 | SSE 4 | 29·90 | 10·1 |
| 13 | WSW 5 | 29·73 | 10·3¾ | 13 | NW 6 | 29·90 | 10 2½ |
| 14 | WbS 3 | 29·85 | 10·2½ | 14 | NbE 3 | 30 02 | 10·0 |
| 15 | — 4 | 30·13 | 10·1 | 15 | NE 4 | 30 02 | 10 3 |
| 16 | WbN 2 | 30·14 | 9·10 | 16 | NW 2 | 30·00 | 10·3 |
| 17 | SE 3 | 30·02 | 9·10 | 17 | North 1 | 29·86 | 9·11 |
| 18 | WbN 5 | 29·53 | 10·1 | 18 | — 1 | 29·86 | 9·11 |
| 19 | NW 5 | 29·60 | 10·2½ | 19 | — 1 | 29·86 | 9·11 |
| 20 | SW 4 | 29·62 | 10·4½ | 20 | SW 3 | 29·74 | 10·7 |
| 21 | NW 3 | 29·60 | 10·5½ | 21 | North 3 | 29·73 | 9·10½ |
| 22 | — 2 | 29·69 | 10·0¼ | 22 | SW 6 | 29·68 | 10·5 |
| 23 | North 3 | 29·92 | 9·8 | 23 | WSW 2 | 29·79 | 10·4 |
| 24 | WbE | 29·92 | 9·8 | 24 | SSE 5 | 29·68 | 10·4 |
| 25 | North 3 | 29·95 | 9·7 | 25 | SW 4 | 29·59 | 10·5 |
| 26 | SW 5 | 29·85 | 10·1½ | 26 | SSE 2 | 29·67 | 10·4 |
| 27 | NW 2 | 29·40 | 10 2½ | 27 | NNW 5 | 29·72 | 10·4 |
| 28 | SE 6 | 29·77 | 10 2 | 28 | SWbS 5 | 29·79 | 11·2 |
| 29 | SW 2 | 29·77 | 10·6 | 29 | SSW 6 | 29·55 | 11 3 |
| 30 | EbS 2 | 29·70 | 10·8 | 30 | SW 6 | 29·50 | 11 3½ |
| 31 | EbN 3 | 29·69 | 10·11 | 31 | — 6 | 29·74 | 10·10½ |
| Mean | | 29·84 | 10·2¾ | | | 29·79 | 10·4 |

The above results are deduced from the day tides. as the height of the Barometer

Figure C.2. In a letter to William Whewell, dated 10 July 1837, William Walker, the dockmaster at Plymouth Dockyard, wrote: "I entirely agree with you that the level of the sea at low water cannot be determined and is consequently an uncertain and fluctuating level and cannot be used correctly for hydrographical uses." He then enclosed his own observations and reductions of the mean tide level from the tide gauge at Plymouth Dockyard. Whewell Papers, R.6.20/318.

synchronize their chronometers, part of the larger attempt to establish Greenwich as zero longitude and standard mean time.[23] That Britain was also attempting to standardize the vertical dimension should come as no surprise. The work of Lubbock, Whewell, Airy, and Colby led directly to a vertical standardized datum plane, one that fit the imperial quest to standardize measurements and bring distant possessions to rule.

The quest to flatten the sea for surveyors and civil engineers was part of a larger vertical consciousness that engulfed science in the early nineteenth century. "There was a growing perception of volume," noted

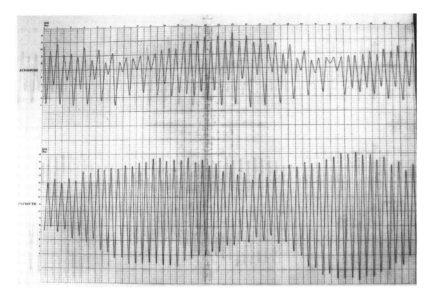

Figure C.3. Whewell's graphical representation of the mean level of the sea at both Singapore and Plymouth. "Researches on the Tides," 7th ser., "On the Diurnal Inequality of the Height of the Tide, Especially at Plymouth and at Singapore; and on the Mean Level of the Sea," *Philosophical Transactions* 127 (1837): 75–87.

Alain Corbin in his recent study of the popular fascination with the sea, "and a three-dimensional interpretation became common, . . . a vision more capable of taking layers into account."[24] Humboldt's emphasis on the vertical orientation of the earth's biota, which he described through vivid maps relating altitude to plant life, had a profound influence on nineteenth-century scientists. The role of vertical zonation in Darwin's theory of coral reef formation, for instance, which relied heavily on the renewed interest in the changing levels of the sea, related directly to Humboldt's vertical vision.[25] Following Humboldt, scientists used both a horizontal and a vertical orientation in understanding the oceanic environment, one that extended from the heavens through the earth's atmosphere all the way down to the tidal oscillations on its surface and eventually to submarine tides in the ocean's depths.[26]

## Science, Ocean Space, and Empire

Study of the oceans had begun in the celestial sphere, with the aim of improving shipping. The national observatories at Greenwich and Paris were both founded to address questions of navigation, particularly the vexing problem of finding longitude. Throughout the late eighteenth and early nineteenth centuries, scientists solved the longitude problem by accounting for the motion of the moon. The perfection of lunar theory, in turn, opened the prospect of answering most other questions of navigation. The moon was known to be responsible for the ocean's tides and thought to cause the fluctuations in the earth's magnetic field and atmospheric pressure. By the second quarter of the nineteenth century, then, scientists hoped their new vertical orientation could help them understand changes in the atmosphere (including meteorology and terrestrial magnetism) and the movement of the oceans (including tides and ocean currents). Moreover, much of the world's landmasses had been mapped, charted, and mostly explored, but the ocean, which covers 70 percent of the earth, remained relatively unknown. Marine scientists felt as if they were approaching a massive and largely uncharted geographical area—on the threshold, as one historian aptly put it, of an unfamiliar world.[27] This new attention to verticality in understanding the physical characteristics of the globe complemented the Admiralty's horizontal expansion throughout the world's oceans.

As maritime commerce moved outward, from the coastal trade of the eighteenth century to the transoceanic trade of the age of imperialism, scientists seized the opportunity to bring rule and rationality to this new frontier. They chose to use the isomap, a synoptic tool that visually illustrated the lawlike nature of the ocean. Whewell's own research had produced isotidal maps of the Atlantic and Pacific, and though innovative for tidal studies, his charts closely resembled contemporary efforts. Humboldt's emphasis on collecting massive amounts of interconnected observations combined with his insistence on graphical representation embodied a powerful new force for industrial nations with imperial designs. The French, Germans, Americans, and British all participated in coordinated global efforts to understand the earth's properties, from its changing magnetic and atmospheric tides and waves to the physical properties of the ocean itself, including temperature, salinity, and depth.

Like Whewell in his work on the tides, researchers in other fields of nautical science began with data confined to the littoral, then tentatively extended their studies to cover the open ocean, a pattern obvious in the Magnetic Crusade, the international quest to discover the laws of the earth's changing magnetic field. Like the study of the tides, determining the earth's magnetism—a project that had begun with Edmond Halley in the seventeenth century—expanded into an international collaboration including Russia, Prussia, France, Britain, and the United States. In

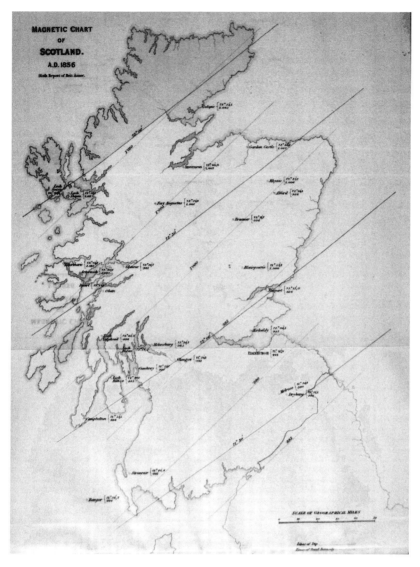

Figure C.4. Sabine's magnetic chart of Scotland. Edward Sabine, "Observations on the Direction and Intensity of the Terrestrial Magnetic Force in Scotland," *British Association Report* 1836 (1837): 97–119.

Britain, after a "false start" before Whewell's international tide experiments, Edward Sabine of the Royal Artillery proved to be the most ardent evangelical magneticist.[28] An Irishman educated at the Royal Military Academy in Woolwich, Sabine joined John Augustus Lloyd and James Clark Ross in studying the direction and intensity of the magnetic force in Ireland using the methods outlined earlier by Humboldt and Carl Friedrich Gauss. Sabine then continued the work to form a chart of the magnitude, intensity, and direction of the magnetic force throughout the entire British Isles.

Like Whewell in his earlier work on the tides, Sabine used his initial studies of terrestrial magnetism around Britain as a template for a magnetic chart of the entire globe. Also like Whewell's work, this entailed a substantial change in the funding and implementation of the research. Sabine suggested that the British Association ask the government to establish fixed magnetic observatories in all its possessions, thus adding to the international cooperation in the field begun earlier in Prussia, Russia, and France. But he also suggested that determining the position of the South Magnetic Pole would require an Antarctic voyage by a ship serving as a traveling observatory. According to Sabine, only through a large number of fixed observatories funded by the government, along with a roaming magnetic observatory traveling the southern seas, could a correct map be produced to help mariners determine the changing magnetic field over the world's oceans.

At the 1838 meeting of the British Association, Whewell brought the subject of an Antarctic expedition and fixed magnetic observatories to the attention of the committee in charge of the mathematical-physical sciences (Section A of the Association), and John Herschel, recently returned from the Cape of Good Hope and now serving as president of Section A, brought the petition before government.[29] With the full support of the British Association, the Royal Society, and the Hydrographic Office, the magnetic lobby convinced government to appropriate £100,000 for fixed colonial observatories and to have officers of the Royal Engineers collect the data. Permanent magnetic stations were established in England, Ireland, Canada, New Zealand, and Africa and on the islands of St. Helena and Jamaica. The East India Company added observatories in India and Singapore, while Russia maintained stations in its own territory, including eleven in Siberia. In the end, over forty permanent observatories registered the earth's magnetism from all over the globe, truly a worldwide survey and one that far surpassed any previous study. The effort included a mobile magnetic observatory under the command of James Clark Ross from 1839 to 1843. And of course, it culminated in isomagnetic charts of the world's oceans. The analysis of previous large-scale tidal research raises intriguing questions about the internal workings of the Magnetic Crusade. The massive amount of computing needed to reduce data on such a grand scale, taken along with other global projects, spotlighted the role of the human computer and brought additional funding. Moreover, the global practice of terrestrial magnetism, like tidology, helped distinguish the scientists (such as Herschel) from the associate laborers (such as the Royal Engineers) within an international and government-funded venture with strong ties to Britain's military and commercial interests.

With some success in both tidology and terrestrial magnetism, scientists hoped a similar global effort could advance the fledgling science of meteorology. Meteorology had long required sustained cooperation from people living in dispersed areas and seemed especially suited to gain from a systematized and coordinated effort.[30] Individuals throughout Great Britain had taken measurements of rainfall and barometric pressure for some time, while government-sponsored voyages often included similar assignments for naturalists and captains on board British vessels. The rooms of the Royal Society and Hydrographic Office were overflowing with meteorological registers, as were the desk drawers of many of the local gentry. Before the nineteenth century, however, these observations were scattered and useless. The domestication of the ocean during the first few decades of the century offered scientists new hope. "There is no branch of physical science," John Herschel exclaimed, "which can be advanced more materially by observations made during sea voyages than meteorology."[31] Owing to the homogeneous nature of the ocean's surface as well as its relatively more stable temperature, observations made at sea had far less disturbing influences than those made on land. The area of the sea also far exceeded that of the land, and thus "a much wider field of observation is laid open, calculated thereby to offer a far more extensive basis for the deduction of general conclusions." The sea's accessibility, geographic extent, and natural uniformity made it ideal for global investigations.

Aided by a grant from the British Association, Herschel worked closely with William Radcliff Birt to reduce the barometric and weather observations reaching Britain from around the world. They studied more than three hundred sets of observations made between 1835 and 1838 to trace the magnitude and direction of what they termed "atmospheric waves."[32] And of course they created isothermal charts of the world's oceans and continents. Further advances were made not in Britain, but in the large landmass of the United States, where American researchers finally surpassed

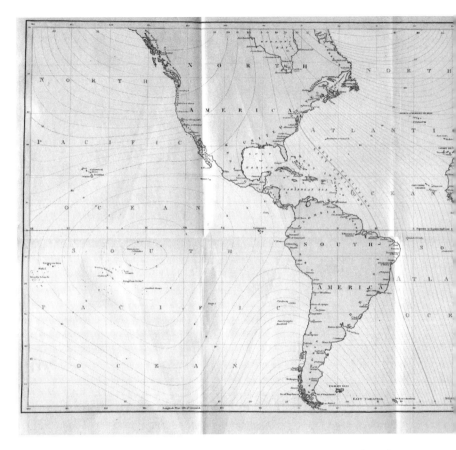

Figure C.5. Maps depicting magnetic lines of equal variation were all the vogue in the early Victorian era. Reproduced here is a map accompanying a paper by Peter Barlow in the *Philosophical Transactions* for 1833, the same volume in which Whewell's first isotidal map of the world appeared. Peter Barlow, "On the Present Situation of the Magnetic Lines of Equal Variation, and Their Changes on the Terrestrial Surface," *Philosophical Transactions* 123 (1833): 667–73.

their British counterparts. Joseph Henry at the Smithsonian Institution, Alexander Dallas Bache of the U.S. Coast Survey, and Matthew Fontaine Maury of the Naval Observatory and the U.S. Hydrographic Office all viewed meteorology as a science the United States could advance.[33]

Both Henry and Bache set up large networks of meteorological observations in the United States, coordinating hundreds of volunteers to observe the changing weather as it moved across the continent. But it was Matthew Fontaine Maury who became the international spokesman for meteorological research. Appointed superintendent of the Depot of

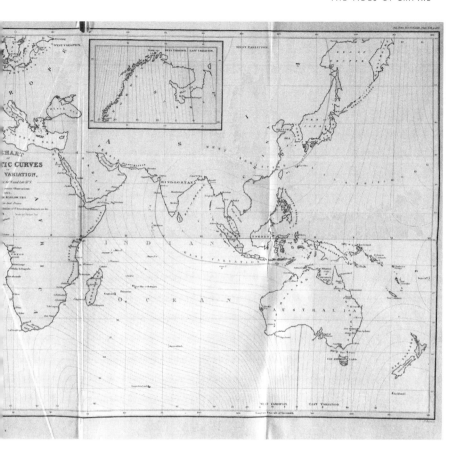

Charts and Instruments for the Navy Department in 1842, Maury assembled meteorological reports he found in the Depot's archives in Washington, DC. From these logs he produced global charts of wind and weather currents throughout the oceans, which he then augmented as new logs and new observations were submitted to his office. His surface temperature and wind charts attracted international attention, and by the time he had published his influential *Physical Geography of the Seas*, a textbook that synthesized his work on the oceans, his system had been adopted worldwide.

From the beginning, Maury viewed meteorology as an international project requiring coordinated readings. He proposed sending blank forms to all surveying vessels, a plan soon adopted by most European maritime nations, and he also suggested holding an international conference to establish a worldwide system covering both land and sea. The conference took place in Brussels in 1853, and Maury was appointed the United States representative to discuss his international system of data collection and reduction. One result of the Brussels conference was the founding of the British Meteorological Department the next year to coordinate meteorological registers, especially at sea. Robert Fitzroy was elected its first director. Fitzroy joined the navy at age fourteen and sailed on the first voyage of HMS *Beagle* at age twenty-three. He became famous for the *Beagle's* second voyage, which circumnavigated the globe, carrying the aspiring young naturalist Charles Darwin. Fitzroy, however, was a respected natural philosopher in his own right, a fellow of numerous scientific societies, including the prestigious Royal Society of London. As head of the British Meteorological Department, he outlined a comprehensive plan of data collection that focused on the global nature of meteorological phenomena and the importance of standardized instruments. He allocated instruments to the navy and the merchant marine, along with detailed instructions on how to measure temperature, barometric pressure, wind, rain, humidity, and other meteorological phenomena, including auroras and magnetic storms. Before his tragic death at his own hands, he spurred the growth of European global meteorology.[34]

As the head of the Meteorological Department, Fitzroy was determined to represent the world's weather in graphical form. As Duncan Agnew has recently revealed, Fitzroy's adaptation of Maury's earlier charts produced the first maps on which the "Marsden squares" appeared, a graphical method of organizing meteorological data by breaking up the ocean into ten-degree squares of latitude and longitude.[35] As Fitzroy viewed it, "the locality of each frequented square may soon become fixed in the mind of the navigator, and serve (like provinces on land) to recall spaces to the mind, rather than points indicated only by latitude and longitude."[36] Fitzroy was intent on transforming the boundless ocean into tidy squares, easily referred to on charts and recalled as "spaces to the mind."

Besides lines of latitude and longitude, by the mid-nineteenth century the world's oceans had been zoned, squared, graphed, and charted into rational and useful categories that proved especially appealing to the Admiralty as it attempted to extend its operations throughout the globe. The quest for general laws went hand-in-hand with the scientist's use of visual representations to represent ocean space. Graphs and charts depicting the distribution of winds, tides, currents, magnetic variation, and barometric pressure became the preferred means for representing the myriad forces at work. Their power to create a synoptic view of a largely unexplored area made the scientists' work exceptionally useful to the expansionist program of maritime nations.

The renewed assault on the earth's magnetism and meteorology produced a flurry of activity in the 1840s, including the appointment of a committee "for conducting the cooperation of the British Association in the system of simultaneous Magnetical and Meteorological Observations."[37] The actual magnetic and meteorological data, however, like Whewell's earlier tidal data, were primarily confined to the littoral zone. Scientists then uneasily expanded their analysis outward over the ocean. "Sea as well as land observations are, however, equally required," Herschel urged in his opening address to the British Association meeting held in Cambridge in 1845.[38] Herschel noted that Ross's expedition on the *Erebus* and the *Terror*, the culmination of the Magnetic Crusade, brought an added dimension to the study of the physical characteristics of the sea. "A ship is an itinerant observatory," Herschel argued, "and, in spite of its instability, one which enjoys several eminent advantages—in the uniform level and nature of the surface, which eliminate a multitude of causes of disturbance and uncertainty, to which land observations are liable." Ross's voyage harked back to Halley's earlier magnetic survey of the Atlantic in its attempt to search out the magnetic variation over an entire ocean. After midcentury, itinerant laboratories fanned out from the trading nations of the world to take soundings of the ocean floor, gather biological specimens, measure temperature and salinity, amass measurements of the tides and currents on the ocean's surface, and account for the tides and waves in the atmosphere above.[39]

Ross's four-year voyage signaled an important shift in the study of the world's oceans. Rather than his traveling *through* the ocean, for Ross the ocean *itself* became the main object of study, a laboratory where scientists could search out physical laws and overarching patterns. This was not lost on Whewell. While helping to lobby for funds in 1838 for Ross's Antarctic voyage, he had begun thinking about a tide expedition, and he continued throughout the 1840s and early 1850s to pursue what he called a "Great Atlantic Tide Expedition" to search out the tides in the uncharted sea. Together with James Clark Ross, he presented a paper on the subject to the British Association meeting held in Oxford in 1847.[40] "The knowledge which we possess of the tides," they began their report, "looking at the connexion of the phaenomenon over the whole surface of the ocean, is extremely imperfect at present, and not at all likely to be completed in any material degree in any finite time, by the observations which voyages mainly directed to other objects will supply." Tidology required an expedition specifically intended to search out the tides, and they suggested that the British Association press government for just such a voyage.

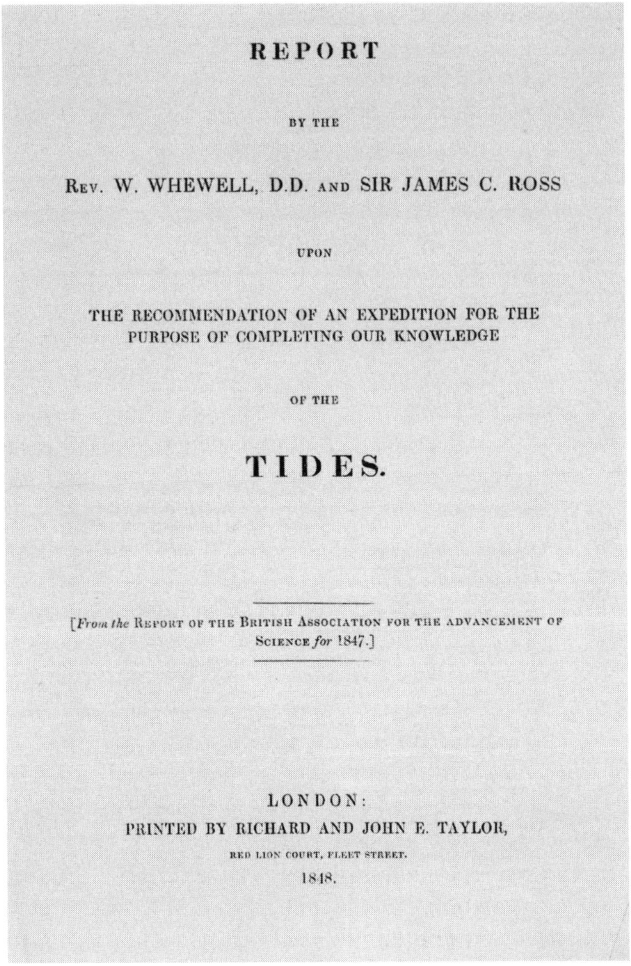

REPORT

BY THE

REV. W. WHEWELL, D.D. AND SIR JAMES C. ROSS

UPON

THE RECOMMENDATION OF AN EXPEDITION FOR THE
PURPOSE OF COMPLETING OUR KNOWLEDGE

OF THE

# TIDES.

[*From the* REPORT OF THE BRITISH ASSOCIATION FOR THE ADVANCEMENT OF
SCIENCE *for* 1847.]

LONDON:
PRINTED BY RICHARD AND JOHN E. TAYLOR,
RED LION COURT, FLEET STREET.
1848.

Figure C.6. William Whewell and James Clark Ross's printed circular
recommending an expedition for searching out the tides of the deep ocean.
Reproduced from the Whewell Papers, R.6.20/5.

Whewell's last publication on his own research in tidology, subtly
entitled "On Our Ignorance of the Tides," was far from an admission of
failure.[41] Rather, Whewell aimed to convince the Admiralty of the need
for his Atlantic tidal expedition. The "ignorance" arose from a lack of
knowledge of the tides in the large ocean basins. "The border of our knowledge," according to Whewell, "ends with the shores of Europe," and
only by means of an expedition would Britain be able to extend that

knowledge outward. "By means of given connected observations (the *connexion* in the mode of making the observations is the essential condition of success), the course of the littoral tide-wave on all the extensive shores of the ocean might be determined, and at the islands; and these results being obtained, the motion of the tidal movements of the ocean on the larger scale would probably be brought into view."[42] Whewell wanted to literally connect the observations by means of a voyage to determine the "movements of the ocean on a larger scale."

Despite the efforts of Whewell, Ross, and Beaufort, the tide expedition never set sail, one of the very few enterprises backed by the hydrographer to the Admiralty that failed to be funded.[43] "I still want a Tidal expedition," Whewell wrote to James David Forbes as late as October 1856, still hoping the Admiralty's interest in the tides would end with an expedition "to *hunt* them."[44] The men behind the proposed tide expedition constituted a powerful force, consisting of the master of Trinity College, Cambridge (Whewell), the hydrographer to the Admiralty (Beaufort), the nation's most revered Arctic explorer (Ross), its major magneticist (Sabine), and one of its top surveyors (Fitzroy). Even in failure, the expedition embodied the renewed effort to understand the physics of the world's oceans, mirroring Ross's earlier voyage to the Antarctic in 1839 and foreshadowing the equally successful voyage of the *Challenger* in 1872. As the use of the ocean intensified after midcentury, these voyages increasingly focused on the ocean itself.

From its roots in similar geophysical initiatives that required massive amounts of data covering large areas of the globe, ocean science was from the beginning expensive and expansive.[45] The financial and practical difficulties of studying the sea made it entirely dependent on financial support from national governments. The new science was on such a massive scale that not even an extremely rich maritime nation like Britain could afford to develop it alone. By midcentury, the study of the ocean had become one of the most fertile areas of investigation, funded by the state and incorporating an international cast of scientists. These early multinational efforts in the geophysical sciences focused the nascent scientific community on shared research topics, helped to strengthen the relationship between national governments and natural philosophers, and led to the standardization of both methods and instruments.

Lubbock, Airy, and Whewell in tides, Sabine, Lloyd, and Ross in magnetism, Herschel and Fitzroy in meteorology—virtually all the major British scientists in this period worked at some point in their careers on projects that demanded observations from around the globe, research that was directly funded, if not actually undertaken, by government. Researchers like Herschel and Whewell used these global projects to advance physical astronomy and to assert their own authority within the scientific community. And though they questioned government's role in directly funding science, they seemed more than willing to accept money to extend the reach of science beyond Britain. Indeed, they expected further support for reducing the data and complained when they failed to receive it.

Empire thus subtly transformed science: in the research pursued, the questions asked, and the theories adopted. The close union between scientists and the British government was exceedingly fruitful for certain kinds of scientific research, especially in natural history and geology, both highly dependent at that time on observations from terrestrial possessions overseas.[46] Extending the reconnaissance of the great age of discovery in the late eighteenth century, aspiring naturalists moved ever inland, following Humboldt's insistence on the need to explore large landmasses. Darwin always referred to the voyage of the *Beagle* as the single most important event in his life as a naturalist, not because he traveled the oceans, but because he witnessed the extravagant abundance of nature and the awesome variety of its flora and fauna on landmasses separated *by* the oceans. Darwin traveled hundreds of miles into Patagonia, Joseph Dalton Hooker scaled the glaciers of the Himalayas, Thomas Henry Huxley explored the interior of New Guinea, and Alfred Russel Wallace boldly followed the course of the Amazon. Their descriptions of the world's diversity, and the collections they brought home to the European botanical gardens and museums, forever transformed Europeans' conception of their own place in nature.[47]

By midcentury, physical scientists had extended the naturalists' approach to cover the world's oceans as well, incorporating both a vertical and a horizontal viewpoint into their study of the sea. Their gains, however, were far more modest. In the study of the tides, by midcentury the Hydrographic Office produced tide tables for over one hundred ports throughout the world (compared with none in 1830). Lubbock and Whewell also made additions to tidal theory, including the empirical laws for the effects of the parallax and declination of the sun and moon and the importance of the diurnal inequality. Yet their insistence on comparing their results with the Newton-Bernoulli equilibrium theory, a decision conditioned by Britain's imperial reach, hampered their ultimate success. The way Whewell approached the study of the tides—a decidedly spatial approach, extending from the coasts of Britain outward to the oceans—came from a detailed analysis of the history of the relationship between science and the state. In his *History*, he outlined how questions in physical astronomy, such as the shape of the earth, had been solved in the eighteenth century only through the "liberal encouragement" of European governments willing to invest in the systematic coordination of large-scale global observations.[48] Only after discussing similar problems in the history of astronomy did Whewell introduce his own research in tidology. "We come, finally, to that result, in which most remains to be done for the verification of the general law of attraction— the subject of the Tides."[49]

For Whewell, the tides remained a question that could be answered only through similar "liberal encouragement" from governments, which had the power and finances to institute global observations. In the end, however, tide prediction is based on the hydrodynamics of waves on local coasts and estuaries. Though the initial forces that produce the tides are celestial, their rise and fall on particular coasts depends more on the shape of the ocean basin, the slope of the coast, the direction of the tidal stream versus the prevailing winds, and the width and depth of the estuary or channel. Local topography rather than celestial dynamics ultimately rules the tides. This was glaringly obvious on the coasts of England and Scotland, where the same astronomical forces operating over the same latitude and longitude produce decidedly different tides.

Whewell therefore had to *make* the tides global, a process that affected his choices of theories and methods of observations. That is, in Whewell's work, one can pinpoint exactly how empire shrewdly shaped the practice of science. He practiced spatial science because the British Empire was not good at solving local problems. It excelled at the global level. With the ability to set up permanent stations around the globe equipped with the latest instruments to measure and ultimately compare observations on a worldwide scale, Whewell could use empire to both test the equilibrium theory and extend the reach of science. But he could not use empire to solve the recalcitrant differential equations required for a hydrodynamic approach, which merely demonstrated his—and the scientists'—limitations. The British Admiralty was the department within the state that spent the most money.[50] It was also the largest patron of science in Britain in the first half of the nineteenth century. And of course it was the institution that projected British power across space. This proved a powerful stimulus for science, which also was aspiring to a global reach. When studying the relationship between science and empire in the Victorian era, the important questions are cultural, not scientific. The science that scientists produced over the world's oceans reflected the culture of empire: it literally and graphically followed the flag.

## Liquid Imperialism

Scientific interest in the ocean followed directly from the growing popular fascination in the late eighteenth century with the natural history and physical topography of the seashore. Beginning in the intertidal zone and then widening their gaze, by the mid-nineteenth century scientists had rendered the oceans not only comprehensible and controllable but also usable for commerce and imperialism. Historians of science have explored how science and technology extended the reach of imperial powers across the globe, facilitating the exploitation of distant natural resources. Yet questions of science and empire have unfortunately continued to focus on land, with the ocean invisibly linking peripheral colonies to the metropolis. This book has attempted to counteract that terrestrial bias. The scientists' three-dimensional efforts to comprehend the overarching physics of the ocean formed part of a broader effort to extend cultural and political control to the far regions of the globe.

It is no coincidence that since the sixteenth century, the world's most powerful countries have been sea powers.[51] Nor is it a coincidence that the spatial turn in science occurred simultaneously with Britain's move toward imperial expansion. Leadership of the modern world required access to, understanding of, and control of the world's maritime frontier, a "liquid imperialism" that heightened the status of science and the new scientist.[52] The ocean became known in the nineteenth century—constructed, mapped, outlined, and defined—largely through science, and by the end of the century its practitioners were charged with the responsibility of interpreting and defining the physical properties of the sea. This stewardship has had dramatic consequences for the geopolitics of nations and for the ocean itself.

Through the marriage of science and navigation, Britain began to control the waves and many of the world's colonial possessions. The British used their knowledge and control to ensure an open system of trade and commerce. They did not claim the sea by fiat as they increasingly claimed land or demand legal sovereignty over it. Rather, the ocean was owned by those who could master its tides and waves, deal with its magnetic and atmospheric undulations, and traverse its waters unimpeded. The ability to define the ocean—to outline its boundaries and contours through isomaps of all types—gave the British dominion over the last part of the natural world still left unclaimed. Whewell and others inscribed the ocean with order and meaning, translated it into easily transferable visuals, then offered the products to eager national governments. The most potent force to emerge from this endeavor was the central position of science in translating the unknown for the good of the state. The intellectual mastery such knowledge conferred led to actual physical mastery of the coastlines, bays, and estuaries on the ocean's outer rim.

Vessels searching the oceans to determine the tides, trade winds, and changing magnetic fields combined to form a powerful statement of British hegemony. By the end of Whewell's work on the tides, Britain had the world's largest naval and commercial fleets and almost complete mastery of the seas. Science, in turn, expanded from a relatively isolated undertaking relying on small amounts of incremental funding to large-scale research involving teams of scientists in different specialties working together with increasingly larger budgets and standardized instruments. Victorians reconceptualized the oceans through the practice of science and thus carved out a powerful intellectual space for the modern scientist. Modern conceptions of the scientist are thereby inextricably linked to the politics of imperialism and the economics of worldwide trade. As Raleigh long ago surmised, control of the ocean translated directly into control of the world's riches and hence the world itself.

# Glossary: Terms Applied to Tidal Theory in the Early Victorian Era

AGE OF THE TIDE: Because most ports are on channels, estuaries, and inland rivers, the time of the tide does not correspond directly to the tide-generating forces of the sun and moon at the port, but corresponds to tide-generating forces somewhere in the ocean, which are then conveyed up the estuaries and rivers. The age of the tide is the delay in the highest tides of the month after the corresponding full or new moon.

CORRECTED ESTABLISHMENT: The vulgar establishment corrected for the age of the tide and the semimenstrual inequality.

COTIDAL (OR ISOTIDAL) LINES: Lines connecting places on the coasts and in the ocean that have high tide at the same time.

DISCUSSION OF DATA: The process whereby calculators formed tables and graphs comparing tidal observations with other variables, such as the position of the sun and moon, to find lawlike relationships.

DIURNAL INEQUALITY: The difference in high water and low water heights and times between two consecutive tides of the same day.

EBB TIDE: The period between high water and the following low water. Often used as a synonym for low tide.

EPHEMERIS (PL. EPHEMERIDES): A table showing the assigned places of celestial bodies at regular intervals.

EPOCH: An arbitrary interval of time used by calculators to match observations to theory. For instance, calculators would tabulate tidal observations and compare them with the position of the sun and moon at thirty-minute intervals, or epochs.

ESTABLISHMENT: Term often used as a synonym for "vulgar establishment," the time of high water on the day of the new or full moon.

FLOOD TIDE: The period between low water and the following high water. Often used as a synonym for high tide.

LUNAR CYCLE: The time needed for the moon to go through its full cycle with respect to the sun and earth and return on the same days of the month– 18.6 years. Thus tide tables based purely on observational data must include almost nineteen years of tide observations to correct for the anomalies associated with the positions of the moon, sun, and earth.

LUNAR DECLINATION: The angle between the sun and moon, between conjunction and quadrature. The tides are highest when the moon is either in conjunction or in opposition with the sun and lowest when the moon is in quadrature.

LUNAR PARALLAX: The effect on the tides caused because the moon does not travel around the earth in an exact circle but is sometimes closer and sometimes farther away. Also termed the lunar parallax correction.

LUNAR TRANSIT: The passage of the moon across the meridian.

LUNITIDAL INTERVAL: The difference in time between high tide and the previous lunar transit.

MOON'S AGE: The number of days that have elapsed since the moon's southing or the time when it passed the meridian.

REDUCTION OF DATA: The process whereby calculators reduced observations to a standard so that they could be compared. For instance, calculators reduced tidal observations to a standard zero, taking into account the position of the tide gauge and the geography of the port.

SEMIMENSTRUAL INEQUALITY: The inequality affecting the tides owing to lunar declination. It depends on the moon's distance from the sun, and it goes through its period twice in one month.

VULGAR ESTABLISHMENT: The time of high water on the day of the new or full moon; the period by which the time of high water follows the moon's transit or meridian passage on the day of the new and full moon. The time of high water is nearly the same time on all days when the moon is new or full.

# Notes

INTRODUCTION: THE LITTORAL IN SCIENCE AND HISTORY

1.  Rachel Carson, *The Sea Around Us* (New York: Oxford University Press, 1989), 97–98.
2.  See Stuart Gilbert and Ray Horner, *The Thames Barrier* (London: Thomas Telford, 1984); Jonathan Schneer, *The Thames* (New Haven, CT: Yale University Press, 2005), 249.
3.  Anthony Dale, *Fashionable Brighton, 1820–1860* (London: Country Life Limited, 1947), 13–15.
4.  Alain Corbin, *The Lure of the Sea: The Discovery of the Seaside in the Western World: 1750–1840*, trans. Jocelyn Phelps (Berkeley and Los Angeles: University of California Press, 1994), 61–69, 96.
5.  Helen M. Rozwadowski, *Fathoming the Ocean: The Discovery and Exploration of the Deep Sea* (Cambridge, MA: Harvard University Press, 2005); David Elliston Allen, *The Naturalist in Britain: A Social History* (Princeton, NJ: Princeton University Press, 1994), especially chap. 11.
6.  The friction produced by the oceanic tides slows the rotation of the earth and lengthens our day. Eventually the earth will no longer turn on its axis.
7.  Charles Lyell, *Principles of Geology,* vol. 1 (Chicago: University of Chicago Press, 1990), 257ff.; J. Hardisty, *The British Seas: An Introduction to the Oceanography and Resources of the North West European Continental Shelf* (London: Routledge, 1990), 95.
8.  Rick Szostak, *The Role of Transportation in the Industrial Revolution: A Comparison of England and France* (Montreal: McGill-Queen's University Press, 1991), 55, 59. See also T. S. Willan, *The English Coasting Trade, 1600–1750* (Manchester: Manchester University Press, 1938), xii.

9. Frederick Wallace Morgan, *Ports and Harbours* (London: Hutchinson's University Library, 1952), 45.

10. Memorandum, "Report Relative to Grant on Harbour, 1834," Admiralty Records, Public Records Office, Kew, England, 1.4862 (hereafter cited as PRO, ADM).

11. James Anderson, "Some Observations on the Peculiarity of the Tides between Fairleigh and the North Foreland," *Philosophical Transactions* 109 (1819): 217.

12. For Humboldt's role in the geographical turn in science, see David N. Livingstone, *The Geographical Tradition: Episodes in the History of a Contested Enterprise* (Oxford: Blackwell, 1992).

13. For a history of tidal theory, see David Edgar Cartwright, *Tides: A Scientific History* (Cambridge: Cambridge University Press, 1999), and Margaret Deacon, *Scientists and the Sea, 1650–1900* (London: Academic Press, 1971).

14. For a discussion of the "silent majority," see the introduction to *Cultures of Natural History*, ed. N. Jardine, J. A. Secord, and E. C. Spary (Cambridge: Cambridge University Press, 1996). Also see the contributions in *Science in Victorian Context*, ed. Bernard Lightman (Chicago: University of Chicago Press, 1997).

15. Edward Said, *Culture and Imperialism* (New York: Vintage, 1993); Matthew H. Edney, *Mapping an Empire: The Geographical Construction of British India, 1765–1843* (Chicago: University of Chicago Press, 1997).

16. This point is ably argued in Philip E. Steinberg, *The Social Construction of the Ocean* (Cambridge: Cambridge University Press, 2001).

17. John D. Milliman and Bilal U. Haq, eds., *Sea Level Rise and Coastal Subsidence: Causes, Consequences, and Strategies* (Dordrecht: Kluwer, 1996), 15.

18. John Brewer, *The Sinews of Power: War, Money and the English State, 1688–1783* (Cambridge, MA: Harvard University Press, 1988), xvi.

19. See for instance, M. Campbell-Kelly et al., eds., *The History of Mathematical Tables: From Sumer to Spreadsheets* (Oxford: Oxford University Press, 2003); and David Alan Grier, *When Computers Were Human* (Princeton, NJ: Princeton University Press, 2005).

20. James Scott, *Seeing Like a State: How Certain Schemes to Improve the Human Condition Have Failed* (New Haven, CT: Yale University Press, 1999).

21. For recent accounts, see especially Richard Yeo, *Defining Science: William Whewell, Natural Knowledge, and Public Debate in Early Victorian Britain* (Cambridge: Cambridge University Press, 1993); Menachem Fisch, *William Whewell, Philosopher of Science* (Oxford: Clarendon Press, 1991); and Menachem Fisch and Simon Schaffer, eds., *William Whewell: A Composite Portrait* (Oxford: Clarendon Press, 1991).

22. Margaret Deacon has devoted most of chapter 12 in her *Scientists and the Sea* to Whewell's tidal studies, linking his work with other maritime investigations of the mid-nineteenth century. Isaac Todhunter offers detailed

summaries of Whewell's tidal research in *William Whewell, D.D., Master of Trinity College Cambridge*, vol. 2 (New York: Johnson Reprint, 1970). Several of the entries in Fisch and Schaffer, eds., *William Whewell: A Composite Portrait,* mention Whewell's tidal studies. See especially Michael Ruse, "William Whewell: Omniscientist," 87–116. See also Paul Hughes, "A Study in the Development of Primitive and Modern Tidetables" (PhD diss., John Moores University, Liverpool, 2005).

23. David Philip Miller, "The Revival of the Physical Sciences in Britain, 1815–1840," *Osiris* 2 (1986): 124.

24. See, for instance, *Journal of Historical Geography* 32, no. 3 (July 2006), and *American Historical Review* 111, no. 3 (June 2006). See also W. Jeffrey Bolster, "Opportunities in Marine Environmental History," *Environmental History* 11, no. 3 (July 2006): 567–97; John McNeill, "The Nature and Culture of Environmental History," *History and Theory* 42 (2003): 5–43; Bernhard Klein and Gesa Mackenthun, eds., *Sea Changes: Historicizing the Ocean* (New York: Routledge, 2003); and Helen Rozwadowski, *Fathoming the Ocean: The Discovery and Exploration of the Deep Sea* (Cambridge, MA: Harvard University Press, 2005).

25. For the spatial turn in history, see Henri Lefebvre, *The Production of Space* (Cambridge: Blackwell, 1991), and David Harvey, *Justice, Nature and the Geography of Difference* (Cambridge: Blackwell, 1996). Work on spatial versus temporal history is growing. See David N. Livingstone, *Putting Science in Its Place: Geographies of Scientific Knowledge* (Chicago: University of Chicago Press, 2003); David N. Livingstone, "The Spaces of Knowledge: Contributions toward a Historical Geography of Science," *Environment and Planning D: Society and Space* 13 (1995): 5–34; Martin W. Lewis and Karen E. Wigen, *The Myth of Continents: Critique of Metageography* (Berkeley and Los Angeles: University of California Press, 2000); and Robert E. Kohler, *All Creatures: Naturalists, Collectors, and Biodiversity, 1850–1950* (Princeton, NJ: Princeton University Press, 2006).

CHAPTER ONE: PHILOSOPHERS, MARINERS, TIDES

1. Alain Corbin, *The Lure of the Sea: The Discovery of the Seaside in the Western World: 1750–1840*, trans. Jocelyn Phelps (Berkeley and Los Angeles: University of California Press, 1994), 2.

2. For the role of the recovery narrative in Western thought, see Carolyn Merchant, "Reinventing Eden: Western Culture as a Recovery Narrative," in *Uncommon Ground: Rethinking the Human Place in Nature,* ed. William Cronon (New York: W. W. Norton, 1996), 132–70.

3. William Shakespeare, *Richard II*, act 2, scene 1.

4. Hydrographic offices were founded in France in 1720 and Denmark in 1784. See Archibald Day, *The Admiralty Hydrographic Service (1785–1919)* (London: Her Majesty's Stationery Office, 1967), 12, 20.

5.  For a discussion of tidal science in the sixteenth and seventeenth centuries, see E. J. Aiton, "Galileo's Theory of the Tides," *Annals of Science* 10 (1954): 44–57, "Descartes' Theory of the Tides," *Annals of Science*, 11 (1955): 337–48, and "The Contributions of Newton, Bernoulli and Euler to the Theory of the Tides," *Annals of Science* 11 (1955): 206–23; Federico Bonelli and Lucio Russo, "The Origin of Modern Astronomical Theories of Tides: Chrisogono, de Dominis and Their Sources," *British Journal for the History of Science* 29 (1996): 385–401; Deacon, *Scientists and the Sea*.

6.  Robert Moray, "A Relation of Some Extraordinary Tydes in the West-Isles of Scotland," *Philosophical Transactions* 1 (1665): 53–55. Oldenburg had by this time received numerous letters on the tides meant for publication. See Henry Oldenburg, *Correspondence*, vol. 3, ed. and trans. A. Rupert Hall and Marie Boas Hall (Madison: University of Wisconsin Press, 1966), no. 510, 94; no. 514, 105–6; no. 518, 116–17; and no. 525, 131. Moray's interest in the tides dates back to the 1650s. See Margaret Deacon, *Scientists and the Sea* (London: Academic Press, 1971), chap. 4.

7.  John Wallis, "An Essay of Dr. John Wallis, Exhibiting His Hypothesis about the Flux and Reflux of the Sea," *Philosophical Transactions* 1 (1666): 263–81.

8.  John Wallis, "Some Inquiries and Directions concerning Tides, Proposed by Dr. Wallis, for the Proving or Disproving of His Lately Publish't Discourse concerning Them," *Philosophical Transactions* 1 (1666): 297–98.

9.  Robert Moray, "Considerations and Enquiries concerning Tides, by Sir Robert Moray; Likewise for a Further Search into Dr. Wallis's Newly Publish't Hypothesis," *Philosophical Transactions* 1 (1666): 298–301; Robert Moray, "Patterns of the Tables Proposed to Be Made for Observing of Tides," *Philosophical Transactions* 1 (1666): 311.

10. Samuel Colepresse, "An Account of Some Observations Made by Mr. Samuel Colepresse at and nigh Plimouth, Anno 1667, by Way of Answer to Some of the Quaeries concerning Tydes," *Philosophical Transactions* 3 (1668): 632–34; Joseph Childrey, "Animadversions on Wallis's Hypothesis about the Flux and Reflux of the Sea," *Philosophical Transactions* 5 (1670): 2061–68; [Henry Oldenburg], "An Account of Several Engagements for Observing of Tydes," *Philosophical Transactions* 2 (1666–67): 378–79; David Edgar Cartwright, *Tides: A Scientific History* (Cambridge: Cambridge University Press, 1999), 52–56.

11. Henry Philips, "A Letter Written to Dr. John Wallis by Mr. Henry Philips, Containing His Observations about the True Time of the Tides," *Philosophical Transactions* 3 (1668): 656.

12. Edmond Halley, "An Account of the Course of the Tides at Tonqueen in a Letter from Mr. Francis Davenport, July 15, 1678, with the Theory of Them at the Barr of Tonqueen by the Learned Edmond Halley," *Philosophical Transactions* 14 (1684): 677–88.

13. Mr. Flamstead [John Flamsteed], "A Correct Tide Table, Shewing the True Times of the High-Waters at London-Bridge, to Every Day in the Year 1683," *Philosophical Transactions* 13 (1683): 12.
14. For the dispute between Newton, Halley, and Flamsteed, see Richard Westfall, *Never at Rest: A Biography of Isaac Newton* (Cambridge: Cambridge University Press, 1980), 541–50, 655–67, 686–97; Victor E. Thoren, "John Flamsteed," in *Dictionary of Scientific Biography,* ed. Charles Gillispie (New York: Charles Scribner's Sons, 1976), 5:22–26; Cartwright, *Tides: A Scientific History,* 55–57.
15. Westfall, *Never at Rest,* chap. 10, esp. 404ff.
16. For Halley's work on the tides, especially the voyages on the *Paramore,* I relied heavily on Alan H. Cook, *Edmond Halley: Charting the Heavens and the Seas* (New York: Clarendon, 1998). See also Norman J. Thrower, ed., *The Three Voyages of Edmond Halley in the "Paramore," 1698–1701* (London: Hakluyt Society, 1981).
17. For a full discussion of Newton's treatment of the tides, see I. Bernard Cohen and Anne Whitman, trans., *Isaac Newton: The "Principia"* (Berkeley and Los Angeles: University of California Press, 1999), esp. 238ff; Deacon, *Scientists and the Sea,* especially chap. 5; and David Edgar Cartwright, *Tides: A Scientific History* (Cambridge: Cambridge University Press, 1999), chaps. 4 and 5.
18. R. S. Westfall, "Newton and the Fudge-Factor," *Science* 179 (1973): 751–58.
19. For a discussion of Newton's reasoning, see N. Kollerstrom, "Newton's Two 'Moon-Tests,'" *British Journal for the History of Science* 24 (1991): 369–72.
20. Edmond Halley, "The True Theory of the Tides, Extracted from That Admired Treatise of Mr. Isaac Newton," *Philosophical Transactions* 19 (1696): 445–57.
21. Edmond Halley, "Discourse Tending to Prove at What Time and Place, Julius Cesar Made His First Descent upon Britain," *Philosophical Transactions* 16 (1686): 499.
22. Roy Porter, *London: A Social History* (Cambridge, MA: Harvard University Press, 1994), 66.
23. Cook, *Halley,* 231.
24. A pink is a ship with a narrow overhanging stern, made to sail swiftly.
25. James Burney, *Chronological History of the Voyages and Discoveries in the South Sea* (London, 1816); *Correspondence and Papers of Edmond Halley,* ed. Eugene Fairfield MacPike (Oxford: Clarendon Press, 1932; reprint 1975), app. 12, 244–47.
26. Halley to Burchett, 23 June 1699, in MacPike, *Correspondence,* 107–8.
27. Halley to Burchett, 29 July 1701, in MacPike, *Correspondence,* 118–19.
28. Cook, *Halley,* 236.
29. Ibid., 290.
30. He published his chart as "A New and Correct Chart of the Channel between England and France with Considerable Improvements Not Extant in Any

Draughts Hitherto Publish'd." See J. Proudman, "Halley's Tidal Chart," *Geographical Journal* 100 (1941): 174–76; Cook, *Halley*, 283.

31. Edmond Halley, "An Historical Account of the Trade Winds, and Monsoons, Observable in the Seas between and near the Tropicks, with an Attempt to Assign the Physical Cause of the Said Winds," *Philosophical Transactions* 16 (1686–92): 153–68.

32. Colin A. Ronan, "Edmund Halley," in *Dictionary of Scientific Biography*, ed. Charles Gillispie (New York: Charles Scribner's Sons, 1976), 14:67–72.

33. Daniel Bernoulli, Leonard Euler, Colin Maclaurin, and Antoine Cavalleri, *Pièces qui ont remporté le prix de L'Académie Royale des Sciences en 1740* (Paris, 1741); Aiton, "Contributions of Newton, Bernoulli and Euler to the Theory of the Tides." See also John Leonard Greenberg, "Interlude II: The Paris Academy's Contest on the Tides (1740)," in his *The Problem of the Earth's Shape from Newton to Clairaut* (Cambridge: Cambridge University Press, 1995).

34. This description was taken largely from John William Lubbock, *Account of the "Traité sur le flux et reflux de la mer," of Daniel Bernoulli* (London, 1830), 5ff.

35. J. W. Norie, *A Complete Epitome of Practical Navigation and Nautical Astronomy, Containing All Necessary Instructions for Keeping a Ship's Reckoning at Sea* (London, 1798), and John Hamilton Moore, *The Practical Navigator* (London, 1778).

36. Norie, *Practical Navigation*, 165. The moon's lunar cycle is almost nineteen years, the time needed for the moon to go through its full cycle with respect to the sun and earth. Mariners referred to it as the golden number. The moon's age is the number of days elapsed since the moon's southing, or the time when it passed the meridian. See the glossary for terms used in nineteenth-century tidology.

37. Norie, *Practical Navigation*, vii.

38. As quoted in Philip Woodworth, "William Hutchinson–Local Hero," *Ocean Challenge* 8, no. 3 (1998): 48.

39. William Hutchinson, *A Treatise on Practical Seamanship* (1777; London: Scholar Press, 1979). The book went through four editions before the turn of the century. Quotations appear on 140, 93, and 99.

40. George Innes to John William Lubbock, 3 March 1832, John William Lubbock Papers, Royal Society Manuscripts, London, England (hereafter cited as Lubbock Papers).

41. The work of Leonard Euler, Daniel Bernoulli, and Pierre Simon Laplace stands out. Many historical works describe their theoretical advances in detail. See especially Cartwright, *Tides*, chap, 5, and Deacon, *Scientists and the Sea*, chap, 5. Deacon admitted that little progress was made in Britain, but she gave no reason.

42. J. R. Rossiter, "The History of Tidal Predictions in the United Kingdom Before the Twentieth Century," *Proceedings of the Royal Society of Edinburgh*, ser. B, 73 (1971–72): 14.

43. Murdoch Mackenzie, "The State of the Tides in Orkney," *Philosophical Transactions* 46 (1749): 149–60, as quoted in Deacon, *Scientists and the Sea*, 252.

44. Patricia Fara has argued a similar point concerning terrestrial magnetism in the eighteenth century. See Patricia Fara, *Sympathetic Attractions: Magnetic Practices, Beliefs and Symbolism in 18th Century England* (Cambridge: Cambridge University Press, 1994), 101ff.

45. Nevil Maskelyne, "Observations of the Tides at the Island of St. Helena," *Philosophical Transactions* 52 (1761): 586–91.

46. Ibid., 590.

47. Henry More, "Observations of the Tides in the Straits of Gibraltar," *Philosophical Transactions* 52 (1761): 447–53.

48. Ibid., 453.

49. James Cook, "An Account of the Flowing of the Tides in the South Seas, Made by Capt. James Cook, at the Request of Mr. Nevil Maskelyne," *Philosophical Transactions* 62 (1772): 357–58.

50. James Cook, "Of the Tides in the South Seas," *Philosophical Transactions* 66 (1776): 447–49.

51. Charles Vallancey to H. Parker, Secretary to the Commissioner of Longitude, 25 October 1789, Board of Longitude Papers, 14.51, Miscellaneous Tide and Trade Winds, Royal Greenwich Observatory Manuscripts, Cambridge University Library, Cambridge, England (hereafter cited as Board of Longitude Papers).

52. Marquis of Buckingham to H. Parker, Secretary to the Commissioner of Longitude, 6 January 1789, Board of Longitude Papers.

53. John Abram to the Right Honorable the Lords Commissioners of the Admiralty, 9 July 1824; Lazarus Cohen to Board of Longitude, 16 May 1822; and Mr. Edward Dean to Board of Longitude, 3 December 1804, all in Board of Longitude Papers.

54. Eliza Maria O'Shea to Board of Longitude, 7 January 1817, Board of Longitude Papers.

55. Eliza Maria O'Shea to Board of Longitude, 1 March 1817, Board of Longitude Papers.

56. Walter Forman, *A New Theory of the Tides Shewing What Is the Immediate Cause of the Phenomenon; and Which Has Hitherto Been Overlooked by Philosophers* (Bath, Eng.: Richard Cruttwell, 1822).

57. Walter Forman to Thomas Young, 20 June 1822, Board of Longitude Papers. One of the more ingenious reasons he gave for the moon's not governing the tides was that the corollary would also be true; that is, the earth would cause tides on the moon's seas. The tides would be so great on the moon that no life could exist, which according to Forman appeared contrary to fact. Forman, *New Theory*, 12.

58. For the advancement of tidal science in France, I have relied heavily on Cartwright, *Tides*, especially chap, 7, and Charles Coulston Gillispie,

*Pierre-Simon Laplace, 1749–1827: A Life in Exact Sciences* (Princeton, NJ: Princeton University Press, 1997).

59. Martha Ornstein, *The Role of Scientific Societies in the Seventeenth Century* (New York: Arno Press Preprint, 1975), 146. See also Roger Hahn, *The Anatomy of a Scientific Institution: The Paris Academy of Sciences, 1666–1803* (Berkeley and Los Angeles: University of California Press, 1971).

60. Cartwright, *Tides*, 59. For a history of the geodetic survey, see Josef W. Konvitz, *Cartography in France, 1660–1848: Science, Engineering, and Statecraft* (Chicago: University of Chicago Press, 1987).

61. Gillispie, *Laplace*, 162.

62. Cartwright, *Tides*, 68.

63. William Whewell, "Essay towards a First Approximation to a Map of Co-tidal Lines," *Philosophical Transactions* 123 (1833): 147.

64. For the changing focus of Laplace's research, and his ultimate fall from power, see Robert Fox, "The Rise and Fall of Laplacian Physics," *Historical Studies in the Physical Sciences* 4 (1974): 89–136.

65. Laplace's difficulties with the mathematics involved led him to characterize the problem of the tides as "la plus épineuse de l'Astronomie Physique (the thorniest in physical astronomy). As quoted by John William Lubbock, "Report on the Tides," *British Association Report* 1832 (1833): 190.

66. As quoted in Michael Shortland, "'On the Connexion of the Physical Sciences': Classification and Organization in Early Nineteenth century Science," *Historia Scientiarum* 41 (1990): 29.

67. See Jed Buchwald, *The Rise of the Wave Theory of Light: Optical Theory and Experiment in the Early Nineteenth Century* (Chicago: Chicago University Press, 1989).

68. Alexander Wood and Frank Oldham, *Thomas Young, Natural Philosopher, 1773–1829* (Cambridge: Cambridge University Press, 1954), 265.

69. The age of the tide is the amount of time that elapses between the moon's southing and high tide.

70. See H. A. Marmer, "On Cotidal Lines," *Geographical Review* 18, no. 1 (1928): 129–43.

71. As quoted in Wood and Oldham, *Thomas Young*, 269.

72. Ibid.; italics added.

73. R. K. Webb, *Modern England: From the Eighteenth Century to the Present,* 2nd ed. (New York: Harper and Row, 1980), 1.

CHAPTER TWO: THE BOUNDED THAMES

1. "Wreck of the *Leeds* Packet from New York," *London Times*, 27 December 1828, 2, col. f.

2. Alison Winter, "'Compasses All Awry': The Iron Ship and the Ambiguities of Cultural Authority in Victorian Britain," *Victorian Studies* 38 (1994): 69–98.

3.  Richard White, "The Nationalization of Nature," *Journal of American History* 86 (1999): 976–86; Sara B. Pritchard, "Reconstructing the Rhone: The Cultural Politics of Nature and Nation in Contemporary France, 1945–1997," *French Historical Studies* 27 (Fall 2004): 765–99; Paul R. Josephson, *Industrialized Nature: Brute Force Technology and the Transformation of the Natural World* (Washington, DC: Shearwater Books, 2002); John Broich, "Engineering and Empire: English Water Supply Systems and Colonial Societies, 1850–1900," *Journal of British Studies* 46 (2007): 346–65.

4.  Dale H. Porter, *The Thames Embankment: Environment, Technology, and Society in Victorian London* (Akron: University of Ohio Press, 1998), 11.

5.  For the relation between environmental change and the practice of science, see Edmund Russell, *War and Nature: Fighting Humans and Insects with Chemicals from World War I to Silent Spring* (Cambridge: Cambridge University Press, 2001); Edmund Russell, "Evolutionary History: Prospectus for a New Field," *Environmental History* 8 (April 2003): 204–28; Robert E. Kohler, "American Museums and Natural History Collecting," paper presented at the History of Science Society annual meeting, Austin, TX, 17 November 2004.

6.  Hugh Clout, ed., *The Times London History Atlas* (London: HarperCollins, 1991), 22; Porter, *Thames Embankment*, 42.

7.  *The Port of London and the Thames Barrage, a Series of Expert Studies and Reports, on the Conditions Prevailing in the Tidal River and Estuary of the Thames, Dealing Especially with Its Geological, Engineering, Navigation...* (London, 1907), 5ff.

8.  Joseph G. Broodbank, *History of the Port of London,* 2 vols. (London: Daniel O'Connor, 1921), 1:29.

9.  George Rennie, "Report on the Progress and Present State of Our Knowledge of Hydraulics as a Branch of Engineering, Part II," *British Association Report* 1833 (1834): 487.

10. Dale H. Porter, *The Thames Embankment: Environment, Technology and Society in Victorian London* (Akron: University of Ohio Press, 1998), 44.

11. Paul Hughes and Alan D. Wall, "The Admiralty Tidal Predictions of 1833: Their Comparison with Contemporary Observations and with a Modern Synthesis," *Journal of Navigation* 57 (2004): 206.

12. Porter, *Thames Embankment*, 45.

13. Clout, *Times London History Atlas*, 81, 162.

14. Porter, *Thames Embankment*, 109.

15. John Armstrong and Philip S. Bagwell, "Coastal Shipping," in *Transport in the Industrial Revolution*, ed. Derek H. Aldcroft and Michael J. Freeman (Manchester, UK: Manchester University Press, 1983), 169.

16. For the importance of the coal trade, see T. S. Willan, *The English Coasting Trade, 1600–1750* (Manchester, UK: Manchester University Press, 1938).

17. Charles P. Kindleberger, *Mariners and Markets* (New York: Harvester, 1992), 15.

18. Willan, *English Coasting Trade*, 31; David J. Starkey, "War and the Market for Seafarers in Britain, 1736–1792," in *Shipping and Trade, 1750–1950: Essays*

*in International Maritime Economic History,* ed. Lewis R. Fischer and Helge W. Nordvik (Pontefract, UK: Lofthouse 1990), 25–42.

19. Alan W. Cafruny, *Ruling the Waves: The Political Economy of International Shipping* (Berkeley and Los Angeles: University of California Press, 1987), 39; also see Judith Blow Williams, *British Commercial Policy and Trade Expansion, 1750–1850* (Oxford: Clarendon Press, 1972).

20. R. J. B. Knight, "The Convulsion of Europe: The Naval Conflict during the Revolutionary and Napoleonic Wars," in *Maritime History,* vol. 2, *The Eighteenth Century and the Classic Age of Sail,* ed. John B. Hattendorf (Malabar, FL: Krieger, 1997), 243–54, esp. 251.

21. Edgar Gold, *Maritime Transport: The Evolution of International Marine Policy and Shipping Law* (Lexington, MA, Lexington Books, 1981), 77; Simon Ville, *English Shipowning during the Industrial Revolution: Michael Henly and Son, London Shipowners, 1770–1830* (Manchester, UK: Manchester University Press, 1987), 12 n. 36.

22. For the convoy system and British shipping interests, see Ville, *English Shipowning.*

23. For a discussion of these different types of thievery, see Roy Porter, *London: A Social History* (Cambridge, MA: Harvard University Press, 1995), 153ff.

24. Broodbank, *History of the Port of London,* 1:83.

25. Porter, *London,* 153.

26. Ibid., 139.

27. Gordon Jackson, "The Ports," in *Transport in the Industrial Revolution,* ed. Derek H. Aldcroft and Michael J. Freeman (Manchester, UK: Manchester University Press, 1983), 203.

28. Knight, "Changing Technologies and Materials," 234.

29. Broodbank, *History of the Port of London,* 1:78.

30. Porter, *London,* 139–40.

31. Broodbank, *History of the Port of London,* 1:156.

32. Porter, *London,* 23.

33. Armstrong and Bagwell, "Coastal Shipping," 165.

34. "London, Floods through the High Tide in the Thames," *London Times,* 27 December 1806, 3, col. a; "London, Floods on Account of the High Tide," *London Times,* 30 December 1814, 2, col. d.

35. "Accidents at the West India Docks, through the High Tide," *London Times,* 15 October 1802, 2, col. b.

36. [Letter to Editor], *London Times,* 2 November 1827, 2, col. f.

37. George Rennie, "Report on the Progress and Present State of Our Knowledge of Hydraulics as a Branch of Engineering, Part II," *British Association Report 1833* (1834): 473.

38. Porter, *Thames Embankment,* 39.

39. Rennie, "Report on Hydraulics" 501.

40. *Port of London and the Thames Barrage,* 30.

41. Thomas Winter, Minutes of Evidence, Sessional Papers, House of Commons, Select Committee on Thames Embankment, 1840 (554), 12.

42. Joseph Robinson, Minutes of Evidence, Sessional Papers, House of Commons, Select Committee on Thames Embankment, 1840 (554), 12:894–95.

43. James Walker, Minutes of Evidence, Sessional Papers, House of Commons, Select Committee on Thames Embankment 1840 (554), 12:279.

44. *The Port of London and the Thames Barrage*, 71.

45. A Sufferer, "To the Editors of the Times," *London Times*, 31 December 1828, 3, col. b.

46. Armstrong and Bagwell, "Coastal Shipping," 168.

47. Patrick K. O'Brien, "Central Government and the Economy, 1688–1815," in *The Economic History of Britain since 1700,* vol. 1, *1700–1815,* ed. R. Floud and D. McCloskey, 2nd ed. (Cambridge: Cambridge University Press, 1994), 205.

48. [Letter to the Editor], *London Times*, 15 November 1828, 3, col. a.

49. "The *British Almanack*," *London Times*, 15 November 1828, 3, col. a. For a discussion of the reading habits of the working class, see R. K. Webb, *The British Working Class Reader, 1790–1848: Literacy and Social Tension* (London: George Allen and Unwin, 1955).

50. Boyd Hilton, *Cash, Corn and Commerce: The Economic Policies of the Tory Governments*, 1815–1830 (Oxford: Oxford University Press, 1977), 20.

51. As quoted in Webb, *British Working Class Reader*, 28.

52. Henry Brougham, *Practical Observations upon the Education of the People, Addressed to the Working Classes and Their Employers* (London, 1825). The treatise, dedicated to George Birkbeck, president of the London Mechanics' Institute, was published separately as a pamphlet that went through many editions.

53. Steven Shapin and Barry Barnes, "Science, Nature and Control: Interpreting Mechanics' Institutes," *Social Studies of Science* 7 (1977): 33; Robert Stewart, *Henry Brougham, 1778–1868: His Public Career* (London: Bodley Head, 1986), 186.

54. Abraham D. Kriegel, "Biography and the Politics of the Early Nineteenth Century," *Journal of British Studies* 29 (1990): 287.

55. Brougham, "Practical Observations," 23.

56. Monica Grobel, "The Society for the Diffusion of Useful Knowledge, 1826–46" (Ph.D. diss., University College, London, 1932), 59.

57. H. Smith, *The Society for the Diffusion of Useful Knowledge, 1826–1846: A Social and Bibliographic Evaluation,* Occasional Paper no. 8 (Halifax, NS: Dalhousie University Library, 1974), 11.

58. Grobel, "Society for the Diffusion of Useful Knowledge," 20.

59. J. N. Hays, "Science and Brougham's Society," *Annals of Science* 20 (1964): 240.

60. Henry Brougham, "On the Objects, Advantages, and Pleasure of Scientific Pursuits," in *Natural Philosophy*, Library of Useful Knowledge, vol. 1 (London, 1829).

61. Webb, *British Working Class Reader*, 29.
62. Charles Knight, *Passages of a Working Life*, 3. vols. (London, 1864), 2: 62.
63. Francis Beaufort to Henry Coates, May 1827, SDUK.
64. Sub-committee Minute Book, Almanac Committee, 27 November 1827, SDUK.
65. "The Tides," *British Almanack* (London, 1828), 44.
66. Stewart, *Henry Brougham*, 192.
67. [Henry Brougham], "New Almanack," *London Times*, 3 January 1828, 3, col. b.
68. Vindex, "The Almanack of the Stationers' Company," *London Times*, 11 January 1828, 3, cols. e and f; 4, col. a.
69. Vindex, "To Mr. Buckingham," *London Times*, 14 January 1828, 3, col. c; "To Vindex, the Apologist of the Stationers' Company in the *Times*," *London Times*, 14 January 1828, 3, col. f. J. S. Buckingham was the editor of the *Athenaeum*.
70. This was figured out only later by Lubbock. See the preface to his *Account of the "Traité sur le flux et reflux de la mer," of Daniel Bernoulli* (London, 1830).
71. Detector, "To the Editors of the Times," *London Times*, 31 December 1828, 3, col. b.
72. Ibid.
73. T. W., "To the Editors of the *Times*," *London Times*, 4 January 1829, 4, col. b.
74. The Stationers' Company was forced to undertake two new, updated almanacs. See Grobel, "Society for the Diffusion of Useful Knowledge," chap. 5, app. 6.
75. "Effects of the *British Almanack*," *London Times*, 2 March 1829, 6, col. e.
76. Edward Maltby to Henry Brougham, as quoted in Grobel, " Society for the Diffusion of Useful Knowledge," 73. Lubbock was seconded by James Mill, the father of John Stuart Mill. At a meeting of the committee on 8 December 1828, with Sir Francis Beaufort in the chair, the committee elected Lubbock a member. General Committee Minute Book 1, SDUK.
77. Coates to Lubbock, 9 December 1828, Lubbock Papers.
78. Sub-committee Minute Book, Almanac Committee, 10 December 1828, SDUK.
79. Ibid.
80. Sub-committee Minute Book, Almanac Committee, 29 December 1828, SDUK.
81. Frederick Wallace Morgan, *Ports and Harbours* (London: Hutchinson's University Library, 1952), 158–59.
82. See, for example, Fernand Braudel, *The Mediterranean and the Mediterranean World in the Age of Philip II*, trans. Siân Reynolds (Berkeley and Los Angeles: University of California Press, 1995), 1:112, fig. 8.
83. Winter, "'Compasses All Awry;'" Kindleberger, *Mariners and Markets*, 35.

84. Sir John Barrow, Bart, to G. Whittam, Esq., Clerk to the Select Committee on Shipwrecks, Admiralty, 10 April 1842, in Parliamentary Papers, 1842, 9:519, app. 6.

### CHAPTER THREE: DESSIOU'S CLAIM

1. Charles Dickens, *David Copperfield* (1849–50; Harmondsworth, UK: Penguin Classics, 1985), 77. The chapter epigraph appears on 90. Yarmouth is on the east coast of England, above the Thames estuary, and experiences exceedingly large tides. In *Great Expectations,* Pip was forced to consult a tide table to determine when the tide was ebbing so he could escape with his benefactor, Magwich.

2. [Paul Hughes], "Dessiou, Joseph Foss," in *Dictionary of Nineteenth-Century British Scientists*, ed. Bernard Lightman (Chicago: University of Chicago Press, 2004), 1:577–78.

3. Mary Croarken, "Mary Edwards: Computing for a Living in 18th-Century England," *IEEE Annals of the History of Computing* 25 (2003): 9–15.

4. Mary Croarken, "Tabulating the Heavens: Computing the Nautical Almanack in 18th-Century England," *IEEE Annals of the History of Computing* 25 (2003): 48–61.

5. See, for instance, George Innes to Lubbock, Lubbock Papers, I8–13.

6. As quoted in Charles Knight, *Passages of a Working Life,* 3 vols. (London, 1864), 2:62.

7. Francis Beaufort to Henry Coates, 18 April 1833. Papers of the Society for the Diffusion of Useful Knowledge, Science Library, University College, London (hereafter cited as SDUK).

8. Joseph Foss Dessiou to John William Lubbock, 27 November 1851, Lubbock Papers, D170.

9. G. B. Airy to the Editors of the *Cambridge Independent Press,* 28 November 1832, Lubbock Papers.

10. John William Lubbock, *Account of the "Traité sur le flux et reflux de la mer," of Daniel Bernoulli* (London, 1830). Thomas Young had resurrected the theoretical study of the tides in England from relative oblivion at the turn of the nineteenth century, but he published his last essay on the tides in 1811, and after his death in 1829 Lubbock was the sole investigator on the theoretical aspect of the tides in England.

11. Lubbock to Coates, 18 December 1829, SDUK.

12. Lubbock to Coates, 1829, SDUK.

13. Isaac Solly to Lubbock, 7 May 1829, Lubbock Papers.

14. Solly to Lubbock, 1829, Lubbock Papers.

15. Coates to Dessiou, 15 June 1829, SDUK.

16. Sub-committee Minute Book, Almanac Committee, 6 July 1829, SDUK.

17. For a more detailed analysis of the computations, see Michael S. Reidy, "The Flux and Reflux of Science: The Study of the Tides and the Organization of

Early Victorian Science" (PhD diss., University of Minnesota, 2000), and
J. W. Lubbock, "The Tides," in *The Companion to the Almanac of the* SDUK *for the Year 1830* (London: Charles Knight, 1830), esp. 57–58.

18. Paul Hughes and Alan D. Wall, "The Admiralty Tidal Predictions of 1833: Their Comparison with Contemporary Observations and with a Modern Synthesis," *Journal of Navigation* 57 (2004): 206.
19. J. W. Lubbock, "On the Tides on the Coast of Great Britain," *Philosophical Magazine* 9 (1831): 334.
20. Dessiou, Statement of the Time Employed on the London Dock Tides, and the British Almanack, SDUK.
21. Lubbock to Wrottesley, 1829, SDUK.
22. Knight to Coates, 5 October 1829, SDUK.
23. Lubbock to Coates, October 1829, SDUK.
24. J. W. L[ubbock], "Explanation of the Columns 'High Water at London,'" in *British Almanac* (London: Charles Knight, 1830), 5.
25. Ibid., 6.
26. J. W. L[ubbock], "The Tides," in *Companion to the Almanac for the Year 1830* (London, Charles Knight, 1830), 62.
27. Thomas Coates, 14 September 1830, Out-letters 19, SDUK.
28. January 1830, General Committee Minute Book 2, SDUK.
29. Lubbock to Coates, 1829, SDUK.
30. Knight to Lubbock, 7 May 1830, Lubbock Papers, K38.
31. Ibid.
32. Knight to Coates, 8 May 1830, SDUK.
33. Almanac Committee, Sub-committee Minute Book, 28 May 1830, SDUK.
34. Lubbock to Coates, June 1830, SDUK.
35. "Nautical Miscellany: Nautical Almanack," *Nautical Magazine* 1 (1832): 40. The *Nautical Magazine was* a monthly journal published out of the Hydrographic Office for mariners stationed throughout Britain's vast possessions.
36. Ibid., 40–41.
37. Columbus, "Nautical Almanack," *London Times,* 30 March 1830, 3, col. d.
38. [Henry Brougham], "Remarks on Almanacs," in *British Almanac* (London, 1829).
39. Beaufort to Pond, 6 August 1830, Royal Hydrographic Office, Taunton, UK (hereafter cited as RHO).
40. Beaufort to Pond, 6 August 1830, RHO.
41. "Nautical Miscellany: *Nautical Almanack,*" *Nautical Magazine* 1 (1832): 40–41.
42. Pond to Lubbock, 7 February 1831, enclosed in Lubbock to Coates, 1 November 1833, SDUK.
43. Lubbock to Whewell, Whewell Papers, Add. MS.a.208/80.
44. Lubbock to Beaufort, 17 June 1831, RHO, Misc. file 12, folder 1, letter 3; Lubbock's emphasis.
45. Beaufort to Lubbock, 20 January 1831, RHO.

46. Beaufort to Lubbock, 22 February 1832, Lubbock Papers. Beaufort wrote this letter only after Lubbock had complained that Beaufort was not doing enough for tidal studies, but it concerned a conversation Beaufort had had with Gilbert in May 1831.
47. Admiralty Records [PRO], ADM.1.12, digest 277 (1831), flag 68.
48. Navy Board (R. Dundas and J. D. Thomson) to Barrow, 19 August 1831, PRO, ADM.106.2296.
49. Navy Board to Lords Commissioners of the Admiralty, 17 October 1831, PRO, ADM.106.2296; 20 October 1831, ADM.1.5006, Pro. R. 289.
50. Beaufort to John Washington, 18 June 1832, RHO.
51. Thomas Brown to John Barrow, 30 October 1832, PRO, ADM.1.3385.
52. Commander Bullock to Beaufort, 27 March 1832, PRO, ADM.1.1580.
53. John Washington to Admiralty, 8 November 1831, Lubbock Papers.
54. Lubbock to Beaufort, 17 March 1831, Lubbock Papers.
55. Beaufort to Lubbock, 20 June 1831, Lubbock Papers.
56. Pond also resigned as astronomer royal, the first of the king's astronomers not to die in office.
57. William Pierce to Lubbock, 22 February 1832, Lubbock Papers.
58. Lubbock to Whewell, 17 March 1831, Whewell Papers.
59. Lubbock to Beaufort, 17 March 1831, Lubbock Papers, L431.
60. Lubbock to Coates, 1833. Marked "Private," in bundle of papers titled "Mr. Lubbock's Correspondence with Mr. Stratford on Mr. Dessiou's Claim," SDUK.
61. Stratford to Lubbock, 6 April 1832, Lubbock Papers, S484.
62. Beaufort to Stratford, 21 October 1833, SDUK.
63. Stratford to Beaufort, 4 November 1833, SDUK.
64. Ibid.; Stratford's emphasis.
65. Ibid.
66. Lubbock to Whewell, Whewell Papers, R.6.20/226.
67. Whewell to Lubbock, 2 November 1833, Whewell Papers, O.15.47/212.
68. Coates to Lubbock, 7 November 1833, Lubbock Papers, C318.
69. Lubbock to Whewell, Whewell Papers, R.6.20//225.
70. Lubbock to Whewell, Whewell Papers, R.6.20/229.
71. Coates to Lubbock, 5 November 1833, Lubbock Papers, C317.
72. Beaufort to Lubbock, 12 November 1833, RHO.
73. Coates to Lubbock, 19 November 1833, Lubbock Papers, C319.
74. Lubbock to Beaufort, 13 November 1833, SDUK.
75. Ibid.
76. J. W. Lubbock, "Note on the Tides," *Philosophical Transactions* 123 (1833): 19.
77. Lubbock to Whewell, 12 November 1833, Whewell Papers, R.6.20/225.
78. Lubbock to Whewell, 23 November 1833, Whewell Papers, R.6.20/227.
79. Stratford to Whewell, 18 November 1833, Whewell Papers, R.6.20/307; Stratford to Whewell, 21 November 1833, Whewell Papers, R.6.20/308; and Stratford to Whewell, 5 December 1833, Whewell Papers, R.6.20/309.

80. Whewell to Stratford, 22 November 1833, Whewell Papers, O.15.47/211.
81. As quoted in Jack Morrell and Arnold Thackray, *Gentlemen of Science: Early Years of the British Association for the Advancement of Science* (Oxford: Clarendon Press, 1981), 515.
82. John William Lubbock, "Discussion of Tide Observations Made at Liverpool," *Philosophical Transactions* 125 (1835): 275–99.
83. Stratford to Lubbock, 26 December 1833, Lubbock Papers, S504.
84. Stratford to Lubbock, 24 August 1835, Lubbock Papers, S519.
85. Stratford to Lubbock, 24 September 1835, Lubbock Papers, S520.
86. Stratford to Lubbock, 16 October 1835, Lubbock Papers, S521.
87. Stratford to Lubbock, Lubbock Papers, S525.
88. Lubbock to Whewell, 25 September 1835, Whewell Papers, R.6.20/232.
89. "Resolution" enclosed in Beaufort to Lords Commissioners of the Admiralty, 1 January 1834, PRO, ADM.1.4282.
90. Roget to Lords Commissioners of the Admiralty, 17 December 1833, PRO, ADM.1.4282.
91. George Elliot to Mr. Roget, 6 February 1834, PRO, ADM.1.4282.
92. Dessiou to Knight, 29 November 1832, SDUK.
93. Ibid. Knight's remarks are written on the back of the letter.
94. David Cartwright, *Tides: A Scientific History* (Cambridge: Cambridge University Press, 1999), 92.
95. Dessiou, "Statement of the Time Employed on the London Dock Tides, and the British Almanack," SDUK.
96. Coates to Lubbock, 23 April 1833, Lubbock Papers, C313; and Coates to Lubbock, 29 October 1833, Lubbock Papers, C316.
97. William Whewell, "Essay towards a First Approximation to a Map of Cotidal Lines," *Philosophical Transactions* 123 (1833): 148.
98. Robert W. Smith, "A National Observatory Transformed: Greenwich in the Nineteenth Century," *Journal for the History of Astronomy* 222 (1991): 5.
99. Lubbock to Whewell, 2 November 1833, Whewell Papers, R.6.20/223.
100. Lubbock to Whewell. Whewell Papers, R.6.20/226; Lubbock's emphasis.
101. Dessiou to Lubbock, enclosed in Lubbock to Coates, 3 August 1837, SDUK.
102. Dessiou to Coates, 20 September 1837, SDUK.
103. Dessiou to Lubbock, 22 February 1838, SDUK.
104. Stratford to Lubbock, 27 March 1837, Lubbock Papers, S527.

CHAPTER FOUR: "TIDOLOGY"

1. George Innes to Lubbock, 17 April 1830, Lubbock Papers. Lubbock's text of 1839 would be the first textbook on the tides used at Cambridge.
2. William Whewell, "Essay towards a First Approximation to a Map of Co-tidal Lines," *Philosophical Transactions* 123 (1833): 147.
3. Beaufort to Lubbock, 22 February 1832, Lubbock Papers.
4. Whewell, "First Approximation," 148.

5. Vladimir Jankovic has used a similar argument to describe John Herschel's interest in meteorology. See Vladimir Jankovic, "Ideological Crests versus Empirical Troughs: John Herschel's and William Radcliffe Birt's Research on Atmospheric Waves, 1843–50," *British Journal for the History of Science* 31 (1998): 21–40. Both approaches owe a great deal to the work of Bruno Latour.

6. Jack Morrell and Arnold Thackray, *Gentlemen of Science: Early Years of the British Association for the Advancement of Science* (Oxford: Clarendon Press, 1981), 515.

7. As quoted in Isaac Todhunter, *William Whewell, D.D., Master of Trinity College, Cambridge: An Account of His Writings with Selections from His Literary and Scientific Correspondence,* 2 vols. (New York: Johnson Reprint, 1970), 1:1.

8. Robert E. Butts, "Whewell, William," in *Dictionary of Scientific Biography,* ed. Charles Gillispie (New York: Charles Scribner's Sons, 1976), 14:292–95; M. J. S. Hodge, "The History of the Earth, Life, and Man: Whewell and Palaetiological Science," in *William Whewell: A Composite Portrait,* ed. Menachem Fisch and Simon Schaffer (Oxford: Clarendon Press, 1991), 255–88; Harvey W. Becher, "William Whewell and Cambridge Mathematics," *Historical Studies in the Physical Sciences* 11 (1980–81): 1–48; Walter F. Cannon, "William Whewell, F.R.S. (1794–1866), 2. Contributions to Science and Learning," *Notes and Records of the Royal Society of London* 19 (1964): 176–91.

9. Robert Peel to Whewell, 17 October 1841, British Library Manuscripts, 40,492.155, British Library, London.

10. Perry Williams, "Passing on the Torch: Whewell's Philosophy and the Principles of English University Education," in *William Whewell: A Composite Portrait,* ed. Menachem Fisch and Simon Schaffer (Oxford: Clarendon Press, 1991), 117.

11. Menachem Fisch, *William Whewell: Philosopher of Science* (Oxford: Clarendon Press, 1991); Richard Yeo, *Defining Science: William Whewell, Natural Knowledge, and Public Debate in Early Victorian Britain* (Cambridge: Cambridge University Press, 1993); Menachem Fisch and Simon Schaffer, eds., *William Whewell: A Composite Portrait* (Oxford: Clarendon Press, 1991).

12. Jack Morrell, "William Whewell: Rough Diamond," *History of Science* 32 (1994): 353.

13. Michael Ruse, for instance, has directly confronted the relationship between Whewell's scientific work and his philosophy. But he spent more time on geology and biology, sciences that Whewell did not directly participate in, than on his research in tidology. Michael Ruse, "William Whewell: Omniscientist," in Fisch and Schaffer, *William Whewell: A Composite Portrait,* 87–116; Michael Ruse, "The Scientific Methodology of William Whewell," *Centaurus* 20 (1976): 227–57.

14. Harvey Becher, "William Whewell's Odyssey: From Mathematics to Moral Philosophy," in Fisch and Schaffer, *Whewell: A Composite Portrait,* 8.

15. Yeo, *Defining Science,* 56. Yeo argued that Whewell's role "made his situation different from that of Herschel or Brewster, who could comment on issues

of method or theory, often in relation to their own original investigations. Whewell could do this too, and we should not forget the work he did on mineralogy and tides; but he seemed to doubt whether this was an adequate basis from which to speak" (56). Other historians have used Whewell's own doubts as evidence that he did not think of himself as a scientist. Yet every time Whewell made such claims, he was attempting to get out of becoming the next president of a scientific body, be it the Geological Society or the British Association. See Michael S. Reidy, "The Flux and Reflux of Science: The Study of the Tides and the Organization of Early Victorian Science" PhD diss., University of Minnesota, 2000), 304. Yeo's recent sixteen-volume collection of Whewell's works makes no mention of his research on the tides. See Richard Yeo, ed., *The Collected Works of William Whewell*, 16 vols. (Sterling, VA: Thoemmes Press, 2000).

16. Fisch, *Whewell: Philosopher of Science*, 41.

17. Joan Richards, "Observing Science in Early Victorian England: Recent Scholarship on William Whewell," *Perspectives on Science* 4 (1996): 235.

18. Morrell and Thackray, *Gentlemen of Science*, 479. Cannon, "Whewell: Contributions to Science and Learning," 177.

19. Becher, "Whewell and Cambridge Mathematics," 15; William Whewell, *A Treatise on Dynamics, Containing a Considerable Collection of Mechanical Problems* (London, 1823); see, for example, 365–72; 378ff.

20. Harvey W. Becher, "Radicals, Whigs and Conservatives: The Middle and Lower Classes in the Analytical Revolution at Cambridge in the Age of Aristocracy." *British Journal for the History of Science* 28 (1995): 421.

21. Whewell to Herschel, 15 October 1823, Whewell Papers.

22. William Whewell, "A General Method of Calculating the Angles Made by Any Planes of Crystals, and the Laws according to Which They Are Formed," *Philosophical Transactions* 115 (1825): 87–130.

23. Ibid., 87. Ten years later Whewell would reduce the mathematical portion of tides to mathematical formulas that could be applied generally.

24. H. Deas, "Crystallography and Crystallographers in England in the Early Nineteenth Century: A Preliminary Survey," *Centaurus* 6 (1959): 135. See Ruse, "Scientific Methodology," 237–40, for further historical significance of Whewell's paper.

25. Whewell to Herschel, 3 April 1828, Whewell Papers.

26. Cannon, "Contributions to Learning," 180; Ruse, "Scientific Methodology," 238–39.

27. Harvey W. Becher, "Voluntary Science in Nineteenth Century Cambridge University to the 1850s," *British Journal for the History of Science* 19 (1986): 61. For Whewell's work in mineralogy, see Deas, "Crystallography and Crystallographers," 129–48; Ruse, "Scientific Methodology," 237–40.

28. Whewell to Richard Jones, 1 June 1827, in William Ashworth, "Calendar of William Whewell's Correspondence," unpublished manuscript, used courtesy of the author.

29. Morrell and Thackray, *Gentlemen of Science,* 431, 480.

30. David B. Wilson, "The Educational Matrix: Physics Education at Early-Victorian Cambridge, Edinburgh and Glasgow," in *Wranglers and Physicists: Studies on Cambridge Physics in the 19th Century,* ed. P. M. Harman (Manchester, UK: Manchester University Press, 1985), 13.

31. Crosbie Smith, "'Mechanical Philosophy' and the Emergence of Physics in Britain: 1800–1850," *Annals of Science* 33 (1976): 25–26; Maurice Crosland and Crosbie Smith, "The Transmission of Physics from France to Britain: 1800–1840," *Historical Studies in the Physical Sciences* 9 (1978): 17.

32. Whewell to Forbes, 16 December 1831, Todhunter, *Whewell,* 2:136. Whewell referred Forbes to numerous texts, including Humboldt's two-volume *Géologie et climatologie asiatiques,* published that year (1831) in Paris.

33. Whewell to Richard Jones, 23 July 1831, in Ashworth, "Correspondence."

34. Forbes to Whewell, 10 January 1832, Whewell Papers, Add.Ms.a.204/5 (1).

35. Lubbock to Whewell, 28 October 1829, Whewell Papers.

36. Lubbock to Whewell, 17 November 1829, Whewell Papers.

37. Whewell to Lubbock, 7 December 1829, Whewell Papers.

38. Richard Sheepshanks to Lubbock, 4 May 1830, Lubbock Papers, S107. Sheepshanks was writing on Whewell's behalf.

39. Whewell to Lubbock, 30 January 1831, Lubbock Papers. Whewell transformed this to "tidology," which he and Lubbock then used throughout their researches.

40. Whewell to Lubbock, 30 January 1831, Lubbock Papers.

41. Fisch, *Whewell: Philosopher of Science,* esp. 57–63.

42. Whewell to Lubbock, 7 December 1829 and 2 April 1832, Whewell Papers.

43. Whewell to Lubbock, 1 April 1832, Lubbock Papers. As I will show below, he had told his sister his intentions a month earlier.

44. Lubbock to Whewell, 7 May 1832, Whewell Papers.

45. Whewell to Lubbock, 8 May 1832, Lubbock Papers.

46. Whewell to Lubbock, 6 July 1832, Whewell Papers, Add.Ms.a.216.

47. Beaufort to Coates, October 1833, SDUK.

48. Ruth Kark, "The Contributions of Nineteenth Century Protestant Missionary Societies to Historical Cartography," *Imago Mundi* 45 (1993): 112–19. For the myriad ways missionaries in the Pacific used science beyond their educational institutions, see Sujit Sivasundaram, *Nature and the Godly Empire: Science and Evangelical Mission in the Pacific, 1795–1850* (New York: Cambridge University Press, 2005).

49. Kenneth Latourette, *A History of the Expansion of Christianity* (Grand Rapids, MI: Zondervan, 1970), and Kark, "Protestant Missionary Societies," 112.

50. Whewell to his sister, 13 March 1832, in [Janet Mary] Stair Douglas, *The Life and Selections from the Correspondence of William Whewell, D.D.* (London, 1881), 143–44.

51. Stair Douglas, *Life,* 144. At Tahiti, high tides occur around noon and midnight, as they should if they were influenced only by the sun. According

to modern tidal theory, a few points exist in midocean, nodal points of the ocean's standing waves, where the moon's influence is negated completely and the solar tide dominates. Tahiti happens to be near one of these nodes. Certain aspects of the Tahitian tides, however, remain unexplained. See William R. Corliss, *Earthquakes, Tides, Unidentified Sounds and Related Phenomena: A Catalog of Geophysical Anomalies* (Glen Arm, MD: Sourcebook Project, 1983), 32.

52. As reported by David Tomlinson, "On the Tides," *American Journal of Science* 34 (1838): 81–85.

53. Captain William Waldegrave to Whewell, 8 December 1833, Whewell Papers, R.6.20/314.

54. "Williams, John," in *Dictionary of National Biography,* ed. Leslie Stephen and Sidney Lee (London: Oxford University Press, 1921–22), 21:423–25.

55. Whewell to his sister, 19 December 1834, in Stair-Douglas, 162.

56. Ibid.

57. Whewell Papers, R.6.20/33(13).

58. "Jowett, William," in *Dictionary of National Biography,* ed. Sir Leslie Stephen and Sir Sidney Lee (London: Oxford University Press, 1967–68), 10:1103–4.

59. Whewell to his sister, 13 February 1835, in Stair-Douglas, 170–71.

60. For an in-depth history of the British Association for the Advancement of Science, see Morrell and Thackray, *Gentlemen of Science.*

61. William Vernon Harcourt to Lubbock, 9 November 1831, Lubbock Papers.

62. Scoresby himself had invited both the Liverpool Institution and the Athenaeum to the York meeting; see Morrell and Thackray, *Gentlemen of Science,* 67.

63. William Vernon Harcourt to Lubbock, 9 November 1831, Lubbock Papers.

64. Morrell and Thackray, *Gentlemen of Science,* 514.

65. Whewell to Vernon Harcourt, 2 September 1832, Whewell Papers.

66. J. W. Lubbock, "Report on the Tides," *British Association Report* 1832 (1833): 192.

67. Lubbock to Phillips, 30 October 1834, Phillips Papers, Oxford Museum of Natural History, Oxford, England; Morrell and Thackray, *Gentlemen of Science,* 320, 515.

68. Whewell to Lubbock, 1 April 1832, Lubbock Papers.

69. Whewell to Lubbock, 9 January 1833, Whewell Papers.

70. Whewell to Lubbock, 3 February 1833, Lubbock Papers.

71. Admiralty to Dr. Thomas Young, Capt. Edwin Sabine, and Michael Faraday, Esq., 7 January 1829, PRO, ADM.1.3469.

72. Alfred Friendly, *Beaufort of the Admiralty: The Life of Sir Francis Beaufort, 1774–1857* (New York: Random House, 1977), 14. See also Nicholas Courtney, *Gale Force 10: The Life and Legacy of Admiral Beaufort, 1774–1857* (London: Review, 2003), and Randolph Cock, "Sir Francis Beaufort and the Co-ordination of British Scientific Activity, 1829–1855" (PhD diss., University of Cambridge, 2003).

73. Friendly, *Beaufort*, 288–89.
74. David McLean, *Education and Empire: Naval Tradition and England's Elite Schooling* (London: British Academic Press, 1999), 24; Janet Browne, "Biogeography and Empire," in *Cultures of Natural History*, ed. N. Jardine, J. A. Secord, and E. C. Spary (Cambridge: Cambridge University Press, 1996), 309.
75. McLean, *Education and Empire*, 24.
76. Henry Palmer, "Description of a Graphical Register of Tides and Winds," *Philosophical Transactions* 121 (1831): 210.
77. Henry Palmer to Lubbock, 10 February 1831, Lubbock Papers.
78. "The Tide Gauge at Sheerness," *Nautical Magazine* 1 (1832): 4014.
79. Ibid., 403.
80. Ibid.
81. Ibid., 404.
82. John Washington to Admiralty, 8 November 1831, Lubbock Papers.
83. Sir Francis Beaufort, *General Instruction for the Hydrographic Surveyors of the Admiralty* (London, 1850). Though it was published only in 1850 as a time-saving mechanism, Beaufort explained that this publication was similar to instructions he had been giving to individual surveyors for twenty years.
84. Navy Board to Master Attendants, PRO, ADM.106.2296. For the expenses and use of tide poles, see Lubbock to Beaufort, Miscellaneous Folder 12, letter no. 2, RHO.
85. Beaufort to Captain Woolmore, 6 September 1833, RHO.
86. Beaufort to Commander Mudge, 1 July 1831, RHO.
87. Beaufort to Johnson, 20 October 1831, RHO.
88. Beaufort to Captain Horsburgh, 1 November 1832, PRO, ADM.1.3478. A formal letter to the "Honourable Court of Directors" was sent nine days later.
89. Beaufort to Sir Edward Parry, 15 April 1833, RHO.
90. Whewell to Lubbock, 7 November 1833, Whewell Papers, O.15.47/208.
91. Whewell to Lubbock, 15 November 1833, Whewell Papers, O.15.47/209.
92. Whewell, "First Approximation."
93. Ibid., 150.
94. See the glossary for scientific terms used in tidology. Whewell repeated this discussion for mariners in several entries in the *Nautical Magazine* in 1833 and 1834. His emphasis on precise terminology confirms Simon Schaffer's argument that Whewell believed "the continuity of knowledge was guaranteed by the permanence of past work embodied in right language." See Simon Schaffer, "Whewell's Politics of Language," in Schaffer and Fisch, *Whewell: A Composite Portrait*, 208.
95. Whewell, "First Approximation," 227.
96. Whewell, *History of the Inductive Sciences, from the Earliest to the Present Time*, 1st ed., 3 vols. (London, 1837), 2:248–49.
97. Todhunter, *Whewell*, 2:172; Ruse, "Scientific Methodology," 227.

98. William Whewell, "On the Empirical Laws of the Tides in the Port of London, with Some Reflexions on the Theory," *Philosophical Transactions* 124 (1834): 17.
99. Ibid., 40.
100. Ibid., 45.
101. Todhunter, *Whewell*, 2:169; Whewell to Lubbock, 31 October 1833, Whewell Papers.
102. Dessiou to Whewell, 14 October 1834, Whewell Papers.
103. John William Lubbock, "Discussion of Tide Observations Made at Liverpool," *Philosophical Transactions* 125 (1835): 275–300.
104. William Whewell, "Researches on the Tides," 4th ser., "On the Empirical Laws of the Tides in the Port of Liverpool," *Philosophical Transactions* 126 (1836): 1.
105. Ibid., 2.
106. Ibid., 3.
107. William Whewell, *Philosophy of the Inductive Sciences, Founded upon Their History* (London, 1840), 2:228.
108. Whewell, "Researches on the Tides," 4th ser., "On the Empirical Laws of the Tides in the Port of Liverpool."
109. Fisch, *Whewell*, 41.
110. See for example, Andrew D. Lambert, "Preparing for the Long Peace: The Reconstruction of the Royal Navy, 1815–1830," *Mariner's Mirror* 82 (1996): 41–54; David K. Brown, "Wood, Sail and Cannonballs to Steel, Steam and Shells: 1815–1895," in *The Oxford Illustrated History of the Royal Navy,* ed. J. R. Hill (Oxford: Oxford University Press, 1995).

CHAPTER FIVE: THE TIDE CRUSADE

1. Whewell to Lubbock, 1 April 1832, Lubbock Papers, W261. The *Beagle* sailed into Rio de Janeiro the first week of April.
2. Charles Darwin to J. S. Henslow, 18 May 1832, in Frederick Burkhardt and Sydney Smith, eds., *The Correspondence of Charles Darwin* (Cambridge: Cambridge University Press, 1985), 1:237.
3. For the importance of Humboldt and the geographical turn, see Susan Faye Cannon, *Science in Culture: The Early Victorian Period* (New York: Science History Publications, 1978); David N. Livingstone, *The Geographical Tradition* (Oxford: Blackwell, 1992); and Mary Louise Pratt, *Imperial Eyes: Travel Writing and Transculturation* (New York: Routledge, 1992).
4. Charles Darwin, *On the Origin of Species* (Harmondsworth, UK: Penguin, 1968), 65.
5. Nicholas Jardine and Emma Spary, "The Natures of Cultural History," in *Cultures of Natural History*, ed. N. Jardine, J. A. Secord, and E. C. Spary (Cambridge: Cambridge University Press, 1996), 8.

6. "On the Advantage Possessed by Naval Men, in Contributing to General Science," *Nautical Magazine* 1 (1832): 180.
7. David Philip Miller, "The Revival of the Physical Sciences in Britain, 1815–1840," *Osiris* 2 (1986): 108–12.
8. See for example, Sir John Robison to Whewell, 22 January 1836, Whewell Papers, Add.a.201.149; and John Wells to John William Lubbock, 5 April 1836, Lubbock Papers, W201.
9. Helen M. Rozwadowski, "Internationalism, Environmental Necessity, and National Interest: Marine Science and Other Sciences," *Minerva* 42 (2004): 127–49.
10. For the internationalization of the economy during the Victorian era, see Marc R. Brawley, *Liberal Leadership: Great Powers and Their Challengers in Peace and War* (Ithaca, NY: Cornell University Press, 1994), and Bernard Semmel, *The Rise of Free Trade Imperialism: Classical Political Economy, the Empire of Free Trade, and Imperialism, 1750–1850* (Cambridge: Cambridge University Press, 2003).
11. [Richard Cobden], *England, Ireland, and America* (London, 1835), 45.
12. Donald Read, *Cobden and Bright: A Victorian Political Partnership* (London: Edward Arnold, 1967), 32.
13. Ibid.
14. William Whewell, "Essay towards a First Approximation to a Map of Cotidal Lines," *Philosophical Transactions* 123 (1833): 148.
15. Nicolaas Rupke, "Humboldtian Distribution Maps: The Spatial Ordering of Scientific Knowledge," in *The Structure of Knowledge: Classifications of Science and Learning since the Renaissance,* ed. Tore Frängsmyr (Berkeley: Office for History of Science and Technology, University of California, 2001), 94.
16. Matthew H. Edney, M*apping an Empire: The Geographical Construction of British India, 1765–1843* (Chicago: University of Chicago Press, 1997); Graham D. Burnett, *Masters of All They Surveyed: Exploration, Geography, and a British El Dorado* (Chicago: University of Chicago Press, 2000).
17. William Snow Harris to Rev. I. H. Macauley, 7 January 1832, Whewell Papers, R.6.20.201.
18. William Whewell, "Essay towards a First Approximation to a Map of Cotidal Lines," 147.
19. Ibid., 148.
20. Ibid., 163.
21. Ibid., 165.
22. Ibid., 165.
23. Ibid., 227.
24. William Whewell, "Memoranda and Directions for Tide Observations," *Nautical Magazine* 2 (1833): 664.
25. Ibid.; Whewell's emphasis.
26. Ibid., 665.

27. Richard Spencer to Whewell, 26 September 1832, Whewell Papers, R.6.20/301.
28. Michael Lewis, *The Navy in Transition 1814–1864: A Social History* (London: Hodder and Stoughton, 1965), 69.
29. Ibid., 81.
30. Circular, George Elliot, "Regulations for the Admission of Seamen into the Coast Guard Service; Their Pay, Allowances, and Rewards," 13 June 1831, PRO, ADM.1.3876.
31. Lewis, *Navy in Transition,* 90.
32. Beaufort to Bowles, 30 July 1833, Beaufort Papers, Letter Book 5.
33. Sparshott to Beaufort, 8 October 1833, RHO, S405.
34. Bowles to the Respective Inspecting Commanders, UK, 17 May 1834, RHO, B456.
35. Bowles to Beaufort, 15 May 1834, RHO, B453.
36. Becher to Whewell, 24 September 1833, Whewell Papers, R.6.20/62.
37. Examples abound. See for instance, Bullock to Beaufort, 27 March 1832, PRO, ADM.1.1580, and Master Attendant at Malta to Secretary of the Admiralty, 22 April 1832, PRO, ADM.1.4988.
38. Whewell to Becher, 24 September 1833, Whewell Papers, R.6.20/62.
39. Beaufort to Whewell, 23 March, 1833, RHO, Letter Book 4.
40. Beaufort to Commander Mudge, 25 March 1833, RHO, Letter Book 4.
41. Beaufort to Commander Mudge, 15 April 1833, RHO, Letter Book 4.
42. Memorandum, "Circular to Surveyors," 26 May 1834, RHO.
43. Rear Admiral Parker to George Elliot, for the Hydrographer, 28 May 1834, PRO, ADM.1.1365.
44. Beaufort to Officers of the Kent and Sussex Coast Guard, 5 June 1834, RHO, Letter Book 5.
45. Ibid.
46. Beaufort to Lubbock, 5 June 1834, Lubbock Papers, B174.
47. William Whewell, "On the Results of Tide Observations Made in June 1834 at the Coast Guard Stations in Great Britain and Ireland," *Philosophical Transactions* 125 (1835): 84.
48. Beaufort to Sparshott, 9 September 1834, RHO, Letter Book 5.
49. Whewell to Lubbock, 4 August 1834, Whewell Papers, O.15.47/213b.
50. Beaufort to Lubbock, 30 July 1834, Lubbock Papers, B175.
51. Whewell to Lubbock, 7 September 1834, Whewell Papers, O.15.47/214.
52. Dessiou to Whewell, 14 October 1834, Whewell Papers, R.6.20.169.
53. Whewell to Beaufort, 26 November 1834, RHO.
54. William Whewell, "On the Results of Tide Observations Made in June 1834," 84.
55. Ibid., 89.
56. Whewell to Lubbock, 22 April 1835, Whewell Papers, O.15.47/218.
57. Beaufort to Whewell, 7 February 1835, RHO, Letter Box 6.
58. Whewell to Beaufort, 8 February 1835, RHO, W554.

59. Customs [Commander Bowles] to Francis Beaufort, 13 February 1835, PRO, ADM.1. 3876.

60. Beaufort to Lords Commissioners of the Admiralty, 14 February 1835, PRO, ADM.1.3485.

61. See, for example, Foreign Office to HM Minister at Paris, 19 May 1834, PRO, ADM.1.4254; Archibald Day, *The Admiralty Hydrographic Service, 1795–1919* (London: Her Majesty's Stationary Office, 1967), 42ff.

62. Memorandum, "Circular to Ministers and Consuls Abroad," 13 March 1835, PRO, FO.83.84.

63. Beaufort to Bowles, 19 February 1835, RHO, Letter Book 6.

64. Thomas Spark to Beaufort, 1 June 1835, RHO, S379.

65. Thomas Spark to Beaufort, 25 May 1835, RHO, S378.

66. Home Office to the Governors of the Channel Islands, 16 May 1835, PRO, ADM.1.3485. Beaufort's letter to the Lords Commissioners is enclosed.

67. See, for example, Thomas Spark to Beaufort, 25 May 1835, 28 May 1835, 1 June 1835, and 4 June 1835, RHO, S376–S380.

68. Beaufort to Whewell, 16 February 1835, RHO, Letter Book 6.

69. Memorandum, "Dispatches from HM Ministers in Sweden, and from HM Consul General at Hamburg," 13 April 1835, PRO, ADM.1.4257.

70. G. S. S. Jermingham, The Hague, to Lord [Palmerston], 26 August 1835, PRO, ADM.1.4259.

71. Forbes to Whewell, 26 June 1835, Whewell Papers, Add.Ms.a.204/21.

72. G. Moll to Whewell, 4 June 1835, Whewell Papers, Add.Ms.a.209.

73. Charles R. Vaughan to Lord Duke [of Wellington], 28 April 1835, PRO, ADM.1.4258.

74. Bache to Whewell, 29 October 1834, Whewell Papers, Add.Ms.a.200/195.

75. Mr. Dickerson to Navy Department, 29 October 1835, PRO, ADM.1.4260. See Hugh Richard Slotten, *Patronage, Practice, and the Culture of American Science: Alexander Dallas Bache and the U.S. Coast Survey* (Cambridge: Cambridge University Press, 1994), esp. chap, 6.

76. Whewell to J. D. Forbes, 23 October 1856, Whewell Papers.

77. William Whewell, "Researches on the Tides," 6th ser., "On the Results of an Extensive System of Tide Observations Made on the Coast of Europe and America in June 1835," *Philosophical Transactions* 126 (1836): 291.

78. Ibid.

79. Admiralty Minutes, 18 July 1835, PRO, ADM.3.231.

80. William Whewell, *The Philosophy of the Inductive Sciences, Founded upon Their History,* 1st ed., 2 vols. (London: John W. Parker, 1840), 2:542.

81. Ibid., 2:543–44.

82. Ibid., 2:546; Whewell's emphasis.

83. Whewell, "Tide Observations Made on the Coast of Europe and America in June 1835," 297.

84. Becher to Whewell, 14 October 1836, Whewell Papers, R.6.20/63.

85. Whewell to Becher, 19 November 1836, RHO.

86. "Orders to Commander Hewett," 27 March 1833, PRO, ADM.1.3479.
87. Beaufort to Hewett, 20 April 1835, RHO, Letter Book 6.
88. Beaufort to Hewett, 8 April 1840, RHO, Letter Book 9.
89. William Hewett, "Tide Observations in the North Sea—Verification of Professor Whewell's Theory," *Nautical Magazine* 9 (1841): 180–83.
90. Beaufort to Whewell, 4 September 1840, RHO, Letter Book 9.
91. "H.M.S. *Fairy*," *Nautical Magazine* 9 (1841): 72; "Loss of the *Fairy*," *Nautical Magazine* 9 (1841): 116–17; "Memorial of Mrs. Hewett," *Nautical Magazine* 9 (1841): 425–27.
92. John Washington, "Tide Observations—North Sea—Professor Whewell's Theory," *Nautical Magazine* 10 (1842): 566–68.
93. George Biddell Airy, "Tides and Waves," in *Encyclopaedia Metropolitana*, vol. 5 (London, 1845), article 525, 376; Cartwright, *Tides: A Scientific History*, 113–16, 168–73.
94. *The Admiralty Manual of Tides* (London: Her Majesty's Stationer, 1941), 200.
95. Ibid., 209.
96. Airy, "Tides and Waves," articles 492, 369.
97. Ibid., articles 497, 370.
98. James C. Scott, *Seeing Like a State: How Certain Schemes to Improve the Human Condition Have Failed* (New Haven, CT: Yale University Press, 1998). Scott noted that a high modernist ideology, which he argued began in Europe in the 1830s (88), "tended to see rational order in remarkably visual aesthetic terms" (4).
99. See for example, John Tempter to Whewell, 31 January 1840, Whewell Papers, R.6.20/313.
100. Whewell, *Philosophy*, 2:546; Whewell, "Tide Observations Made on the Coast of Europe and America in June 1835," 290.
101. Whewell, "Tide Observations Made on the Coast of Europe and America in June 1835," 290–92, 294.
102. On the material hidden by scientific publications, see Alan Gross, *The Rhetoric of Science* (Cambridge, MA: Harvard University Press, 1990), 5.
103. Bruno Latour, "Drawing Things Together," in *Representation in Scientific Practice*, ed. Michael Lynch and Steve Woolgar (Cambridge: MIT Press, 1990), 19–68.
104. Bruno Latour, "The Force of Reason and Experiment," in *Experimental Inquiries: Historical, Philosophical and Social Studies of Experimentation in Science*, ed. H. E. Le Grand (Boston: Kluwer Academic Publishers, 1990), 58.
105. For a discussion of scientific writing reenacting the process of knowledge creation, see Alan G. Gross, Joseph E. Harmon, and Michael Reidy, *Communicating Science: The Scientific Article from the 17th Century to the Present* (Oxford: Oxford University Press, 2002).
106. Whewell coined the term only in 1844, but the antithesis between fact and theory was a central doctrine of his as early as 1833 in his "Address" to the

British Association for the Advancement of Science in Cambridge, and he focused on just this "fundamental antithesis" to open both his *History* and his *Philosophy*. Menachem Fisch, "Necessary and Contingent Truth in William Whewell's Antithetical theory of Knowledge," *Studies in the History and Philosophy of Science* 16 (1985): 279.

107. Whewell, *Philosophy*, 2:202.

108. Whewell, "Tide Observations Made on the Coast of Europe and America in June 1835," 289.

109. Thomas L. Hankins and Robert J. Silverman, *Instruments and the Imagination* (Princeton, NJ: Princeton University Press, 1995), esp. chap. 6.

110. J. F. W. Herschel, "On the Investigation of the Orbits of Revolving Double Stars; Being a Supplement to a Paper Entitled 'Micrometrical Measures of 364 Double Stars,'" *Memoirs of the Royal Astronomical Society* 5 (1833): 171–222.

111. Rupke, "Humboldtian Distribution Maps," 96. See also Eric L. Mills, "*De Motu Marium*: Understanding the Oceans Before the Second Scientific Revolution," Stillman Drake Lecture, Canadian Society for History and Philosophy of Science, June 1999, esp. 2–3.

112. For geology, see Martin Rudwick, *The Great Devonian Controversy: The Shaping of Scientific Knowledge among Gentlemanly Specialists* (Chicago: University of Chicago Press, 1985).

113. Darwin, *Origin of Species*, 121.

114. Alan Gross, "Darwin's Diagram," paper presented at the annual meeting of the History of Science Society, Milwaukee, WI, 7 November 2002.

115. Latour, "Drawing Things Together," 45; italics in original.

116. "Address of His Royal Highness the President," 30 November 1837, *Abstracts of Papers in the "Philosophical Transactions," and, the "Proceedings of the Royal Society of London"* (London, 1838); my emphasis.

117. Cannon, *Science in Culture*, 227.

CHAPTER SIX: CALCULATED COLLABORATIONS

1. Peter Roget to Lords Commissioners of the Admiralty, 17 December 1833, PRO, ADM.1.4282.

2. James C. Scott, *Seeing Like a State: How Certain Schemes to Improve the Human Condition Have Failed* (New Haven, CT: Yale University Press, 1998), 310ff.

3. William Whewell, "On the Empirical Laws of the Tides in the Port of London, with Some Reflexions on the Theory," *Philosophical Transactions* 124 (1834): 15.

4. Ibid., 16.

5. For an in-depth discussion of the Cambridge Network and scientific servicemen, see David Philip Miller, "The Revival of the Physical Sciences in Britain, 1815–1840," *Osiris* 2 (1986): 108–12; Susan Faye Cannon, *Science in Culture: The Early Victorian Period* (New York: Science History Publications, 1978),

esp. chap. 2; and Robert W. Smith, "The Cambridge Network in Action: The Discovery of Neptune," *Isis* 80 (1989): 395–422.

6. William Whewell, "Address," *British Association Report* 1833 (1834): xii. Whewell also used the term "scientist" in this address, and in the next chapter I will focus on his distinction between the two.

7. Anne Secord, "Science in the Pub: Artisan Botanists in Early Nineteenth-Century Lancashire," *History of Science* 32 (1994): 269–315; Anne Secord, "Corresponding Interests: Artisans and Gentlemen in 19th-Century Natural History," *British Journal for the History of Science* 27 (1994): 383–408.

8. See, for example, Vladimir Jankovic, "Ideological Crests versus Empirical Troughs: John Herschel's and William Radcliffe Birt's Research on Atmospheric Waves, 1843–50," *British Journal for the History of Science* 31 (1998): 21–40.

9. See Dorinda Outram, "New Spaces in Natural History," in *Cultures of Natural History,* ed. N. Jardine, J. A. Secord, and E. C. Spary (Cambridge: Cambridge University Press, 1996), esp. 149–50.

10. Whewell to Phillips, 10 August 1836, Whewell Papers.

11. Anne Secord, "Corresponding Interests: Artisans and Gentlemen in Nineteenth-Century Natural History," *British Journal for the History of Science* 27 (1994): 383–408; Bernard Lightman, ed., *Victorian Science in Context* (Chicago: University of Chicago Press, 1997); James Secord, *Victorian Sensation* (Chicago: University of Chicago Press, 2003).

12. W. Bennis and P. W. Biederman, *Organising Genius: The Secrets of Creative Collaboration* (London: Addison-Wesley, 1997).

13. N. C. Russell, E. M. Tansey, P. V. Lear, "Missing Links in the History and Practice of Science: Teams, Technicians, and Technical Work," *History of Science* 38 (2000): 237.

14. As quoted in John Belchem, "Liverpool in 1848: Image, Identity and Issues," *Transactions of the Historic Society of Lancashire and Cheshire* 147 (1997): 2.

15. Thomas Baines, *History of the Commerce and Town of Liverpool, and of the Rise of Manufacturing Industry in the Adjoining Counties* (London, 1852), 749.

16. M. J. Power, "The Growth of Liverpool," in *Popular Politics, Riot and Labour: Essays in Liverpool History, 1790 to 1940,* ed. John Belchem (Liverpool, UK: Liverpool University Press, 1992), 21–37.

17. W. E. Minchinton, ed., *The Growth of English Overseas Trade in the Seventeenth and Eighteenth Centuries* (London: Methuen, 1969), ix; Kenneth Morgan, "Bristol and the Atlantic Trade in the Eighteenth Century," *English Historical Review* 424 (1992): 626.

18. Graham H. Hills, *Navigation of the Approaches to Liverpool, with Some Account of the Tides, Currents, and Soundings of the Irish Sea, with a Chart of Liverpool Bay* (Liverpool, UK, 1871).

19. As quoted in Jos. Brooks Yates, "Memoir on the Rapid and Extensive Changes Which Have Taken Place at the Entrance to the River Mersey," an appendix to William Williams Mortimer, *The History of the Hundred of Wirral,*

*with a Sketch of the City and County of Chester, Compiled from the Earliest Authentic Records* ([London], 1847), 25.

20. Baines, *History of the Commerce and Town of Liverpool*, 648.
21. Ibid., 558, 640.
22. J. F. Smith, Gordon Hemm, and A. Ernest Shennan, *Liverpool, Past, Present, Future* (Liverpool, UK: Northern, 1948), 8.
23. Power, " Growth of Liverpool," 21.
24. Michael S. Reidy, "Masters of Tidology: The Cultivation of the Physical Sciences in Early Victorian Liverpool," *Transactions of the Historic Society of Lancashire and Cheshire* 152 (2003): 45–71.
25. Hutchinson's manuscript is housed in the Liverpool Central Library.
26. I discuss Whewell's work with the Liverpool observations in chapter 4.
27. Holden to Whewell, 19 April 1834, Whewell Papers, R.6.20/205. Holden's methods were not divulged until they were copyrighted in 1867. See J. R. Rossiter, "The History of Tidal Predictions in the United Kingdom Before the Twentieth Century," *Proceedings of the Royal Society of Edinburgh,* ser. B, 73 (1971–72): 20; P. L. Woodworth, "Three Georges and One Richard Holden: The Liverpool Tide Table Makers," *Transactions of the Historic Society of Lancashire and Cheshire* 151 (2002): 19–51.
28. Holden to Whewell, 27 September 1833, Whewell Papers, R.6.20/204.
29. Holden to Whewell, 19 April 1834, Whewell Papers, R.6.20/205.
30. Joseph Yates to Joseph Foss Dessiou, September 1835, Lubbock Papers, Y6.
31. Bywater to Whewell, 21 November 1835, Whewell Papers, R.6.20/147; Wylie to Lubbock, March 1836, Lubbock Papers, W503.
32. E. A. Bryant, "The Lit and Phil–a Retrospective," handwritten monograph housed in the Liverpool Central Library Archives, Hq 1021, Papers, Chiefly on Psychology. Delivered to the Liverpool Literary and Philosophical Society of Liverpool on 22 March 1937.
33. Bywater to Lubbock, 14 November 1835, Lubbock Papers, B627. See also Ken Adler, *The Measure of All Things: The Seven-Year Odyssey and Hidden Error That Transformed the World* (New York: Free Press, 2002), 184.
34. Bywater to Lubbock, 14 November 1835, Lubbock Papers, B627; Bywater to Whewell, 16 December 1835, Whewell Papers, R.6.20/148.
35. Bywater to Whewell, 21 November 1835, Whewell Papers, R.6.20/147.
36. Bywater to Whewell, 16 December 1835, Whewell Papers, R.6.20/148.
37. Whewell to Lubbock, 1 April 1837, Whewell Papers, O.15.47/225.
38. Isaac Todhunter, *William Whewell, D.D., Master of Trinity College, Cambridge: An Account of His Writings with Selections from His Literary and Scientific Correspondence*, 2 vols. (New York: Johnson Reprint, 1970), 1:79.
39. Brown to Lubbock, 12 March 1840, Lubbock Papers, B493.
40. Brown to Lubbock, 21 March 1840, Lubbock Papers, B494.
41. Brown to Lubbock, 12 March 1840 and 21 March 1840, Lubbock Papers, B493 and B494.

42. Whewell, "On the Empirical Laws of the Tides in the Port of London," 16. He noted Bywater's contributions in his "seventh series," 133.
43. "Address by Professor Traill, M.D., *"British Association Report* 1837 (1838).
44. John William Lubbock, *An Elementary Treatise on the Tides* (London, 1839).
45. Whewell to Phillips, 10 August 1836, Whewell Papers.
46. Lubbock to Phillips, 30 October 1834, Phillips Papers, OMNH. For Lubbock's disparagement of the British Association, see Jack Morrell and Arnold Thackray, *Gentlemen of Science: Early Years of the British Association for the Advancement of Science* (Oxford: Clarendon Press, 1981), 320, 515.
47. See, for instance, Bunt to Whewell, 13 February 1836, Whewell Papers, R.6.20.85, and especially Bunt to Whewell, 8 November 1841, Whewell Papers, Add.Ms.a.58/25. See also Todhunter, *Whewell,* 1:84.
48. "The Late Mr. Bunt," *Western Daily Press,* 6 June 1872, 5; [Obituary Notice], *Bristol Mercury,* 8 June 1872, 6.
49. Bunt to Whewell, 7 February 1838, Whewell Papers, R.6.20/113.
50. Bunt to Whewell, 17 April 1840, Whewell Papers, R.6.20/135.
51. Michael Raymond Neve, "Natural Philosophy, Medicine, and the Culture of Science in Provincial England: The Cases of Bristol, 1790–1850, and Bath, 1750–1820" (PhD diss., University of London, 1984), 131.
52. Ibid., 158.
53. Bunt to Whewell, 12 October 1835, Whewell Papers, R.6.20/86; Bunt's emphasis.
54. Stutchbury to Whewell, 12 October 1835, Whewell Papers, R.6.20/311.
55. Bunt to Whewell, 12 October 1835, Whewell Papers, R.6.20/87.
56. Ibid.
57. "The Late Mr. Bunt," *Western Daily Press,* 6 June 1872, 5.
58. Bunt to Whewell, 31 December 1835, Whewell Papers, R.6.20/89.
59. Bunt to Whewell, undated, Whewell Papers; Bunt's emphasis.
60. Bunt to Whewell, 29 October 1835, Whewell Papers, R.6.20/87.
61. Bunt to Whewell, 29 April 1836, Whewell Papers, R.6.20/94.
62. Bunt to Whewell, 9 April 1836, Whewell Papers, R.6.20/93.
63. Ibid.
64. Bunt to Whewell, 2 May 1836, Whewell Papers, R.6.20/29.
65. Society of Merchant Venturers, 12 October 1838, HB 18.
66. Ibid.
67. "Memorial Respecting the Registration of the Tides in the Port of Bristol, Respectfully Addressed to the Mayor, Alderman, and Councilors, of the City and County of Bristol," undated, Whewell Papers, R.6.20/152. Stutchbury wrote the initial draft of the "Memorial," then sent it to Whewell for his additions and corrections.
68. Ibid.
69. William Whewell, "Description of a New Tide-Gauge, Constructed by Mr. T. G. Bunt, and Erected on the Eastern Bank of the River Avon, in Front of the Hotwell House, Bristol, 1837," *Philosophical Transactions* 128 (1838): 249–55.

70. See chapter 3 for the initial invention of the self-registering tide gauge.
71. Bunt to Beaufort, 14 October 1837, Beaufort Papers, B649.
72. Whewell, *The Philosophy of the Inductive Sciences, Founded upon Their History*, 2 vols. (London: John W. Parker, 1840), 2:557.
73. For the importance of a "steady hand," see Whewell, *Philosophy of the Inductive Sciences,* 2:545ff.
74. Bunt to Whewell, February 1839, Whewell Papers, R.6.20/127(1).
75. The word "discussion" in this passage refers to the reduction of the data, though it could easily refer to Bunt's and Whewell's discussion of the best means to proceed. See my clarification below. Bunt to Whewell, February 1839, Whewell Papers, R.6.20/127(1); Bunt's emphasis.
76. Bunt to Whewell, February 1839, Whewell Papers, R.6.20/127(1); Bunt's emphasis. Bunt showed him a description earlier in Bunt to Whewell, 30 September 1837, Whewell Papers, R.6.20/107.
77. William Whewell, "Researches on the Tides," 9th ser., "On the Determination of the laws of the Tides from Short Series of Observations," *Philosophical Transactions* 128 (1838): 231–47. Whewell cites Bunt on 233–35.
78. Bunt to Whewell, 6 January 1837, Whewell Papers, R.6.20/97(1).
79. Secord, "Corresponding Interests," 383–408.
80. Bunt to Whewell, 13 January 1838, Whewell Papers, R.6.20/111.
81. William Whewell, "Report on Discussions of Bristol Tides, Performed by Mr. Bunt under the Direction of the Rev. W. Whewell," *British Association Report* 1841 (1842): 30.
82. Bunt to Whewell, 14 January 1841, Whewell Papers, R.6.20/141.
83. Whewell, "Discussions of Bristol Tides," 31. Bunt reported that every inch of fall in the barometer produced a one-fifteen-inch rise in the tide.
84. Thomas Gamlen Bunt, "Discussion of Tide Observations at Bristol," *Philosophical Transactions* 157 (1867): 1–6.
85. Whewell to Lubbock, undated, Whewell Papers; my italics.
86. See, for example, Ross to Whewell, 31 July 1847, Whewell Papers, R.6.20/271.
87. Ross to Whewell, 5 October 1839, Whewell Papers, R.6.20/266.
88. Ross to Whewell, 8 October 1839, Whewell Papers. R.6.20/267.
89. William Whewell, "Report on the Discussion of Leith Tide Observations, Executed by Mr. D. Ross, of the Hydrographer's Office, Admiralty, under the Direction of the Rev. W. Whewell," *British Association Report* 1841 (1842): 33–36.
90. Ibid., 34.
91. Ibid., 34–35.
92. Ross to Whewell, 4 August 1847, Whewell Papers, R.6.20/273.
93. William Whewell, "Researches on the Tides," 13th ser., "On the Tides of the Pacific, and on the Diurnal Inequality," Bakerian Lecture, *Philosophical Transactions* 138 (1848): 1–29. He acknowledged Ross's contribution on 12.

94. Ross to Whewell, 20 July 1847, Whewell Papers, R.6.20/269. The range of high tide differs each month. "Spring tides" are the highest high tides each year, "neap tides" the lowest high tides.
95. Ibid.
96. Ross to Whewell, 27 July 1847, Whewell Papers, R.6.20/270.
97. Ross to Beaufort, 17 October 1849, RHO.
98. William Whewell, "Researches on the Tides," 14th ser., "On the Results of Continued Tide Observations at Several Places on the British Coasts," *Philosophical Transactions* 140 (1850): 227–33.
99. William Whewell, *History of the Inductive Sciences, from the Earliest to the Present Time*, 1st ed., 3 vols. (London, 1837), 1:430; my italics.
100. Ibid., 434. Notice how similar this process is to Whewell's project in tidology.
101. Whewell, *History*, 2:127.
102. Ibid., 2:216–25.
103. Ibid., 2:246–53.
104. Ibid., 2:247.
105. K. B. Martin, "Tides and Tide Tables," *Nautical Magazine and Naval Chronicle* 9 (1839): 86.
106. Ibid.; Martin's italics.
107. Ibid.
108. Whewell, *History*, 1:457.

CHAPTER SEVEN: CREATING SPACE FOR THE "SCIENTIST"

1. Charles Babbage, *Reflections on the Decline of Science in England, and on Some of Its Causes* (London, 1830; reprint, Westmead, UK: Gregg International Publishers, 1969). Babbage built his machine to produce mathematical tables of all types, but he politicked for funds based on the production of navigation tables. See Anthony Hyman, *Charles Babbage: Pioneer of the Computer* (Princeton: Princeton University Press, 1982), 49. See also Simon Schaffer, "Babbage's Intelligence: Calculating Engines and the Factory System," *Critical Inquiry* 21 (1994): 203–27.
2. Babbage, *Reflections*, 126.
3. William J. Ashworth, "'Labour Harder Than Thrashing': John Flamsteed, Property and Intellectual Labour in Nineteenth Century England," in *Flamsteed's Stars: New Perspectives on the Life and Work of the First Astronomer Royal, 1646–1719*, ed. Frances Willmoth (Rochester, NY: Boydell Press, 1997), 199–201; William J. Ashworth, "The Calculating Eye: Baily, Herschel, Babbage and the Business of Astronomy," *British Journal for the History of Science* 27 (1994): 409–41.
4. Richard Yeo, "Scientific Method and the Rhetoric of Science in Britain, 1830–1917," in *The Politics and Rhetoric of Scientific Method: Historical Studies*, ed.

John A. Schuster and Richard Yeo (Dordrecht: Kluwer Academic Publishers, 1986); Ashworth, "'Labour Harder Than Thrashing'"; Richard Yeo, *Defining Science: William Whewell, Natural Knowledge, and Public Debate in Early Victorian Britain* (Cambridge: Cambridge University Press, 1993), 129–24; and Timothy L. Alborn, "The Business of Induction: Industry and Genius in the Language of British Scientific Reform, 1820–1840," *History of Science* 34 (1996): 91–121.

5. Bernard Lightman, "Introduction," in *Victorian Science in Context,* ed. Bernard Lightman (Chicago: University of Chicago Press, 1997), 10.

6. See Jack Morrell and Arnold Thackray, *Gentlemen of Science: Early Years of the British Association for the Advancement of Science* (Oxford: Clarendon Press, 1981), and Yeo, *Defining Science.*

7. Joost Mertens, "From Tubal Cain to Faraday: William Whewell as a Philosopher of Technology," *History of Science* 38 (2000): 321.

8. David Philip Miller, "The Revival of the Physical Sciences in Britain, 1815–1840," *Osiris* 2 (1986): 124.

9. Though Whewell coined the term at this meeting, he used the word in print for the first time the next year in the *Quarterly Review* in a review of Mary Somerville's *On the Connexion of the Physical Sciences.*

10. Whewell, "Address," *British Association Report* 1833 (1834): iv–xii.

11. Ibid., xiv.

12. Ibid., xx.

13. Ibid., xxv.

14. Whewell to Jones, 21 October 1833, in *William Whewell, D.D., Master of Trinity College, Cambridge: An Account of His Writings with Selections from His Literary and Scientific Correspondence,* ed. Isaac Todhunter, 2 vols. (New York: Johnson Reprint, 1970), 2:171.

15. For a discussion of the history of Whewell's *History,* see Menachem Fisch, *William Whewell, Philosopher of Science.* (Oxford: Clarendon Press, 1991).

16. William Whewell, *History of the Inductive Sciences, from the Earliest to the Present Time,* 1st ed., 2 vols. (London, 1882), 1:4.

17. Todhunter, *Whewell,* 2:186.

18. Menachem Fisch, "A Philosopher's Coming of Age: A Study in Erotetic Intellectual History," in *William Whewell: A Composite Portrait,* ed. Menachem Fisch and Simon Schaffer (Oxford: Clarendon Press, 1991), 32; Whewell, *Philosophy,* 2:212ff.

19. William Whewell, *The Philosophy of the Inductive Sciences, Founded upon Their History,* 2nd ed., 2 vols. (London: John W. Parker, 1842), 2:213.

20. Ibid., 2:220.

21. Whewell, *Philosophy,* 2:260.

22. Ibid., 1:4.

23. Ibid., 2:542.

24. Whewell, "Address" (1833), xxv.

25. See Morrell and Thackray, *Gentlemen of Science*, 256–64; Gerard L'E. Turner, *Nineteenth-Century Scientific Instruments* (Berkeley and Los Angeles: University of California Press, 1983); and W. D. Hackmann, "The Nineteenth Century Trade in Natural Philosophy Instruments in Britain," in *Nineteenth-Century Scientific Instruments and Their Makers: Papers Presented at the Fourth Scientific Instruments Symposium, Amsterdam, 23–26 October 1984,* ed. P. R. de Clercq (Amsterdam, 1985), 53–91.

26. Albert Van Helden and Thomas L. Hankins, "Introduction: Instruments in the History of Science," *Osiris* 9 (1994): 3.

27. See Thomas L. Hankins and Robert J. Silverman, *Instruments and the Imagination* (Princeton, NJ: Princeton University Press, 1995), esp. 113–47.

28. Philip L. Woodworth, *A Study of Changes in High Water Levels and Tides at Liverpool during the Last Two Hundred and Thirty Years with Some Historical Background,* Report 56 (Birkenhead, UK : Proudman Oceanographic Laboratory, 1999), 28. See also G. H. Darwin, "On the Harmonic Analysis of Tidal Observations of High and Low Water," *Proceedings of the Royal Society of London* 48 (1890): 278–339.

29. Jack Morrell, "The Judge and Purifier of All," *History of Science* 30 (1992): 97–114.

30. Ibid., 111; Yeo, *Defining Science,* 227–28.

31. For instrument makers as fellows of the Royal Society, see J. A. Bennett, "Instrument Makers and the 'Decline of Science in England': The Effect of Institutional Change on the Elite Makers of the Early Nineteenth Century," in *Nineteenth-Century Scientific Instruments and Their Makers: Papers Presented at the Fourth Scientific Instruments Symposium, Amsterdam, 23–26 October 1984,* ed. R. de Clercq (Amsterdam, 1985), 13–27.

32. William Whewell, "On the Results of Observations Made with a New Anemometer," *Transactions of the Cambridge Philosophical Society* 6, pt. 2, Whewell Papers, R.6.9/2–3.

33. William Whewell, "Note on the Working of the Anemometer Since the Account Given to the Society May 1, 1837." read 20 May 1839, Whewell Papers. R.6.9/4.

34. Thomas A. Southwood to Whewell, 19 December 1836, Add.Ms.a.212/131, Whewell Papers.

35. Whewell, "Note on the Working of the Anemometer."

36. Whewell to Airy, 17 November 1837, in Todhunter, *Whewell,* 2:262.

37. Airy to Smyth, undated, Airy Papers, RGO.6.499. The correspondence between Bunt and Airy dates the letter to December 1840.

38. Smyth to Airy, undated, Airy Papers, RGO.6.499.

39. Airy to Bunt, 28 December 1840, Airy Papers, RGO.6.499.42.

40. Bunt to Airy, undated, Airy Papers, RGO.6.499.43.

41. Airy to Whewell, 21 January 1841, Whewell Papers, Add.Ms.a.200/35.

42. Airy to Hugh Godfray, 14 August 1843, Airy Papers, RGO.6.499.82; italics added.

43. Hugh James Rose was succeeded as general editor by his brother, Henry John Rose, sometime during the preparation of Airy's manuscript for publication (about 1842).

44. Rose to Airy, 8 June 1842, Airy Papers, RGO.500.71.

45. Airy to Rose, 10 June 1842, Airy Papers, RGO.500.72.

46. Whewell, *Philosophy,* 2:485.

47. Ibid., 2:500–501.

48. Bennett, "Instrument Makers and the 'Decline of Science in England.'"

49. Whewell to Herschel, 3 April 1836, Whewell Papers.

50. William Whewell, "Address," *British Association Report* 1841 (1842): xxxiii.

51. Roy MacLeod, "Introduction: On the Advancement of Science," in *The Parliament of Science: The British Association for the Advancement of Science, 1831–1981,* ed. Roy MacLeod and Peter Collins (Northwood, UK: Science Reviews, 1981), 31.

52. For magnetism, see Patricia Fara, *Sympathetic Attractions: Magnetic Practices, Beliefs and Symbolism in 18th Century England* (Cambridge: Cambridge University Press, 1994).

53. Alison Winter, "'Compasses All Awry': The Iron Ship and the Ambiguities of Cultural Authority in Victorian Britain," *Victorian Studies* 38 (1994): 69–98; Frank James, "Davy in the Dockyard: Humphrey Davy, the Royal Society and the Electro-chemical Protection of the Copper Sheeting of His Majesty's Ships in the Mid-1820s," *Physis* 29 (1992): 217.

54. Lucille Brockway, *Science and Colonial Expansion: The Role of the British Royal Botanical Gardens* (New York: Academic Press, 1979); Robert A. Stafford, *Scientist of Empire: Sir Roderick Murchison, Scientific Exploration, and Victorian Imperialism* (Cambridge: Cambridge University Press, 1989), esp. chap. 8.

55. Olin J. Eggen, "Airy, Sir George Biddell," in *Dictionary of Scientific Biography,* ed. Charles Gillispie (New York: Charles Scribner's Sons, 1976), 1:84–87.

56. Charles H. Cotter, "The Early History of Ship Magnetism: The Airy–Scoresby Controversy," *Annals of Science* 34 (1977): 589–99; J. A. Bennett, "George Biddell Airy and Horology," *Annals of Science* 37 (1980): 269–85.

57. As quoted in Robert W. Smith, "A National Observatory Transformed: Greenwich in the Nineteenth Century." *Journal for the History of Astronomy* 222 (1991): 9.

58. George Biddell Airy, "On the Laws of the Rise and Fall of the Tide in the River Thames," *Philosophical Transactions* 132 (1842): 1–8; Airy, "On the Laws of Individual Tides at Southampton and at Ipswitch," *Philosophical Transactions* 133 (1843): 45–54; Airy, "On the Laws of the Tides on the Coasts of Ireland, as Inferred from an Extensive Series of Observations Made in Connection with the Ordnance Survey of Ireland," *Philosophical Transactions* 135 (1845): 1–123.

59. Wilfrid Airy, ed., *Autobiography of Sir George Biddell Airy* (Cambridge: Cambridge University Press, 1896), 217.

60. Published in George Biddell Airy, *Essays* (London, 1865).

61. Ibid.; italics added.
62. Airy, "On the Laws of the Tides on the Coast of Ireland," 103.
63. Susan Faye Cannon, *Science in Culture: The Early Victorian Period* (New York: Science History Publications, 1978), chap. 2.
64. Robert V. Bruce, *The Launching of Modern American Science: 1846–1876* (Ithaca, NY: Cornell University Press, 1987), 220.
65. For the use of the term "interdependence" and the most promising lines of research in this area, see R. W. Home and Sally Kohlstedt, eds., *International Science and National Scientific Identity: Australia between Britain and America* (Boston: Kluwer Academic Publishers, 1991).
66. Bache to Whewell, 29 October 1834, Whewell Papers, Add.Ms.a.200/195.
67. Charles Henry Davis to Edward Everett, president of Harvard University, Cambridge, Massachusetts, enclosed in Edward Everett to William Whewell, 10 March 1848, Whewell Papers, R.6.20/176a(1–3). See also Bache to Whewell, 13 August 1852, Whewell Papers, R.6.20/53(1).
68. For a discussion of Alexander Dallas Bache's tidal work in relation to his other responsibilities as head of the Coast Survey, see Hugh Richard Slotten, *Patronage, Practice, and the Culture of American Science: Alexander Dallas Bache and the U.S. Coast Survey* (Cambridge: Cambridge University Press, 1994).
69. Davis to Everett, 7 March 1848, Whewell Papers.
70. He also served as the superintendent of the Naval Observatory from 1865 to 1867 and from 1873 to 1877. See Frederick Nebeker, *Astronomy and the Geophysical Tradition in the United States in the 19th Century: A Guide to Manuscript Sources in the Library of the American Philosophical Society* (Philadelphia: American Philosophical Society Library Press, 1991), 30.
71. Everett to Whewell, 10 March 1848, Whewell Papers.
72. Charles Henry Davis, "A Memoir upon the Geological Action of the Tidal and Other Currents of the Ocean," *Memoirs of the American Academy of Arts and Sciences* 4 (1849), 117–56; Davis, "The Law of the Deposit of the Flood Tide: Its Dynamical Action and Office," *Smithsonian Institution Contributions to Knowledge* 3, no. 33, (1852).
73. Davis to Everett, 7 March 1848, Whewell Papers.
74. Bache to Whewell, 13 August 1852, Whewell Papers.
75. Ibid.
76. Ibid.
77. Slotten, *Patronage*, 129.
78. Whewell, "Researches on the Tides," 9th ser., "On the Determination of the Laws of the Tides from Short Series of Observations," *Philosophical Transactions* 128 (1838): 231–48, esp. 233–35; Davis to Everett, 7 March 1848, Whewell Papers.
79. Slotten, *Patronage*, 172.
80. Ibid., 127. Astronomy and hydrography were the top two.
81. Joseph Henry, "Eulogy on Professor Alexander Dallas Bache, Late Superintendent of the United States Coast Survey," in *General Appendix to the*

*Smithsonian Report for 1870* (Washington, DC: Government Printing Office, 1871), 91–116.

82. A. D. Bache, "Preliminary Determination of Cotidal Lines on the Atlantic Coast of the United Sates," *American Journal of Science and Arts* 21 (1856): 14.

83. A. D. Bache, *Tide Tables for the Principal Sea Ports of the United States* (New York: E and G. W. Blunt, 1855); American Philosophical Society, Philadelphia, Pamphlet V 316, no. 3.

84. Slotten, *Patronage,* 124.

85. William Whewell, "On Mr. Superintendent Bache's Tide Observations," *British Association Report* 1854 (1855): 28.

86. For the ways Whewell and Bache attempted to define the practice of science, see Richard Yeo, *Defining Science: William Whewell, Natural Knowledge, and Public Debate in Early Victorian Britain* (Cambridge: Cambridge University Press, 1993), and Hugh Richard Slotten, *Patronage, Practice, and the Culture of American Science: Alexander Dallas Bache and the U.S. Coast Survey* (Cambridge: Cambridge University Press, 1994).

87. Robert K. Merton, "De-gendering 'Man of Science': The Genesis and Epicene Character of the Word 'Scientist,'" in *Sociological Visions,* ed. Kai Erikson (Lanham, MD: Rowman and Littlefield, 1997), 225–53; Sydney Ross, "Scientist: The Story of a Word," *Annals of Science* 18, no. 2 (1962): 65–85.

88. Alexander Dallas Bache, *AAAS Reports* (1851). As quoted in Bruce, *Launching of Modern American Science,* 217.

89. Robert Brain, "Going to the Exhibition," in *The Physics of Empire: Public Lectures,* ed. Richard Staley (Cambridge: Whipple Museum of the History of Science, 1994), 137.

90. William Whewell, "Inaugural Lecture, November 26, 1851: The General Bearing of the Great Exhibition on the Progress of Art and Science," Cam.c.851.17, Rare Books Room, Cambridge Library, 6.

91. Yeo, *Defining Science,* 224.

92. See, for example, Iwan Rhys Morus, *Frankenstein's Children: Electricity, Exhibition, and Experiment in Early 19th-Century London* (Princeton, NJ: Princeton University Press, 1998); Simon Schaffer, "Babbage's Intelligence: Calculating Engines and the Factory System," *Critical Inquiry* 21 (1994): 203–27; Martin J. S. Rudwick, *The Great Devonian Controversy: The Shaping of Scientific Knowledge among Gentlemanly Specialists* (Chicago: University of Chicago Press, 1985).

93. For a discussion of the traveler motif and the relation between science and empire, see Simon Schaffer, "Empires of Physics," in *The Physics of Empires: Public Lectures,* ed. Richard Staley (Cambridge: Whipple Museum of the History of Science, 1994), 97–112.

94. Leah Ceccarelli has termed this rhetorical device a "conceptual chiasmus." See Leah Ceccarelli, *Shaping Science with Rhetoric: The Cases of Dobzhansky, Schrödinger, and Wilson* (Chicago: University of Chicago Press, 2001).

95. Rhetoricians term this a double hierarchy argument based on sequential connections. See Ch. Perelman and L. Olbrechts-Tyteca, *The New Rhetoric*

(Notre Dame: University of Notre Dame Press, 1969), and Jeanne Fahne-stock, *Rhetorical Figures in Science* (Oxford: Oxford University Press, 2002).

96. As quoted in R. G. W. Anderson, "Were Scientific Instruments in the Nine-teenth Century Different? Some Initial Considerations," in *Nineteenth-Century Scientific Instruments and Their Makers: Papers Presented at the Fourth Scientific Instruments Symposium, Amsterdam, 23–26 October 1984*, ed. P. R. de Clercq (Amsterdam, 1985), 6.

97. The medals awarded ranged from honorable mentions to prize medals, to council medals. Britain took sixteen of the fifty-four honorable mentions, forty-three of the ninety-four prize medals, and seventeen of the thirty-one council medals. See W. D. Hackmann, "The Nineteenth-Century Trade in Natural Philosophy Instruments in Britain," in *Nineteenth-Century Scientific Instruments and Their Makers: Papers Presented at the Fourth Scientific Instruments Symposium, Amsterdam, 23–26 October 1984*, ed. P. R. de Clercq (Amsterdam, 1985) 53–91, esp. 62–63.

98. As quoted in Bennett, "Instrument Makers and the 'Decline of Science in England,'" 23.

99. Whewell to Beaufort, 2 November 1836, Beaufort Papers.

100. Whewell to Jones, 6 November 1836, in William Ashworth, "Calendar of William Whewell's Correspondence," unpublished manuscript, used cour-tesy of the author.

101. David Knight, "Tyrannies of Distance in British Science," in *International Sci-ence and National Scientific Identity*, ed. W. R. Home and Sally Gregory Kohlst-edt (Boston: Kluwer Academic Publishers, 1991), 39–54.

102. For the insufficiency of class analysis in the Victorian era, see Gareth Sted-man Jones, *Languages of Class: Studies in English Working Class History 1832–1982* (Cambridge: Cambridge University Press, 1983); Patrick Joyce, Vis*ions of the People: Industrial England and the Question of Class, 1848–1914* (Cam-bridge: Cambridge University Press, 1991); and David Cannadine, *Class in Britain* (London: Penguin, 2000).

CONCLUSION: THE TIDES OF EMPIRE

1. As quoted in Edgar Gold, *Maritime Transport: The Evolution of International Marine Policy and Shipping Law* (Lexington, MA: Lexington Books, 1981), 41.

2. Phillip E. Steinberg, *The Social Construction of the Ocean* (Cambridge: Cam-bridge University Press, 2001).

3. Simon Ville, *English Shipowning during the Industrial Revolution* (Manchester, UK: Manchester University Press, 1987), 11; John Armstrong and Philip S. Bagwell, "Coastal Shipping," in *Transport in the Industrial Revolution*, ed. Derek H. Aldcroft and Michael J. Freeman (Manchester, UK: Manchester University Press, 1983), 145.

4. Nick Harley, "Foreign Trade: Comparative Advantage and Performance," in *The Economic History of Britain since 1700*, vol. 1, ed. Roderick Flood and

Donald McCloskey (Cambridge: Cambridge University Press, 1994), 300–331.

5.   Helen M. Rozwadowski, *Fathoming the Ocean: The Discovery and Exploration of the Deep Sea* (Cambridge, MA: Harvard University Press, 2005); David Elliston Allen, *The Naturalist in Britain: A Social History* (Princeton, NJ: Princeton University Press, 1994), esp. chap. 11.

6.   Andrew D. Lambert, "Preparing for the Long Peace: The Reconstruction of the Royal Navy, 1815–1830," *Mariner's Mirror* 82 (1996): 42.

7.   Michael Lewis, *The Navy in Transition, 1814–1864: A Social History* (London: Hodder and Stoughton, 1965), 12.

8.   Paul M. Kennedy, *The Rise and Fall of British Naval Mastery* (New York: Scribner, 1976), 163.

9.   Barry M. Gough, "The British Reoccupation and Colonization of the Falkland Islands, or Malvinas, 1832–1843," *Albion* 22 (1990): 273.

10.   R. O. Morris, "Surveying Ships of the Royal Navy from Cook to the Computer Age," *Mariner's Mirror* 72 (1986): 389.

11.   Eric L. Mills, "*De Motu Marium*: Understanding the Oceans Before the Second Scientific Revolution," Stillman Drake Lecture, Canadian Society for History and Philosophy of Science, June 1999; L. Kellner, "Alexander von Humboldt and the Organization of International Collaboration in Geophysical Research," *Contemporary Physics* 1 (1959): 35–48.

12.   As quoted in Michael Dettelbach, "Humboldtian Science," in *Cultures of Natural History*, ed. N. Jardine, J. A. Secord, and E. C. Spary (Cambridge: Cambridge University Press, 1996), 298.

13.   Charles Close, *The Early Years of the Ordnance Survey* (Newton Abbot, UK: David and Charles Reprints, 1969), 55.

14.   John Augustus Lloyd to Mr. Kater, October 1830, Lubbock Papers. Official letters concerning Lloyd's commission can be found in PRO, ADM.1.1365.

15.   John William Lubbock, *An Elementary Treatise on the Tides* (London, 1839), 24.

16.   John Augustus Lloyd, "Account of Levellings Carried across the Isthmus of Panama, to Ascertain the Relative Height of the Pacific Ocean at Panama, and of the Atlantic at the Mouth of the River Chagres, Accompanied by Geographical and Topographical Notices of the Isthmus," *Philosophical Transactions* 120 (1830): 50–68.

17.   [Reidy, Michael S.], "Lloyd, John Augustus," in *Dictionary of Nineteenth-Century British Scientists*, ed. Bernard Lightman (Chicago: University of Chicago Press, 2004); see also Ruth Brindze, *The Rise and Fall of the Seas: The Story of the Tides* (London: Harcourt, 1964).

18.   William Whewell, "Researches on the Tides," 7th ser., "On the Diurnal Inequality of the Height of the Tide, Especially at Plymouth and at Singapore; and on the Mean Level of the Sea," *Philosophical Transactions* 127 (1837): 84.

19. William Whewell, "Researches on the Tides," 10th ser., "On the Laws of Low Water at the Port of Plymouth, and on the Permanency of Mean Water," *Philosophical Transactions* 129 (1839): 151–63.
20. G. B. Airy, "On the Laws of the Tides on the Coasts of Ireland, as Inferred from an Extensive Series of Observations Made in Connection with the Ordnance Survey of Ireland, " *Philosophical Transactions* 135 (1845): 1.
21. David Cartwright, *Tides: A Scientific History* (Cambridge: Cambridge University Press, 1999), 116–17. Denham to Lubbock, 7 May 1841, D111, Lubbock Papers; Richard Veevers, "On the Tides and Datums of the Lancashire Coast," *Transactions of the Historic Society of Lancashire and Cheshire* 49 (1897): 171–75; Michael S. Reidy, "Masters of Tidology: The Cultivation of the Physical Sciences in Early Victorian Liverpool," *Transactions of the Historic Society of Lancashire and Cheshire* 152 (2003): 51–77.
22. See especially Ken Alder, *The Measure of All Things: The Seven-Year Odyssey and Hidden Error That Transformed the World* (New York: Free Press, 2002), and Simon Schaffer, "Empires of Physics," in *The Physics of Empire: Public Lectures,* ed. Richard Staley (Cambridge: Whipple Museum of the History of Science, 1994), 97–112.
23. This tradition continues with the dropping of a ball on New Year's Eve.
24. Alain Corbin, *The Lure of the Sea: The Discovery of the Seaside in the Western world: 1750–1840,* Trans. Jocelyn Phelps (Berkeley and Los Angeles: University of California Press, 1994), 106.
25. Alistair Sponsel, "Fathoming the Depth of Charles Darwin's Theory of Coral Reef Formation," paper presented at the History of Science Society annual meeting, Austin, TX, 20 November 2004.
26. See Cartwright, *Tides,* esp. 129–52.
27. Susan Schlee, *The Edge of an Unfamiliar World: A History of Oceanography* (New York: E. P. Dutton, 1973).
28. John Cawood, "The Magnetic Crusade," *Isis* 70 (1979): 500; see also Cawood, "Terrestrial Magnetism and the Development of International Collaboration in the Early Nineteenth Century," *Annals of Science* 34 (1977): 551–87.
29. Jack Morrell and Arnold Thackray, *Gentlemen of Science: Early Years of the British Association for the Advancement of Science* (Oxford: Clarendon Press, 1981), 359.
30. Katharine Anderson, *Predicting the Weather: Victorians and the Science of Meteorology* (Chicago: University of Chicago Press, 2005), esp. chap. 6.
31. John William Herschel, "Meteorology," in *The Admiralty Manual of Scientific Enquiry* (London: His Majesty's Stationer, 1849).
32. See especially Morrell and Thackray, *Gentlemen of Science,* 520, and Vladimir Jankovic, "Ideological Crests versus Empirical Troughs: John Herschel's and William Radcliffe Birt's Research on Atmospheric Waves, 1843–50," *British Journal for the History of Science* 31 (1998): 21–40.
33. See James Rodger Fleming, *Meteorology in America, 1800–1870* (Baltimore: Johns Hopkins University Press, 1990).

34. Anderson, Predicting the Weather.
35. Duncan Carr Agnew, "Robert Fitzroy and the Myth of the 'Marsden Square': Transatlantic Rivalries in Early Marine Meteorology," *Notes and Records of the Royal Society of London* 58 (2004): 21–46.
36. As quoted in ibid., 29.
37. See "Report of the Committee," *British Association Reports* 1841 (1842): xxi. The committee consisted of Herschel, Whewell, Lloyd, the dean of Ely and Sabine.
38. Sir John F. W. Herschel, "Address," *British Association Report* 1845 (1846): xxxv.
39. Margaret Deacon, *Scientists and the Sea, 1650–1900* (London: Academic Press, 1971), esp. chap. 13.
40. W. Whewell and James C. Ross, "Report by the Rev. W. Whewell, D.D., and Sir James C. Ross, upon the Recommendation of an Expedition for the Purpose of Completing Our Knowledge of the Tides," *British Association Report* 1847 (1848): 134–35.
41. William Whewell, "On Our Ignorance of the Tides," *British Association Report* 1850 (1851): 27.
42. Ibid., 28.
43. Nicholas Courtney, *Gale Force 10: The Life and Legacy of Admiral Beaufort, 1774–1857* (London: Review, 2002), 230–31.
44. Whewell to Forbes, 23 October 1856, Whewell Papers; italics added.
45. Eric L. Mills, "The Historian of Science and Oceanography After Twenty Years," *Earth Sciences History* 12, no. 1 (1993): 5–18.
46. See especially Dorinda Outram, "New Spaces in Natural History," in *Cultures of Natural History*, ed. N. Jardine, J. A. Secord, and E. C. Spary (Cambridge: Cambridge University Press, 1996), 249–50; Robert A. Stafford, "Geological Surveys, Mineral Discoveries, and British Expansion, 1835–1871," *Journal of Imperial and Commonwealth History* 12 (1984): 5–32; Lucille Brockway, *Science and Colonial Expansion: The Role of the British Royal Botanic Gardens* (New York: Academic Press, 1979).
47. Roy MacLeod and Fritz Rehbock, eds., *Nature in Its Greatest Extent: Western Science in the Pacific* (Honolulu: University of Hawaii Press, 1988).
48. William Whewell, *History of the Inductive Sciences* (London: John W. Parker, 1837), 457.
49. Ibid.
50. Richard Harding, *The Evolution of the Sailing Navy, 1509–1815* (London: Saint Martin's Press, 1995), 1.
51. George Modelski and William R. Thompson, *Seapower in Global Politics, 1494–1993* (Seattle: University of Washington Press, 1988), 16.
52. "Liquid history" was coined by John Burns to describe the Thames, as quoted in Frank C. Bowen, *Port of London Guide* (London: Coram, 1955), 21.

.

# Bibliography

MANUSCRIPTS CONSULTED

[Airy Papers] George Biddell Airy Papers, Royal Greenwich Observatory Manuscripts [RGO] 6.499–550, University of Cambridge Libraries, Cambridge

American Philosophical Society Manuscript Collection, Philadelphia

[Beaufort Papers] Sir Francis Beaufort Papers, Royal Hydrographic Office, Taunton, UK

[Board of Longitude Papers] Board of Longitude Papers, Miscellaneous Tide and Trade Winds (14.51), Royal Greenwich Observatory Manuscripts [RGO] 6.499–550, University of Cambridge Library, Cambridge

Bristol Institution Archives, reference 32079, Bristol Records Office, Bristol, UK

British Library Manuscripts (40,492), British Library, London

[HB] Merchant Hall Book of Proceedings, Bristol

[Herschel Papers] John William Herschel Papers, Royal Society of London

[Lubbock Papers] Sir John William Lubbock Papers, Royal Society of London, London

[OMNH] Oxford University Museum of Natural History, Oxford

Phillips Papers, Oxford University Museum of Natural History, Oxford, England

[PRO] Admiralty Records, Public Records Office, Kew, UK

[Royal Society Papers] Royal Society of London Manuscripts, Royal Society of London, London

[RHO] Royal Hydrographic Office, Taunton, UK

[SDUK] Society for the Diffusion of Useful Knowledge Manuscripts, Science Library, University College, London

[SMV] Society of Merchant Venturers, Merchant Hall Book of Proceedings, Bristol, UK

[Whewell Papers] William Whewell Papers, Trinity College, University of Cambridge, Cambridge

PUBLISHED SOURCES

"Accidents at the West India Docks, through the High Tide." *London Times*, 15 October 1802, 2, col. b.

"Address of His Royal Highness the President," 30 November 1837. *Abstracts of Papers in the "Philosophical Transactions," and, the "Proceedings of the Royal Society of London."* London, 1838.

Alder, Ken. *The Measure of All Things: The Seven-Year Odyssey and Hidden Error That Transformed the World.* New York: Free Press, 2002.

Agar, Jon. *The Government Machine: A Revolutionary History of the Computer.* Cambridge, MA: MIT Press, 2003.

Agnew, Duncan Carr. "Robert Fitzroy and the Myth of the 'Marsden Square': Transatlantic Rivalries in Early Marine Meteorology." *Notes and Records of the Royal Society of London* 58 (2004): 21–46.

Airy, George Biddell. *Essays.* London, 1865.

———. "On the Laws of Individual Tides at Southampton and at Ipswitch." *Philosophical Transactions* 133 (1843): 45–54.

———. "On the Laws of the Rise and Fall of the Tide in the River Thames." *Philosophical Transactions* 132 (1842): 1–8.

———. "On the Laws of the Tides on the Coasts of Ireland, as Inferred from an Extensive Series of Observations Made in Connection with the Ordnance Survey of Ireland." *Philosophical Transactions* 135 (1845): 1–123.

———. "Tides and Waves." In *Encyclopaedia Metropolitana*, vol. 5 (London, 1845), 241–396.

Airy, Wilfrid, ed. *Autobiography of Sir George Biddell Airy.* Cambridge: Cambridge University Press, 1896.

Aiton, E. J. "Galileo's Theory of the Tides." *Annals of Science* 10 (1854): 44–57.

———. "The Contributions of Newton, Bernoulli and Euler to the Theory of the Tides." *Annals of Science* 11 (1955): 206–23.

———. "Descartes' Theory of the Tides." *Annals of Science* 11 (1955): 337–48.

Alborn, Timothy L. "The Business of Induction: Industry and Genius in the Language of British Scientific Reform, 1820–1840." *History of Science* 34 (1996): 91–121.

Allen, David Elliston. *The Naturalist in Britain: A Social History.* Princeton, NJ: Princeton University Press, 1994.

*American Historical Review* 111, no. 3 (June 2006).

Anderson, James. "Some Observations on the Peculiarity of the Tides between Fairleigh and the North Foreland." *Philosophical Transactions* 109 (1819): 217–33.

Anderson, Katherine. *Predicting the Weather: Victorians and the Science of Meteorology*. Chicago: University of Chicago Press, 2005.

Anderson, R. G. W. "Were Scientific Instruments in the Nineteenth Century Different? Some Initial Considerations." In *Nineteenth-Century Scientific Instruments and Their Makers: Papers Presented at the Fourth Scientific Instruments Symposium, Amsterdam, 23–26 October 1984*, ed. P. R. de Clercq, 1–12. Amsterdam, 1985.

Armstrong, John, and Philip S. Bagwell. "Coastal Shipping." In *Transport in the Industrial Revolution, ed.* Derek H. Aldcroft and Michael J. Freeman, 142–76. Manchester, UK: Manchester University Press, 1983.

Ashworth, William J. "The Calculating Eye: Baily, Herschel, Babbage and the Business of Astronomy." *British Journal for the History of Science* 27 (1994): 409–41.

———. "Calendar of William Whewell's Correspondence." Unpublished manuscript, used courtesy of the author.

———. "'Labour Harder Than Thrashing': John Flamsteed, Property and Intellectual Labour in Nineteenth Century England." In *Flamsteed's Stars: New Perspectives on the Life and Work of the First Astronomer Royal, 1646–1719*, ed. Frances Willmoth. Rochester, NY: Boydell Press, 1997.

Babbage, Charles. *Reflections on the Decline of Science in England, and on Some of Its Causes*. London, 1830; Reprint Westmead, UK: Gregg International Publishers, 1969.

Bache, A. D. "Approximate Cotidal lines of Diurnal and Semidiurnal Tides of the Coast of the United States and the Gulf of Mexico." *American Journal of Science* 23 (1856): 12–14.

———. "Approximate Cotidal Lines of the Pacific Coast." *Coast Survey Report*, 1855, 338.

———. "Comparison of the Diurnal Inequality of the Tides at San Diego, San Francisco, and Astoria, on the Pacific." *American Journal of Science* 21 (1854): 10.

———. "Notice of the Tidal Observations Made on the Coast of the United States on the Gulf of Mexico, with Type Curves at Several Stations, and Their Decomposition into the Curves of the Diurnal and Semidiurnal Tides." *American Journal of Science* 21 (1856): 23.

———. "On the Tides on the Western Coast of the United States—Tides of San Francisco Bay, California, with Two Plates." *American Journal of Science* 21 (1856): 1–4.

———. "Preliminary Determination of Cotidal Lines on the Atlantic Coast of the United Sates." *American Journal of Science and Arts* 21 (1854): 14–18.

———. *Tide Tables for the Principal Sea Ports of the United States*. New York: E and G. W. Blunt, 1855.

Baines, Thomas. *History of the Commerce and Town of Liverpool, and of the Rise of Manufacturing Industry in the Adjoining Counties*. London, 1852.

Beaufort, Francis. *General Instruction for the Hydrographic Surveyors of the Admiralty*. London, 1850.

Becher, Harvey W. "Radicals, Whigs and Conservatives: The Middle and Lower Classes in the Analytical Revolution at Cambridge in the Age of Aristocracy." *British Journal for the History of Science* 28 (1995): 405–26.

———. "Voluntary Science in Nineteenth Century Cambridge University to the 1850s." *British Journal for the History of Science* 19 (1986): 57–87.

———. "The Whewell Story: Essay Review of Menachem Fisch, *William Whewell: Philosopher of Science.*" *Annals of Science* 49 (1992): 377–84.

———. "William Whewell and Cambridge Mathematics." *Historical Studies in the Physical Sciences* 11 (1980–81): 1–48.

Bede, Venerable. *The Reckoning of Time*. Trans. with introduction, notes, and commentary by Faith Wallis. Liverpool, UK: Liverpool University Press, 1999.

Belchem, John. "Liverpool in 1848: Image, Identity and Issues." *Transactions of the Historic Society of Lancashire and Cheshire* 147 (1997): 1–26.

Bennett, J. A. "George Biddell Airy and Horology." *Annals of Science* 37 (1980): 269–85.

———. "Instrument Makers and the 'Decline of Science in England': The Effect of Institutional Change on the Elite Makers of the Early Nineteenth Century." In *Nineteenth-Century Scientific Instruments and Their Makers: Papers Presented at the Fourth Scientific Instruments Symposium, Amsterdam, 23–26 October 1984.* ed. P. R. de Clercq, 13–27. Amsterdam, 1985.

Bennett, Scott. "The Editorial Character and Readership of *The Penny Magazine*: An Analysis." *Victorian Periodicals Review* 17 (1984): 127–41.

Bennis, W., and P. W. Biederman. *Organising Genius: The Secrets of Creative Collaboration*. London: Addison Wesley, 1997.

Bernoulli, Daniel, Leonard Euler, Colin Maclaurin, and Antoine Cavalleri. *Pièces qui ont remporté le prix de L'Académie Royale des Sciences en 1740*. Paris, 1741.

Bolster, W. Jeffrey. "Opportunities in Marine Environmental History." *Environmental History* 11, no. 3 (July 2006): 567–97.

Bonelli, Federico, and Lucio Russo. "The Origin of Modern Astronomical Theories of Tides: Chrisogono, de Dominis and Their Sources." *British Journal for the History of Science* 29 (1996): 385–401.

Bowen, Frank C. *Port of London Guide*. London: Coram, 1955.

Brain, Robert. "Going to the Exhibition." In *The Physics of Empire: Public Lectures*, ed. Richard Staley. Cambridge: Whipple Museum of the History of Science, 1994.

Braudel, Fernand. *The Mediterranean and the Mediterranean World in the Age of Philip II*, trans. Siân Reynolds (Berkeley and Los Angeles: University of California Press, 1995).

Brawley, Marc R. *Liberal Leadership: Great Powers and Their Challengers in Peace and War*. Ithaca, NY: Cornell University Press, 1994.

Brewer, John. *The Sinews of Power: War, Money and the English State, 1688–1783*. Cambridge, MA: Harvard University Press, 1988.

Brewster, David. "Double Refraction and Polarization of Light." In *Natural Philosophy*. Library of Useful Knowledge, vol. 1. London, 1829.

Brindze, Ruth. *The Rise and Fall of the Seas: The Story of the Tides*. London: Harcourt, 1964.

"The *British Almanack* ." *London Times*, 15 November 1828 3, col. a.

Brock, W. H. "British Science Periodicals and Culture: 1820–50." *Victorian Periodicals Review* 21 (1988): 47–55.

_____. "Humboldt and the British: A Note on the Character of British Science." *Annals of Science* 50 (1993): 365–72.

Brockway, Lucille. *Science and Colonial Expansion: The Role of the British Royal Botanic Gardens*. New York: Academic Press, 1979.

Broich, John. "Engineering the Empire: British Water Supply Systems and Colonial Societies, 1850–1900." *Journal of British Studies* 46 (2007): 346–65.

Broodbank, Joseph G. *History of the Port of London*. 2 vols. London: Daniel O'Connor, 1921.

Brougham, Henry. "On the Objects, Advantages, and Pleasures of Science." In *Natural Philosophy*. Library of Useful Knowledge, vol. 1. London, 1829.

_____. *Practical Observations upon the Education of the People, Addressed to the Working Classes and Their Employers*. London, 1825.

_____. "Remarks on Almanacs." In *British Almanac*. London, 1829.

Brown, David K. "Wood, Sail and Cannonballs to Steel, Steam and Shells: 1815–1895." In *The Oxford Illustrated History of the Royal Navy*, ed. J. R. Hill, 200–226. Oxford: Oxford University Press, 1995.

Browne, Janet. "Biogeography and Empire." In *Cultures of Natural History*, ed. N. Jardine, J. A. Secord, and E. C. Spary, 305–21. Cambridge: Cambridge University Press, 1996.

Bruce, Robert V. *The Launching of Modern American Science: 1846–1876*. Ithaca, NY: Cornell University Press, 1987.

Buchwald, Jed Z. *The Rise of the Wave Theory of Light: Optical Theory and Experiment in the Early Nineteenth Century*. Chicago: University of Chicago Press, 1989.

Bunt, Thomas Gamlen. "Description of a New Tide-Gauge, Constructed by Mr. T. G. Bunt, and Erected on the Eastern Bank of the River Avon, in Front of the Hotwell House, Bristol, 1837." *Philosophical Transactions* 128 (1838): 249–51.

_____. "Discussion of Tide Observations at Bristol." *Philosophical Transactions* 157 (1867): 1–6.

Burkhardt, Frederick, and Sydney Smith, eds. *The Correspondence of Charles Darwin*. Vol. 1. Cambridge: Cambridge University Press, 1985.

Burnett, Graham D. *Masters of All They Surveyed: Exploration, Geography, and a British El Dorado*. Chicago: University of Chicago Press, 2000.

Burney, James. *Chronological History of the Voyages and Discoveries in the South Sea*. London, 1816. In *Correspondence and Papers of Edmond Halley*, ed. Eugene Fairfield MacPike, 244–47. Oxford: Clarendon Press, 1932. Reprint 1975.

Butts, Robert E. "Whewell, William." In *Dictionary of Scientific Biography*, ed. Charles Gillispie, 14:292–95. New York: Charles Scribner's Sons, 1976.

Cafruny, Alan W. *Ruling the Waves: The Political Economy of International Shipping*. Berkeley and Los Angeles: University of California Press, 1987.

Campbell-Kelly, M., et al., eds. *The History of Mathematical Tables: From Sumer to Spreadsheets.* Oxford: Oxford University Press, 2003.

Cannadine, David. *Class in Britain.* London: Penguin, 2000.

Cannon, Susan Faye. *Science in Culture: The Early Victorian Period.* New York: Science History Publications, 1978.

Cannon, Walter F. "William Whewell, F.R.S. (1794–1866). 2. Contributions to Science and Learning." *Notes and Records of the Royal Society of London* 19 (1964): 176–91.

Cantor, Geoffrey N. "Between Rationalism and Romanticism: Whewell's Historiography of the Inductive Sciences." In *William Whewell: A Composite Portrait*, ed. Menachem Fisch and Simon Schaffer, 67–86. Oxford: Clarendon Press, 1991.

Cardwell, D. S. L. *Technology, Science and History.* London: Heinemann, 1972.

Carson, Rachel. *The Sea Around Us.* New York: Oxford University Press, 1969.

Cartwright, David Edgar. *Tides: A Scientific History.* Cambridge: Cambridge University Press, 1999.

Cawood, John. "The Magnetic Crusade." *Isis* 70 (1979): 493–518.

_____. "Terrestrial Magnetism and the Development of International Collaboration in the Early Nineteenth Century." *Annals of Science* 34 (1977): 551–87.

Ceccarelli, Leah. *Shaping Science with Rhetoric: The Cases of Dobzhansky, Schrödinger, and Wilson.* Chicago: University of Chicago Press, 2001.

Childrey, Joseph. "A Letter of Mr. Joseph Childrey to the Right Reverend Sith Lord Bishop of Sarum, Containing some Animadversions upon the Reverend Dr. John Wallis's Hypothesis about the Flux and Reflux of the Sea, Publish't No. 16 of These Tracts." *Philosophical Transactions* 5 (1670): 2061–68.

Close, Charles. *The Early Years of the Ordnance Survey.* Newton Abbot, UK: David and Charles Reprints, 1969.

Clout, Hugh, ed. *The Times London History Atlas.* London: HarperCollins, 1991.

[Cobden, Richard]. *England, Ireland, and America.* London, 1835.

Cock, Randolph. "Sir Francis Beaufort and the Co-ordination of British Scientific Activity, 1829–1855." PhD diss., University of Cambridge, 2003.

Cohen, I. Bernard, and Anne Whitman, trans. *Isaac Newton: The "Principia."* Berkeley and Los Angeles: University of California Press, 1999.

Colepresse, Samuel. "An Account of Some Observations Made by Mr. Samuel Colepresse at and nigh Plymouth, Anno 1667, by Way of Answer to Some of the Queries concerning Tydes, Propos'd Numb. 17 and 18." *Philosophical Transactions* 3 (1668): 632–34.

Cook, Alan H. *Edmond Halley: Charting the Heavens and the Seas.* New York: Clarendon, 1998.

Cook, James. "An Account of the Flowing of the Tides in the South Seas, Made by Capt. James Cook, at the Request of Mr. Nevil Maskelyne." *Philosophical Transactions* 62 (1772): 357–58.

_____. "Of the Tides in the South Seas." *Philosophical Transactions* 66 (1776): 447–49.

Corbin, Alain. *The Lure of the Sea: The Discovery of the Seaside in the Western World: 1750–1840*. Trans. Jocelyn Phelps. Berkeley and Los Angeles: University of California Press, 1994.

Corliss, William R. *Earthquakes, Tides, Unidentified Sounds and Related Phenomena: A Catalog of Geophysical Anomalies*. Glen Arm, MD: Sourcebook Project, 1983.

Cotter, Charles H. "The Early History of Ship Magnetism: The Airy–Scoresby Controversy." *Annals of Science* 34 (1977): 589–99.

Courtney, Nicholas. *Gale Force 10: The Life and Legacy of Admiral Beaufort, 1774–1857*. London: Review, 2002.

Croarken, Mary. "Mary Edwards: Computing for a Living in 18th-Century England." *IEEE Annals of the History of Computing* 25 (2003): 9–15.

_____. "Tabulating the Heavens: Computing the *Nautical Almanack* in 18th-Century England." *IEEE Annals of the History of Computing* 25 (2003): 48–61.

Cronon, William. *Nature's Metropolis: Chicago and the Great West*. New York: Norton, 1991.

Crosland, Maurice, and Crosbie Smith. "The Transmission of Physics from France to Britain: 1800–1840." *Historical Studies in the Physical Sciences* 9 (1978): 1–61.

Darwin, Charles. *On the Origin of Species*. Harmondsworth, UK: Penguin, 1968.

Darwin, G. H. "On the Harmonic Analysis of Tidal Observations of High and Low Water." *Proceedings of the Royal Society of London* 48 (1890): 278–339.

Davenport, Francis. "An Account of the Course of the Tides at Tonqueen in a Letter from Mr. Francis Davenport, July 15. 1678." *Philosophical Transactions* 14 (1684): 677–84.

Davis, Charles Henry. "The Law of the Deposit of the Flood Tide: Its Dynamical Action and Office." *Smithsonian Institution Contributions to Knowledge* 3, no. 33 (1852).

_____. "A Memoir upon the Geological Action of the Tidal and Other Currents of the Ocean." *Memoirs of the American Academy of Arts and Sciences* 4 (1849): 117–56.

Day, Archibald. *The Admiralty Hydrographic Service, 1795–1919*. London: Her Majesty's Stationery Office, 1967.

Deacon, Margaret. *Scientists and the Sea, 1650–1900*. London: Academic Press, 1971.

Deas, H. "Crystallography and Crystallographers in England in the Early Nineteenth Century: A Preliminary Survey." *Centaurus* 6 (1959): 129–48.

Dettelbach, Michael. "Humboldtian Science." In *Cultures of Natural History*, ed. N. Jardine, J. A. Secord, and E. C. Spary, 287–304. Cambridge: Cambridge University Press, 1996.

Dickens, Charles. *David Copperfield*. Harmondsworth, UK: Penguin Classics, 1985.

Doodson, A. T., and H. D. Warburg. *Admiralty Manual of Tides*. London: His Majesty's Stationery Office, 1941.

Drouin, Jean-Marc, and Bernadette Bensaude-Vincent. "Nature for the People." In *Cultures of Natural History*, ed. N. Jardine, J. A. Secord, and E. C. Spary, Cambridge: Cambridge University Press, 1996.

Edney, Matthew H. *Mapping an Empire: The Geographical Construction of British India, 1765–1843*. Chicago: University of Chicago Press, 1997.

"Effects of the *British Almanack* ." *London Times*, 2 March 1829 6, col. e.

Eggen, Olin J. "Airy, Sir George Biddell." In *Dictionary of Scientific Biography*, ed. Charles Gillispie, 1:84–87. New York: Charles Scribner's Sons, 1976.

Fahnestock, Jeanne. *Rhetorical Figures in Science*. Oxford: Oxford University Press, 2002.

Fara, Patricia. *Sympathetic Attractions: Magnetic Practices, Beliefs and Symbolism in 18th Century England*. Cambridge: Cambridge University Press, 1994.

Finkelstein, Gabriel. "Headless in Kashgar." Paper presented at the annual meeting of the History of Science Society, Kansas City, MO, October 1998.

Fisch, Menachem. "Necessary and Contingent Truth in William Whewell's Antithetical Theory of Knowledge." *Studies in the History and Philosophy of Science* 16 (1985): 275–314.

———. "A Philosopher's Coming of Age: A Study in Erotetic Intellectual History." In *William Whewell: A Composite Portrait*, ed. Menachem Fisch and Simon Schaffer, 31–66. Oxford: Clarendon Press, 1991.

———. *William Whewell, Philosopher of Science*. Oxford: Clarendon Press, 1991.

Fisch, Menachem, and Simon Schaffer, eds. *William Whewell: A Composite Portrait*. Oxford: Clarendon Press, 1991.

Fischer, Lewis R., and Helge W. Nordvik, eds. *Shipping and Trade, 1750–1950: Essays in International Maritime Economic History*. Pontefract, UK: Lofthouse, 1990.

Fitzroy, Robert. *Narrative of the Surveying Voyages of "H.M.S. Adventure" and "Beagle," between 1826 and 1836*. London, 1839.

[Flamsteed, John.] "A Correct Tide Table, Shewing the True Times of the High-Waters at London-Bridge, to Every Day in the Year 1683." *Philosophical Transactions* 13 (1683): 10–15.

Fleming, James Rodger. *Meteorology in America, 1800–1870*. Baltimore: Johns Hopkins University Press, 1990.

Forman, Walter. *A New Theory of the Tides Shewing What Is the Immediate Cause of the Phenomenon; and Which Has Hitherto Been Overlooked by Philosophers*. Bath, UK: Richard Cruttwell, 1822.

Fox, Robert. "The Rise and Fall of Laplacian Physics." *Historical Studies in the Physical Sciences* 4 (1974): 89–136.

Friendly, Alfred. *Beaufort of the Admiralty: The Life of Sir Francis Beaufort, 1774–1857*. Random House: New York, 1977.

Gilbert, Stuart, and Ray Horner, *The Thames Barrier*. London: Thomas Telford, 1984.

Gillispie, Charles Coulston. *Pierre-Simon Laplace, 1749–1827: A Life in Exact Sciences*. Princeton, NJ: Princeton University Press, 1997.

Gold, Edgar. *Maritime Transport: The Evolution of International Marine Policy and Shipping Law*. Lexington, MA: Lexington Books, 1981.

Golinski, Jan. *Science as Public Culture: Chemistry and Enlightenment in Britain, 1760–1820*. Cambridge: Cambridge University Press, 1992.

Gough, Barry M. "The British Reoccupation and Colonization of the Falkland Islands, or Malvinas, 1832–1843." *Albion* 22, no. 2 (1990): 261–87.

Greenberg, John Leonard. "Interlude II: The Paris Academy's Contest on the Tides (1740)." In his *The Problem of the Earth's Shape from Newton to Clairaut: The Rise of Mathematical Science in Eighteenth-Century Paris and the Fall of "Normal" Science*, chap. 8. Cambridge: Cambridge University Press, 1995.

Grier, David Alan. *When Computers Were Human*. Princeton, NJ: Princeton University Press, 2005.

Grobel, Monica. "The Society for the Diffusion of Useful Knowledge, 1826–46." PhD diss., University College, London, 1932.

Gross, Alan G. "Darwin's Diagram." Paper presented at the annual meeting of the History of Science Society, Milwaukee, WI, 7 November 2002.

———. *The Rhetoric of Science*. Cambridge, MA: Harvard University Press, 1990.

Gross, Alan G., Joseph E. Harmon, and Michael Reidy. *Communicating Science: The Scientific Article from the 17th Century to the Present*. Oxford: Oxford University Press, 2002.

Hackmann, W. D. "The Nineteenth Century Trade in Natural Philosophy Instruments in Britain." In *Nineteenth-Century Scientific Instruments and Their Makers: Papers Presented at the Fourth Scientific Instruments Symposium, Amsterdam, 23–26 October 1984*, ed. P. R. de Clercq., 53–91. Amsterdam, 1985.

Hall, Marie Boas. *All Scientists Now: The Royal Society in the Nineteenth Century*. Cambridge: Cambridge University Press, 1984.

Halley, Edmond. "An Account of the Course of the Tides at Tonqueen in a Letter from Mr. Francis Davenport, July 15, 1678, with the Theory of Them at the Barr of Tonqueen by the Learned Edmund Halley." *Philosophical Transactions* 14 (1684): 677–88.

———. "Discourse Tending to Prove at What Time and Place, Julius Cesar Made His First Descent upon Britain." *Philosophical Transactions* 16 (1686): 495–501.

———. "An Historical Account of the Trade Winds, and Monsoons, Observable in the Seas between and near the Tropicks, with an Attempt to Assign the Physical Cause of the Said Winds." *Philosophical Transactions* 16 (1686–92): 153–68.

———. "The True Theory of the Tides, Extracted from That Admired Treatise of Mr. Isaac Newton, Entitled, *Philosophiae Naturalis Principia Mathematica*; Being a Discourse Presented with That Book to the Late King James, by Mr. Edmund Halley." *Philosophical Transactions* 19 (1695–97): 445–57.

Hamilton, C. I. "The Victorian Navy." *Historical Journal* 25 (1982): 471–87.

Hankins, Thomas L. "A 'Large and Graceful Sinuosity': John Herschel's Graphical Method." *Isis* 97 (2006): 605–33.

Hankins Thomas L., and Robert J. Silverman. *Instruments and the Imagination.* Princeton, NJ: Princeton University Press, 1995.

Harding, Richard. *The Evolution of the Sailing Navy, 1509–1815.* London: Saint Martin's Press, 1995.

Hardisty, J. *The British Seas: An Introduction to the Oceanography and Resources of the North West European Continental Shelf.* London: Routledge, 1990.

Harley, Nick. "Foreign Trade: Comparative Advantage and Performance." In *The Economic History of Britain since 1700*, vol, 1, ed. Roderick Flood and Donald McCloskey, 300–31. Cambridge: Cambridge University Press, 1994.

Harvey, David. *Justice, Nature and the Geography of Difference.* Cambridge: Blackwell, 1996.

Hays, J. N. "Science and Brougham's Society." *Annals of Science* 20 (1964): 227–41.

Henry, Joseph. "Eulogy on Professor Alexander Dallas Bache, Late Superintendent of the United States Coast Survey." In *General Appendix to the Smithsonian Report for 1870*, 91–116. Washington, DC: Government Printing Office, 1871.

Herschel, John F. W. "Address." *British Association Report* 1845 (1896): xxvii–xlv.

———. "On the Investigation of the Orbits of Revolving Double Stars; Being a Supplement to a Paper Entitled 'Micrometrical Measures of 364 Double Stars.'" *Memoirs of the Royal Astronomical Society* 5 (1833): 171–222.

——— ed. *A Manual of Scientific Enquiry.* London, 1849.

Hewett, William. "Tide Observations in the North Sea—Verification of Professor Whewell's Theory." *Nautical Magazine* 9 (1841): 180–83.

Hills, Graham H. *Navigation of the Approaches to Liverpool, with Some Account of the Tides, Currents, and Soundings of the Irish Sea, with a Chart of Liverpool Bay.* Liverpool, 1871.

Hilton, Boyd. *Corn, Cash Commerce: The Economic Policies of the Tory Governments, 1815–1830.* Oxford: Oxford University Press, 1977.

"H.M.S. *Fairy.*" *Nautical Magazine* 9 (1841): 72.

Hodge, M. J. S. "The History of the Earth, Life, and Man: Whewell and Palaetiological Science." In *William Whewell: A Composite Portrait*, ed. Menachem Fisch and Simon Schaffer, 255–88. Oxford: Clarendon Press, 1991.

Home, R. W., and Sally Gregory Kohlstedt, eds. *International Science and National Scientific Identity: Australia between Britain and America.* Boston: Kluwer Academic Publishers, 1991.

Home, Rod. "Humboldtian Science Revisited: An Australian Case Study." *History of Science* 33 (1995): 1–22.

Howarth, Richard J. "Sources for a History of the Ternary Diagram." *British Journal for the History of Science* 29 (1996): 337–56.

[Hughes, Paul]. "Dessiou, Joseph Foss." In *The Dictionary of Nineteenth-Century British Scientists*, ed. Bernard Lightman, 1:577–78. Chicago: University of Chicago Press, 2004.

Hughes, Paul. "A Study in the Development of Primitive and Modern Tidetables." PhD diss., John Moores University, Liverpool, 2005.

Hughes, Paul, and Alan D. Wall. "The Admiralty Tidal Predictions of 1833: Their Comparison with Contemporary Observations and with a Modern Synthesis." *Journal of Navigation* 57 (2004): 203–14.

Hyman, Anthony. *Charles Babbage: Pioneer of the Computer*. Princeton, NJ: Princeton University Press, 1982.

Humboldt, Alexander. *Géologie et climatologie asiatiques*. 2 vols. Paris, 1831.

Hutchinson, William. *A Treatise on Practical Seamanship*. London: Scholar Press, 1979.

Jackson, Gordon. "The Ports." In *Transport in the Industrial Revolution*, ed. Derek H.. Aldcroft and Michael J. Freeman, 177–209. Manchester, UK: Manchester University Press, 1983.

James, Frank. "Davy in the Dockyard: Humphry Davy, the Royal Society and the Electro-chemical Protection of the Copper Sheeting of His Majesty's Ships in the Mid-1820s." *Physis* 29 (1992): 205–25.

Jankovic, Vladimir. "Ideological Crests versus Empirical Troughs: John Herschel's and William Radcliffe Birt's Research on Atmospheric Waves, 1843–50." *British Journal for the History of Science* 31 (1998): 21–40.

Jardine, Nicholas, and Emma Spary. "The Natures of Cultural History." In *Cultures of Natural History*, ed. N. Jardine, J. A. Secord, and E. C. Spary., 3–16. Cambridge: Cambridge University Press, 1996.

Jones, Gareth Stedman. *Languages of Class: Studies in English Working Class History, 1832–1982*. Cambridge: Cambridge University Press, 1983.

Josephson, Paul R. *Industrialized Nature: Brute Force Technology and the Transformation of the Natural World*. Washington, DC: Shearwater Books, 2002.

Jourdin, Michel Mollat du. *Europe and the Sea*. Trans. Teresa Lavender Fagan. Cambridge, MA: Blackwell, 1993.

*Journal of Historical Geography* 32, no. 3 (July 2006).

"Jowett, William." In *Dictionary of National Biography*, ed. Sir Leslie Stephen and Sir Sidney Lee, 10:1103–4. Oxford: Oxford University Press, 1967–68.

Joyce, Patrick. *Visions of the People: Industrial England and the Question of Class, 1848–1914*. Cambridge: Cambridge University Press, 1991.

Kargon, Robert. *Science in Victorian Manchester: Enterprise and Expertise*. Baltimore: Johns Hopkins University Press, 1977.

Kark, Ruth. "The Contributions of Nineteenth Century Protestant Missionary Societies to Historical Cartography." *Imago Mundi* 45 (1993): 112.

Kellner, L. "Alexander von Humboldt and the Organization of International Collaboration in Geophysical Research." *Contemporary Physics* 1 (1959): 35–48.

Kennedy, Paul M. *The Rise and Fall of British Naval Mastery*. New York: Scribner, 1976.

Kindleberger, Charles P. *Mariners and Markets*. New York: Harvester, 1992.

Klein, Bernhard, and Gesa Mackenthun, eds. *Sea Changes: Historicizing the Ocean*. New York: Routledge, 2003.

Knight, Charles. *Passages of a Working Life*. 3 vols. London, 1864.

Knight, David. "Tyrannies of Distance in British Science." In *International Science and National Scientific Identity*, ed. R. W. Home and Sally Gregory Kohlstedt, 39–54. Boston: Kluwer Academic Publishers, 1991.

Knight, R. J. B. "Changing Technologies and Materials." In *Maritime History*. Vol. 2, *The Eighteenth Century and the Classic Age of Sail*, ed. John B. Hattendorf, 233–42. Malabar, FL: Krieger, 1997.

———. "The Convulsion of Europe: The Naval Conflict during the Revolutionary and Napoleonic Wars." In *Maritime History*. Vol. 2, *The Eighteenth Century and the Classic Age of Sail*, ed. John B.. Hattendorf, 243–54. Malabar, FL: Krieger, 1997.

Kohler, Robert E. *All Creatures: Naturalists, Collectors, and Biodiversity, 1850–1950*. Princeton, NJ: Princeton University Press, 2006.

———. "American Museums and Natural History Collecting." Paper presented at the History of Science Society annual meeting, Austin, TX, 17 November 2004.

Kohlstedt, Sally Gregory. *The Formation of the American Scientific Community: The American Association for the Advancement of Science, 1848–60*. Urbana: University of Illinois Press, 1976.

Kollerstrom, N. "Newton's Two 'Moon-Tests.'" *British Journal for the History of Science* 24 (1991): 369–72.

Konvitz, Josef W. *Cartography in France, 1660–1848: Sciences, Engineering, and Statecraft*. Chicago: University of Chicago Press, 1987.

Kriegel, Abraham D. "Biography and the Politics of the Early Nineteenth Century." *Journal of British Studies* 29 (1990): 281–91.

Lambert, Andrew D. "Preparing for the Long Peace: The Reconstruction of the Royal Navy, 1815–1830." *Mariner's Mirror* 82 (1996): 41–54.

Laplace, Pierre Simon. "Recherches sur plusiers points du système du monde." *Mémoires Académie Royale des Sciences* 88 (1776): 177–264.

———. *Traité de mécanique céleste*, vol. 5, bk. 13. Paris, 1825.

"The Late Mr. Bunt." *Western Daily Press*, 6 June 1872 5.

Latour, Bruno. "The Force of Reason and Experiment." In *Experimental Inquiries: Historical, Philosophical and Social Studies of Experimentation in Science*, ed. H. E. Le Grand, 49–80. Boston: Kluwer Academic Publishers, 1990.

———. "Drawing Things Together." In *Representation in Scientific Practice*, ed. Michael Lynch and Steve Woolgar, 19–68. Cambridge, MA: MIT Press, 1990.

Latour, Bruno, and Steve Woolgar. *Laboratory Life: The Construction of Scientific Facts*. 2nd ed. Princeton, NJ: Princeton University Press, 1986.

Latourette, Kenneth. *A History of the Expansion of Christianity*. Grand Rapids, MI: Zondervan, 1970.

Lefebvre, Henri. *The Production of Space*. Cambridge: Blackwell, 1991.

[Letter to Editor]. *London Times*, 2 November 1827 2, col. f.

[Letter to Editor]. *London Times*, 15 November 1828 3, col. a.

Lewis, Martin W., and Karen E. Wigen. *The Myth of Continents: Critique of Metageography*. Berkeley and Los Angeles: University of California Press, 2000.

Lewis, Michael. *The Navy in Transition, 1814–1864: A Social History*. London: Hodder and Stoughton, 1965.

Lightman, Bernard, ed. *Victorian Science in Context*. Chicago: University of Chicago Press, 1997.

Livingstone, David N. *The Geographical Tradition: Episodes in the History of a Contested Enterprise*. Oxford: Blackwell, 1992.

———. *Putting Science in Its Place: Geographies of Scientific Knowledge*. Chicago: University of Chicago Press, 2003.

———. "The Spaces of Knowledge: Contributions toward a Historical Geography of Science." *Environment and Planning D: Society and Space* 13 (1995): 5–34.

Lloyd, Christopher. *Mr. Barrow of the Admiralty: A Life of Sir John Barrow, 1764–1848*. London: Collins, 1970.

Lloyd, John Augustus. "Account of Levellings Carried across the Isthmus of Panama, to Ascertain the Relative Height of the Pacific Ocean at Panama, and of the Atlantic at the Mouth of the River Chagres, Accompanied by Geographical and Topographical Notices of the Isthmus." *Philosophical Transactions* 120 (1830): 50–68.

"London, Floods on Account of the High Tide." *London Times*, 30 December 1814 2, col. d.

"London, Floods through the High Tide in the Thames." *London Times*, 27 December 1806 3, col. a.

"Loss of the *Fairy*."*Nautical Magazine* 9 (1841): 116–17.

"Lowrie, Walter." In *Dictionary of American Biography*, ed. Dumas Malone, 11:476. New York: Charles Scribner's Sons, 1933.

[Lowrie, Walter H.] *A New Theory of the Causes of the Tides, and Oceanic and Atmospheric Currents*. Pittsburgh: Barr and Myers, 1859.

Lubbock, John William. "Account of the Discussions of Observations of the Tides." *British Association Report* 1836 (1837): 31–32.

———. "Account of the Recent Discussions of Observations of the Tides." *British Association Report* 1835 (1836): 285–87.

———, Account of the *"Traité sur le flux et reflux de la mer,"* of Daniel Bernoulli. London, 1830.

———. "Discussion of Tide Observations Made at Liverpool." *Philosophical Transactions* 125 (1835): 275–300.

———. "Discussion of Tide Observations Made at Liverpool." *Philosophical Transactions* 126 (1836): 57–75.

———. *An Elementary Treatise on the Tides*. London, 1839.

———. "Explanation of the Columns 'High Water at London.'" In *British Almanac*, 5–6. London: Charles Knight, 1830.

———. "Note on the Tides in the Port of London." *Philosophical Transactions* 122 (1832): 595–600.

———. "On the Tides." *Philosophical Transactions* 122 (1832): 51–56.

———. "Note on the Tides." *Philosophical Transactions* 123 (1833): 19–22.

———. "On the Tides." *Philosophical Transactions* 124 (1834): 143–66.

_____. "On the Tides in the Port of London." *Philosophical Transactions* 121 (1831): 379–415.

_____. "On the Tides on the Coast of Great Britain." *Philosophical Magazine* 9 (1831): 333–35.

_____. "Report on the Tides." *British Association Report* 1832 (1833): 189–95.

_____. "The Tides." In *The Companion to the Almanac of the SDUK for the Year 1830*. London: Charles Knight, 1830.

_____. "The Tides in the Port of London." *Philosophical Transactions* 122 (1831): 379–416.

Lunteren, Frans van. "Humboldtian Ideals and Practical Benefits: The Foundation of the Royal Dutch Meteorological Institute." Paper presented at the annual meeting of the History of Science Society, Kansas City, MO, October 1998.

Lyell, Charles. *Principles of Geology*. Vol. 1. Chicago: University of Chicago Press, 1990.

Mackay, D. L. "A Presiding Genius of Exploration: Banks, Cook and Empire, 1767–1805." In *Captain James Cook and His Times*, ed. Robin Fischer and Hugh Johnston, 21–39. Seattle: University of Washington Press, 1979.

Mackenzie, Murdoch. "The State of the Tides in Orkney." *Philosophical Transactions* 46 (1749): 149–60.

MacLeod, Roy. "Introduction: On the Advancement of Science." In *The Parliament of Science: The British Association for the Advancement of Science, 1831–1981*, ed. Roy MacLeod and Peter Collins, 17–42. Northwood, UK: Science Reviews, 1981.

_____. "The Royal Society and the Government Grant: Notes on the Administration of Scientific Research, 1849–1914." *Historical Journal* 9 (1971): 323–58.

MacLeod, Roy, and Fritz Rehbock, eds. *Nature in Its Greatest Extent: Western Science in the Pacific*. Honolulu: University of Hawaii Press, 1988.

MacLeod, Roy, and Peter Collins, eds. *The Parliament of Science: The British Association for the Advancement of Science, 1831–1981*. London: Science Reviews, 1981.

Manning, Thomas G. *U.S. Coast Survey vs. Naval Hydrographic Office: A 19th-Century Rivalry in Science and Politics*. Tuscaloosa: University of Alabama Press, 1988.

Marmer, H. A. "On Cotidal Lines." *Geographical Review* 18 (1928): 129–43.

Martin, K. B. "Tides and Tide Tables." *Nautical Magazine and Naval Chronicle* 9 (1839).

Maskelyne, Nevil. "Observations of the Tides at the Island of St. Helena." *Philosophical Transactions* 52 (1761): 586–91.

Matthaus, Wolfgang. "On the History of Recording Tide Gauges." *Proceedings of the Royal Society of Edinburgh*, ser. B, 73 (1971–72): 26–34.

McLean, David. *Education and Empire: Naval Tradition and England's Elite Schooling*. London: British Academic Press, 1999.

McNeill, John. "The Nature and Culture of Environmental History." *History and Theory* 42 (2003): 5–43.

"Memorial of Mrs. Hewett." *Nautical Magazine* 9 (1841): 425–27.

Merchant, Carolyn. "Reinventing Eden: Western Culture as a Recovery Narrative." In *Uncommon Ground: Rethinking the Human Place in Nature*, ed. William Cronon, 132–70. New York: W. W. Norton, 1996.

Mertens, Joost. "From Tubal Cain to Faraday: William Whewell as a Philosopher of Technology." *History of Science* 38 (2000): 321–42.

Merton, Robert K. "De-gendering 'Man of Science': The Genesis and Epicene Character of the Word 'Scientist.'" In *Sociological Visions*, ed. Kai Erikson, 225–53. Lanham, MD: Rowman and Littlefield, 1997.

Miller, David Philip. "Between Hostile Camps: Sir Humphry Davy's Presidency, 1820–1827." *British Journal for the History of Science* 16 (1983): 1–47.

———. "The Revival of the Physical Sciences in Britain, 1815–1840." *Osiris* 2 (1986): 107–34.

———. "Sir Joseph Banks: An Historiographical Perspective." *History of Science* 19 (1981): 284–92.

Milliman, John D., and Bilal U. Haq, eds. *Sea Level Rise and Coastal Subsidence: Causes, Consequences, and Strategies*. Dordrecht: Kluwer Academic Publishers, 1996.

Mills, Eric L. "The Historian of Science and Oceanography After Twenty Years." *Earth Sciences History* 12, no. 1 (1993): 5–18.

Minchinton, W. E., ed. *The Growth of English Overseas Trade in the Seventeenth and Eighteenth Centuries*. London: Methuen, 1969.

Modelski, George, and William R. Thompson. *Seapower in Global Politics, 1494–1993*. Seattle: University of Washington Press, 1988.

Mokyr, Joel. *The Level of Riches: Technological Creativity and Economic Progress*. New York: Oxford University Press, 1990.

Moore, John Hamilton. The *Practical Navigator*. London, 1778.

Moray, Robert. "Considerations and Enquiries concerning Tides, by Sir Robert Moray; Likewise for a Further Search into Dr. Wallis's Newly Publish't Hypothesis." *Philosophical Transactions* 1 (1666): 298–301.

———. "Patterns of the Tables Proposed to Be Made for Observing of Tides." *Philosophical Transactions* 1 (1666): 311.

———. "A Relation of Some Extraordinary Tydes in the West-Isles of Scotland." *Philosophical Transactions* 1 (1665): 53–55.

More, Henry. "Observations of the Tides in the Straits of Gibraltar." *Philosophical Transactions* 52 (1761): 447–53.

Morgan, Frederick Wallace. *Ports and Harbours*. London: Hutchinson's University Library, 1952.

Morgan, Kenneth. "Bristol and the Atlantic Trade in the Eighteenth Century." *English Historical Review* 424 (1992): 626–50.

Morrell, Jack. "The Judge and Purifier of All." *History of Science* 30 (1992): 97–114.

———. "William Whewell: Rough Diamond." *History of Science* 32 (1994): 345–59.

Morrell, Jack, and Arnold Thackray. *Gentlemen of Science: Early Years of the British Association for the Advancement of Science*. Oxford: Clarendon Press, 1981.

Morris, R. O. "Surveying Ships of the Royal Navy from Cook to the Computer Age." *Mariner's Mirror* 72 (1986): 385–414.

Morris, Roger. "200 Years of Admiralty Charts and Surveys." *Mariner's Mirror* 83 (1996): 420–35.

Morton, Alan Q. "Concepts of Power: Natural Philosophy and the Uses of Machines in Mid-Eighteenth-Century London." *British Journal for the History of Science* 28 (1995): 63–78.

Morus, Iwan Rhys. *Frankenstein's Children: Electricity, Exhibition, and Experiment in Early 19th-Century London.* Princeton, NJ: Princeton University Press, 1998.

――――. "Manufacturing Nature: Science, Technology and Victorian Consumer Culture." *British Journal for the History of Science* 29 (1996): 403–34.

Musson, Albert, and Eric Robinson. *Science and Technology in the Industrial Revolution.* Manchester, UK: Manchester University Press, 1969.

"Nautical Almanac." *Nautical Magazine* 1 (1832): 40–41.

Nauticus. "On Tides." *Nautical Magazine* 1 (1832): 119–20.

Nebeker, Frederick. *Astronomy and the Geophysical Tradition in the United States in the 19th Century: A Guide to Manuscript Sources in the Library of the American Philosophical Society.* Philadelphia: American Philosophical Society Library Press, 1991.

Neve, Michael Raymond. "Natural Philosophy, Medicine, and the Culture of Science in Provincial England: The Cases of Bristol, 1790–1850, and Bath, 1750–1820." PhD diss., University of London, 1984.

Nicolson, Malcolm. "Alexander von Humboldt, Humboldtian Science and the Origins of the Study of Vegetation." *History of Science* 25 (1987): 167–94.

――――. "Humboldtian Plant Geography After Humboldt: The Link to Ecology." *British Journal for the History of Science* 29 (1996): 289–310.

Norie, J. W. *A Complete Epitome of Practical Navigation and Nautical Astronomy, Containing All Necessary Instructions for Keeping a Ship's Reckoning at Sea.* London, 1798.

[Obituary Notice]. *Bristol Mercury*, 8 June 1872, 6.

O'Brien, Patrick K. "Central Government and the Econonomy, 1688–1815." In *The Economic History of Britain since 1700.* Vol. 2, *1700–1815.* 2nd ed., ed. R. Floud and D. McCloskey, 205–41. Cambridge: Cambridge University Press, 1994.

[Oldenburg, Henry]. "An Account of Several Engagements for Observing of Tydes." *Philosophical Transactions* 2 (1666–67): 378–79.

Oldenburg, Henry. *Correspondence*, vol. 3. Ed. and trans. A. Rupert Hall and Marie Boas Hall. Madison: University of Wisconsin Press, 1966.

"On the Advantage Possessed by Naval Men, in Contributing to General Science." *Nautical Magazine* 1 (1832): 180–81.

Outram, Dorinda. "New Spaces in Natural History." In *Cultures of Natural History*, ed. N. Jardine, J. A. Secord, and E. C. Spary, 249–65. Cambridge: Cambridge University Press, 1996.

Palmer, Henry. "Description of a Graphical Register of Tides and Winds." *Philosophical Transactions* 121 (1831): 209–13.

Pancaldi, Giuliano, "Scientific Internationalism and the British Association." In *The Parliament of Science: The British Association for the Advancement of Science, 1831–1981*, ed. Roy MacLeod and Peter Collins, 145–69. Northwood, UK: Science Reviews, 1981.

Philips, Henry. "A Letter Written to Dr. John Wallis by Mr. Henry Philips, Containing His Observations about the True Time of the Tides." *Philosophical Transactions* 3 (1668): 656–59.

Perelman Ch., and L. Olbrechts-Tyteca. *The New Rhetoric*. Notre Dame, IN: University of Notre Dame Press, 1969.

Porter, Dale H. *The Thames Embankment: Environment, Technology and Society in Victorian London*. Akron: University of Ohio Press, 1998.

Porter, Roy. *London: A Social History*. Cambridge, MA: Harvard University Press, 1995.

Porter, Theodore M. *The Rise of Statistical Thinking, 1820–1940*. Princeton, NJ: Princeton University Press, 1986.

*The Port of London and the Thames Barrage, a Series of Expert Studies and Reports, on the Conditions Prevailing in the Tidal River and Estuary of the Thames, Dealing Especially with Its Geological, Engineering, Navigation . . .* London, 1907.

Power, M. J. "The Growth of Liverpool." In *Popular Politics, Riot and Labour: Essays in Liverpool History, 1790 to 1940*, ed. John Belchem, 21–37. Liverpool, UK: Liverpool University Press, 1992.

Pratt, Mary Louise. *Imperial Eyes: Travel Writing and Transculturation*. New York: Routledge, 1992.

Pritchard, Sara B. "Reconstructing the Rhone: The Cultural Politics of Nature and Nation in Contemporary France, 1945–1997." *French Historical Studies* 27 (Fall 2004): 765–99.

Proudman, J. "Halley's Tidal Chart." *Geographical Journal* 100 (1941): 174–76.

Read, Donald. *Cobden and Bright: A Victorian Political Partnership*. London: Edward Arnold, 1967.

Redfield, W. C. "Remarks on the Tides and the Prevailing Currents of the Ocean and Atmosphere." *American Journal of Science and Arts* 45 (1843): 132–35.

Rediker, Marcus. *Between the Devil and the Deep Blue Sea: Merchant Seamen, Pirates and the Anglo-American Maritime World, 1700–1750*. Cambridge: Cambridge University Press, 1993.

Reidy, Michael S. "The Flux and Reflux of Science: The Study of the Tides and the Organization of Early Victorian Science." PhD diss., University of Minnesota, 2000.

———, "Gauging Science and Technology in the Early Victorian Era," In *The Machine in Neptune's Garden: Historical Perspectives on the Marine Environment*, ed. Helen M. Rozwadowski and David K. Van Keuren, 1–31. Sagamore Beach, MA: Science History Publications, 2004.

_____. "Masters of Tidology: The Cultivation of the Physical Sciences in Early Victorian Liverpool." *Transactions of the Historic Society of Lancashire and Cheshire* 152 (2003): 45–71.

[Reidy, Michael S.] "Lloyd, John Augustus." In *The Dictionary of Nineteenth-Century British Scientists*, ed. Bernard Lightman. Chicago: University of Chicago Press, 2004.

Reidy, Michael S., Gary Kroll, and Erik M. Conway. *Exploration and Science: Social Impact and Interaction*. Santa Barbara, CA.: ABC-Clio, 2006.

Rennie, George. "Report on the Progress and Present State of Our Knowledge of Hydraulics as a Branch of Engineering. Part II." *British Association Report* 1833 (1834): 486–512.

Richards, Joan. "Observing Science in Early Victorian England: Recent Scholarship on William Whewell." *Perspectives on Science* 4 (1996): 231–47.

Ronan, Colin A. "Edmund Halley." In *Dictionary of Scientific Biography*, ed. Charles Gillispie, 14:67–72. New York: Charles Scribner's Sons, 1976.

Ross, Sydney. "Scientist: The Story of a Word." *Annals of Science* 18, no. 2 (1962): 65–85.

Rossiter, J. R. "The History of Tidal Predictions in the United Kingdom Before the Twentieth Century." *Proceedings of the Royal Society of Edinburgh*, ser. *B*, 73 (1971–72): 13–23.

Rozwadowski, Helen M. *Fathoming the Ocean: The Discovery and Exploration of the Deep Sea*. Cambridge, MA: Harvard University Press, 2005.

_____. "Internationalism, Environmental Necessity, and National Interest: Marine Science and Other Sciences." *Minerva* 42 (2004): 127–49.

Rudwick, Martin J. S. *The Great Devonian Controversy: The Shaping of Scientific Knowledge* among *Gentlemanly Specialists*. Chicago: University of Chicago Press, 1985.

Rupke, Nicolaas. "Humboldtian Distribution Maps: The Spatial Ordering of Scientific Knowledge." In *The Structure of Knowledge: Classifications of Science and Learning since the Renaissance, ed*. Tore Frängsmyr, 93–116. Berkeley: Office for History of Science and Technology, University of California, 2001.

Ruse, Michael. "The Scientific Methodology of William Whewell." *Centaurus* 20 (1976): 227–57.

_____. "William Whewell: Omniscientist." In *William Whewell: A Composite Portrait*, ed. Menachem Fisch and Simon Schaffer, 87–116. Oxford: Clarendon Press, 1991.

Russell, Edmund. "Evolutionary History: Prospectus for a New Field." *Environmental History* 8 (April 2003): 204–28.

_____. *War and Nature: Fighting Humans and Insects with Chemicals from World War I to Silent Spring*. Cambridge: Cambridge University Press, 2001.

Russell, N. C., E. M. Tansey, and P. V. Lear. "Missing Links in the History and Practice of Science: Teams, Technicians and Technical Work." *History of Science* 38 (2000): 237–41.

Said, Edward. *Culture and Imperialism*. New York: Vintage, 1993.

"Saxton, Joseph." In *Dictionary of American Biography*, ed. Dumas Malone, 16:400. New York: Charles Scribner's Sons, 1935.

Schaffer, Simon. "Babbage's Intelligence: Calculating Engines and the Factory System." *Critical Inquiry* 21 (1994): 203–27.

———. "Empires of Physics." In *The Physics of Empire: Public Lectures*, ed. Richard Staley, 97–112. Cambridge: Whipple Museum of the History of Science, 1994.

———. "The History and Geography of the Intellectual World: Whewell's Politics of Language." In *William Whewell: A Composite Portrait*, ed. Menachem Fisch and Simon Schaffer, 201–32. Oxford: Clarendon Press, 1991.

Schlee, Susan. *The Edge of an Unfamiliar World: A History of Oceanography.* New York: E. P. Dutton, 1973.

Schneer, Jonathan. *The Thames.* New Haven: CT: Yale University Press, 2005.

Schofield, Robert. *The Lunar Society of Birmingham: A Social History on Provincial Science and Industry in Eighteenth-Century England.* Oxford: Clarendon Press, 1963.

Schweber, S. S. "Scientists as Intellectuals: The Early Victorians." In *Victorian Science and Victorian Values: Literary Perspectives*, ed. James Paradis and Thomas Postlewait, 1–38. New Brunswick, NJ: Rutgers University Press, 1985.

Scott, James C. *Seeing Like a State: How Certain Schemes to Improve the Human Condition Have Failed.* New Haven, CT: Yale University Press, 1998.

Secord, Anne. "Artisan Botany." In *Cultures of Natural History*, ed. N. Jardine, J. A. Secord, and E. C. Spary, 378–93. Cambridge: Cambridge University Press, 1996.

———. "Corresponding Interests: Artisans and Gentlemen in Nineteenth-Century Natural History." *British Journal for the History of Science* 27 (1994): 383–408.

———. "Science in the Pub: Artisan Botanists in Early Nineteenth-Century Lancashire." *History of Science* 32 (1994): 269–315.

Secord, James. *Victorian Sensations.* Chicago: University of Chicago Press, 2003.

Semmel, Bernard. *The Rise of Free Trade Imperialism: Classical Political Economy, the Empire of Free Trade, and Imperialism, 1750–1850.* Cambridge: Cambridge University Press, 2003.

Shakespeare, William. *Richard II.* New Haven, CT: Yale University Press, 1999.

Shapin, Steven. "The Pottery Philosophical Society, 1819–1835: An Examination of the Cultural Uses of Provincial Science." *Science Studies* 2 (1972): 311–36.

Shapin, Steven, and Barry Barnes. "Science, Nature and Control: Interpreting Mechanics' Institutes." *Social Studies of Science* 7 (1977): 31–74.

Shapin, Steven, and Arnold Thackray. "Prosopography as a Research Tool in History of Science: The British Scientific Community, 1700–1900S." *History of Science* 12 (1974): 1–28.

Shiach, Morag. *Discourse on Popular Culture: Class, Gender and History in Cultural Analysis, 1730 to the Present.* Cambridge: Cambridge University Press, 1989.

Sivasundaram, Sujit. *Nature and the Godly Empire: Science and Evangelical Mission in the Pacific, 1795–1850*. New York: Cambridge University Press, 2005.

Slotten, Hugh Richard. *Patronage, Practice, and the Culture of American Science: Alexander Dallas Bache and the U.S. Coast Survey*. Cambridge: Cambridge University Press, 1994.

Smith, Crosbie. "'Mechanical Philosophy' and the Emergence of Physics in Britain: 1800–1850." *Annals of Science 33* (1976): 3–29.

Smith, H. *The Society for the Diffusion of Useful Knowledge, 1826–1846: A Social and Bibliographic Evaluation*. Occasional Paper no. 8. Halifax, NS: Dalhousie University Library, 1974.

Smith, J. F., Gordon Hemm, and A. Ernest Shennan. *Liverpool, Past, Present, Future*. Liverpool, UK: Northern, 1948.

Smith, Robert W. "The Cambridge Network in Action: The Discovery of Neptune." *Isis* 80 (1989): 395–422.

———. "A National Observatory Transformed: Greenwich in the Nineteenth Century." *Journal for the History of Astronomy* 222 (1991): 1–21.

Spence, Graeme. *A Geographical and Nautical Description, of Scilly; with Suitable Sailing Directions, Leading Marks, Tides and Tide Tables: Adapted to the Maritime Survey Thereof, in Two Charts . . . : in the Years 1790, 1791, and 1792*. London, 1792.

Sponsel, Alistair. "Fathoming the Depth of Charles Darwin's Theory of Coral Reef Formation." Paper presented at the History of Science Society annual meeting, Austin, TX, 20 November 2004.

Stafford, Robert A. "Geological Surveys, Mineral Discoveries, and British Expansion, 1835–1871." *Journal of Imperial and Commonwealth History* 12 (1984): 5–32.

Stafford, Robert A. *Scientist of Empire: Sir Roderick Murchison, Scientific Exploration, and Victorian Imperialism*. Cambridge: Cambridge University Press, 1989.

Stair-Douglas, J. *The Life and Selections from the Correspondence of William Whewell, D.D.* London, 1881.

Starkey, David J. "War and the Market for Seafarers in Britain, 1736–1792." In *Shipping and Trade, 1750–1950: Essays in International Maritime Economic History,* ed. Lewis R. Fischer and Helge W. Nordvik, 25–42. Pontefract, UK: Lofthouse, 1990.

Steinberg, Philip E. *The Social Construction of the Ocean*. Cambridge: Cambridge University Press, 2001.

Stewart, Larry, and Paul Weindling. "Philosophical Threads: Natural Philosophy and Public Experiment among the Weavers of Spitalfields." *British Journal for the History of Science* 28 (1995): 37–62.

Stewart, Robert. *Henry Brougham, 1778–1868: His Public Career*. London: Bodley Head, 1986.

Sufferer. "To the Editors of the *Times*." *London Times*, 31 December 1828, 3, col. b.

Szostak, Rick. *The Role of Transportation in the Industrial Revolution: A Comparison of England and France*. Montreal: McGill-Queen's University Press, 1991.

Thackray, Arnold. "Natural Knowledge in Cultural Context: The Manchester Model." *American Historical Review* 79 (1974): 672–709.

Thoren, Victor E. "John Flamsteed." In *Dictionary of Scientific Biography*, ed. Charles Gillispie, 5:22–25. New York: Charles Scribner's Sons, 1976.

Thrower, Norman J. W., ed. *The Three Voyages of Edmond Halley in the "Paramore," 1698–1701*. London: Hakluyt Society, 1981.

"The Tide Gauge at Sheerness." *Nautical Magazine* 1 (1832): 401–4.

"The Tides." In *British Almanac*, 44. London, 1828.

"To the Editors of the Times." *London Times*, 31 December 1828 3, col. b.

"To Vindex, the Apologist of the Stationers' Company in the *Times*." *London Times*, 14 January 1828 3, col. f.

Todhunter, Isaac. *William Whewell, D.D., Master of Trinity College, Cambridge: An Account of His Writings with Selections from His Literary and Scientific Correspondence*. 2 vols. New York: Johnson Reprint, 1970.

Tomlinson, David. "On the Tides." *American Journal of Science* 34 (1838): 81–85.

Tufte, E. R. *The Visual Display of Quantitative Information*. Cheshire, CT: Graphics Press, 1983.

Turner, Gerard L'E. *Nineteenth-Century Scientific Instruments*. Berkeley and Los Angeles: University of California Press, 1983.

T. W. "To the Editors of the *Times*." *London Times*, 4 January 1829, 4, col. b.

Van Helden, Albert, and Thomas L. Hankins, "Introduction: Instruments in the History of Science," *Osiris* 9 (1994): 1–6.

Veevers, Richard. "On the Tides and Datums of the Lancashire Coast." *Transactions of the Historic Society of Lancashire and Cheshire* 49 (1897): 171–75.

Ville, Simon P. *English Shipowning during the Industrial Revolution: Michael Henly and Son, London Shipowners, 1770–1830*. Manchester, UK: Manchester University Press, 1987.

Vindex. "The Almanack of the Stationers' Company." *London Times*, 11 January 1828, 3, cols. e and f; 4, col. a.

———. "To Mr. Buckingham." *London Times*, 14 January 1828, 3 col. c.

Wallis, John. "An Essay of Dr. John Wallis, Exhibiting His Hypothesis about the Flux and Reflux of the Sea." *Philosophical Transactions* 1 (1666): 263–81.

———. "Some Inquiries and Directions concerning Tides, Proposed by Dr. Wallis, for the Proving or Disproving of His Lately Publish't Discourse concerning Them." *Philosophical Transactions* 1 (1666): 297–98.

Washington, John. "Tide Observations—North Sea—Professor Whewell's Theory." *Nautical Magazine* 10 (1842): 566–68.

Webb, R. K. *The British Working Class Reader, 1790–1848: Literacy and Social Tension*. London: George Allen and Unwin, 1955.

———. *Modern England: From the Eighteenth Century to the Present*. 2nd ed. New York: Harper and Row, 1980.

Westfall, Richard S. *Never at Rest: A Biography of Isaac Newton*. Cambridge: Cambridge University Press, 1980.

———. "Newton and the Fudge-Factor." *Science* 179 (1973): 751–58.

Whewell, William. "Account of a Level Line, Measured from the Bristol Channel to the English Channel, during the Year 1837–38, by Mr. Bunt, under the Direction of a Committee of the British Association." *British Association Report* 1838 (1839): 1–11.

———. "Address." *British Association Report* 1833 (1834): iv–xii.

———. "Address." *British Association Report* 1841 (1842): xv–xxxiii.

———. "Description of a New Tide-Gauge, Constructed by Mr. T. G. Bunt, and Erected on the Eastern Bank of the River Avon, in Front of the Hotwell House, Bristol, 1827." *Philosophical Transactions* 128 (1838): 249–51.

———. "Essay towards a First Approximation to a Map of Co-tidal Lines." *Philosophical Transactions* 123 (1833): 147–236.

———. "A General Method of Calculating the Angles Made by Any Planes of Crystals, and the Laws according to Which They Are Formed." *Philosophical Transactions* 115 (1825): 87–130.

———. *History of the Inductive Sciences, from the Earliest to the Present Time.* 3rd ed., 2 vols. New York: Appleton, 1882.

———. *Inaugural Lecture, November 26, 1851: The General Bearing of the Great Exhibition on the Progress of Art and Science.* London, 1851.

———. "Memoranda and Directions for Tide Observations." *Nautical Magazine* 2 (1833): 662–65; 3 (1834): 41–43, 98–102, 170–71, 532–37.

———. "On Mr. Superintendent Bache's Tide Observations." *British Association Report* 1854 (1855): 28.

———. "On Our Ignorance of the Tides." *British Association Report* 1850 (1851): 27–28.

———. "On the Empirical Laws of the Tides in the Port of London, with Some Reflexions on the Theory." *Philosophical Transactions* 124 (1834): 15–46.

———. "On the Results of Observations Made with a New Anemometer." *Transactions of the Cambridge Philosophical Society* 6, pt. 2 (1837): 301–11.

———. "On the Results of Tide Observations Made in June 1834 at the Coast-Guard Stations in Great Britain and Ireland." *Philosophical Transactions* 125 (1835): 83–90.

———. *The Philosophy of the Inductive Sciences, Founded upon Their History.* 1st ed. London: John W. Parker, 1840.

———. "Report on Discussions of Bristol Tides, Performed by Mr. Bunt under the Direction of the Rev. W. Whewell." *British Association Report* 1841 (1842): 30–33.

———. "Report on the Discussion of Leith Tide Observations, Executed by Mr. D. Ross, of the Hydrographer's Office, Admiralty, under the Direction of the Rev. W. Whewell." *British Association Report* 1841 (1842): 33–36.

———. "Researches on the Tides." 4th ser. "On the Empirical Laws of the Tides in the Port of Liverpool." *Philosophical Transactions* 125 (1835): 1–16.

———. "Researches on the Tides." 5th ser. "On the Solar Inequality, and on the Diurnal Inequality of the Tides at Liverpool." *Philosophical Transaactions* 126 (1836): 131–48.

_____. "Researches on the Tides." 6th ser. "On the Results of an Extensive System of Tide Observations Made on the Coast of Europe and America in June 1835." *Philosophical Transactions* 126 (1836): 289–342.

_____. "Researches on the Tides." 7th ser. "On the Diurnal Inequality of the Height of the Tide, Especially at Plymouth and at Singapore; and on the Mean Level of the Sea."*Philosophical Transactions* 127 (1837): 75–86.

_____. "Researches on the Tides." 9th ser. "On the Determination of the Laws of the Tides from Short Series of Observations." *Philosophical Transactions* 128 (1838): 231–48.

_____. "Researches on the Tides." 10th ser. "On the Laws of Low Water at the Port of Plymouth, and on the Permanency of Mean Water." *Philosophical Transactions* 129 (1839): 151–63.

_____. "Researches on the Tides," 13th ser. "On the Tides of the Pacific, and on the Diurnal Inequality," Bakerian Lecture, *Philosophical Transactions* 138 (1848): 1–29.

_____. "Researches on the Tides." 14th ser. "On the Results of Continued Tide Observations at Several Places on the British Coasts."*Philosophical Transactions* 140 (1850): 227–34.

_____. A *Treatise on Dynamics, Containing a Considerable Collection of Mechanical Problems*. Cambridge, 1823.

Whewell, W., and James C. Ross. "Report by the Rev. W. Whewell, D.D., and Sir James C. Ross, upon the Recommendation of an Expedition for the Purpose of Completing Our Knowledge of the Tides." *British Association Report* 1847 (1848): 134–35.

White, Richard. "The Nationalization of Nature." *Journal of American History* 86 (1999): 976–86.

Willan, T. S. *The English Coasting Trade, 1600–1750*. Manchester, UK: Manchester University Press, 1938.

"Williams, John." In *Dictionary of National Biography*, ed. Leslie Stephen and Sidney Lee, 21:423–25. London: Oxford University Press, 1921–22.

Williams, Judith Blow. *British Commercial Policy and Trade Expansion, 1750–1850*. Oxford: Clarendon Press, 1972.

Williams, Perry. "Passing on the Torch: Whewell's Philosophy and the Principles of English University Education." In *William Whewell: A Composite Portrait*, ed. Menachem Fisch and Simon Schaffer, 117–47. Oxford: Clarendon Press, 1991.

Wilson, Adrian, ed. *Rethinking Social History: English Society 1570–1920 and Its Interpretation*. Manchester, UK: Manchester University Press, 1993.

Wilson, David B. "Convergence: Metaphysical Pleasure versus Physical Constraint." In *William Whewell: A Composite Portrait*, ed. Menachem Fisch and Simon Schaffer, 233–54. Oxford: Clarendon Press, 1991.

_____. "The Educational Matrix: Physics Education at Early-Victorian Cambridge, Edinburgh and Glasgow." In *Wranglers and Physicists: Studies on*

*Cambridge Physics in the 19th Century*, ed. P. M. Harman. Manchester, UK: Manchester University Press, 1985.

Winter, Alison. "'Compasses All Awry': The Iron Ship and the Ambiguities of Cultural Authority in Victorian Britain." *Victorian Studies* 38 (1994): 69–98.

Wood, Alexander, and Frank Oldham. *Thomas Young, Natural Philosopher, 1773–1829*. Cambridge: Cambridge University Press, 1954.

Woodworth, Philip L. *A Study of Changes in High Water Levels and Tides at Liverpool during the Last Two Hundred and Thirty Years with Some Historical Background*. Report 56. Liverpool: Proudman Oceanographic Laboratory, 1999.

———. "Three Georges and One Richard Holden: The Liverpool Tide Table Makers." *Transactions of the Historic Society of Lancashire and Cheshire* 151 (2002): 19–51.

———. "William Hutchinson—Local Hero." *Ocean Challenge* 8, no. 3 (1998): 47–51.

"Wreck of the Leeds Packet from New York." *London Times*, 27 December 1828 2, col. f.

Wrottesley, Lord. "Navigation." In *Natural Philosophy*. Library of Useful Knowledge, vol. 3. London, 1831.

Yates, Jos. Brooks. "Memoir on the Rapid and Extensive Changes Which Have Taken Place at the Entrance to the River Mersey." An appendix to William Williams Mortimer, *The History of the Hundred of Wirral, with a Sketch of the City and County of Chester, Compiled from the Earliest Authentic Records*. London, 1847.

Yeo, Richard. *Defining Science: William Whewell, Natural Knowledge, and Public Debate in Early Victorian Britain*. Cambridge: Cambridge University Press, 1993.

———. "Scientific Method and the Image of Science." In *The Parliament of Science: The British Association for the Advancement of Science, 1831–1981*, ed. Roy MacLeod and Peter Collins, 65–88. Northwood, UK: Science Reviews, 1981.

———. "Scientific Method and the Rhetoric of Science in Britain, 1830–1917." In *The Politics and Rhetoric of Scientific Method: Historical Studies*, ed. John A. Schuster and Richard R. Yeo, 259–97. Dordrecht: Kluwer Academic Publishers, 1986.

———, ed. *The Collected Works of William Whewell*. 16 vols. Sterling, VA: Thoemmes Press, 2000.

# Index

Note: Page numbers in italics indicate illustrations.

cotidal lines; curves, graphic method of; isomaps
Gravesend, 67
Great Atlantic Tide Expedition, 289–91, *290*
Great Britain: access to ports of, 5; astronomic studies of, 234, 281–82; class antagonism with industrialization, 79–80; coastline of, 19; control of the seas, 273, 274–75; dangerous ports of, 4–5; as dominant maritime power, 7–10; exploration and study of world's lands and oceans, 8; history of tidal research in, 10; as imperialistic power, 10, 37, 163–64, 177, 240; as an island, 56; level line for England, 277, *279*; mapping of magnetic forces in, 284; naval damage in Revolutionary and Napoleonic Wars, 67, 113; participation in international marine experiments, 160, 180–81, 282; significance of littoral of, 2; significance of tides to commercial and military success of, 19; status following Napoleonic Wars, 7, 169, 192, 195, 275; study of tides on coast of, 169–75, 191, 240, 242, 316n23; system of navigation, 4; tides of, 19, 293; transformation in social organization of, 15–16; war with Spain, 272. *See also* British Admiralty; British Association for the Advancement of Science; Preventive Coast Guard; Royal Navy; Royal Society of London; Society for the Diffusion of Useful Knowledge; *and specific cities*
*Great Britain's Coasting Pilot* (Collins), 35
Great Exposition on the Progress of the Arts and Sciences (1851), 252, 262–69, 336n97
Great Slump, 169, 192, 275
Greeks, 19
Greenwich, as prime meridian, 63, 278, 280. *See also* Royal Greenwich Observatory
Gregory, Olinthus Gilbert, 78, 110, 111

Halifax, 146
Hall, Fitzedward, 261
Halley, Edmond: as astronomer royal, 40–41; beginning of tidal analysis, 30–35; calculation of Caesar's landing, 34, 256; charting of Thames estuary and English Channel, 35, 37; as commander of *Paramore, 30,* 34–46, 57; controversy

over lunar observations, 29; criticism of Flamsteed's tables, 27; diplomatic missions of, 37, 40; graphical techniques of, 36, 37, *38–39,* 42; Newton's theory of universal gravitation and, 29; presentation of Newton's theory to James II, 31–32, 56–57; publication of *Principia,* 30–31; research on tides, 10, 20; rule for calculating the tides, 42; study of terrestrial magnetism, 35, 36, 37, *38–39,* 289; Whewell's use of work of, 162
harbormasters, 9, 200, 276, 277
Harcourt, Vernon, 137, 138, 211
Harris, Rollin A., 188
Harris, William Snow, 164, 248
Hartley, Jesse, 207
Harvey, David, 157, 196
Henry, Joseph, 286
Henslow, John, 157
Herschel, John: approach of, 128; article in *Admiralty Manual of Scientific Enquiry,* 254; Bunt's introduction to, 224; Bunt's invention shown to, 212; exclusion of associate labor from science, 238; in leadership of British Association, 254, 258; meteorological studies of, 16, 285, 289; multinational research of, 8, 291–92; on *Nautical Almanac* committee, 103; original investigations of, 315n15; as science evaluator, 126; use of graphic techniques, 194; and Whewell, 125, 128; work with associate labor, 201
Hewett, William, 186–88, 193
Hills, Graham H., 204
*Historia Coelestis Britannica* (Flamsteed), 24
history: as basis of Whewell's approach to science, 124; of command of the sea, 272–74; distortion of in sciences, 12–13; of induction, 230–31, 243; instruction in scientific method in, 231–32; as process nurturing specialized fields, 16, 125; reduction to laws of mechanics, 33–34; of science, 6–10, 13, 23; scientists' knowledge of their place in, 254, 270; of state support of science, 292–93; of tide studies, 2, 10–17, 23, 41, 231; Whewell's study of, 13, 124, 125, 132, 234, 243–45, 292–93
*History of the Commerce and Town of Liverpool* (Baines), 204

international tidal experiments);
advancement through global studies,
196, 291–92; Beaufort's contribution to,
147; collaborative nature of research in,
15, 159–60; commitment to study of
tides, 19; comparing observation and
theory, 93; contribution of calculators
to, 13, 92, 116–17, 119, 200–201, 203,
232; extension of science geographically
and intellectually, 255; graphical
techniques for, 36; lessons from history
of, 151; Lubbock's work in, 108;
Maskelyne's interest in, 47; need for
increase in data, 41; position in
hierarchy of sciences, 254; reliance on
broad base of interest, 155–56; temporal
approach to, 158; Whewell's study of
history of, 124, 132; work of associate
labor in, 200–201. *See also* tidology
*Physical Geography of the Seas* (Maury),
287
physical science, 11–12, 159. *See also specific
fields of study*
Picard, Jean, 51
Pierce, Benjamin, 260
Pierce, William, 78, 106, 108
pink, 36, 303n24
pirates, 68
Playfair, William, 194, 268–69
Plymouth Port, 105, 164, 276–77, *280*
Poisson, Simeon-Denis, 53
Pond, John: agreement to pay Dessiou,
103–4; as astronomer royal, 98, 103,
119; chronometer synchronization and,
278, 280; on *Nautical Almanac*
committee, 103; resignation of, 106
Pool, The, 63, *65*, 76–77, 78
*Poor Robin* almanac, 79, 85
Pope, Alexander, 59, 60
Port Royal, 146
Portsmouth Port, 105
Port Stevens, 147
Portugal, 180
port watchmen, 45–46
positional astronomy, 92
Powle, Henry, 23
*Practical Navigator* (Moore), 43–44, 153
*Practical Observations concerning Sea-Bathing*
(Buchan), 2
*Practical Observations upon the Education of
the People, Addressed to the Working*

*Classes and Their Employers* (Brougham),
80
Preventive Coast Guard: observations by,
169–72, 191, 242; suggestions to
Whewell, 159, 172; Whewell's maps
sent to, 191; work with Whewell, 261
prime meridian, 63, 278, 280
Prince of Wales Island, 146
Princess Dock, 87
*Principia* (Newton), 31–32
*Principles of Geology* (Lyell), 157, 158
prize competitions, 7, 42, 49, 51
Proudman, Joseph, 188
Prussia, 282, 284. *See also* Humboldt,
Alexander von

Raleigh, Sir Walter, 272, 273
Ramsgate, 146, 233
Rankine, J., 248, 249
reduction analysis: of Australian data, 227;
Bach's system for, 259, 288; of Coast
Guard observations, 172, 173, 175;
control of Thames through, 2; cost and
payment for, 13, 58, 100–102, 106–13,
116, 117–21, 123–24, 138, 139, 237; for
creation of tide tables, 92; with curves,
189, 191; Flamsteed's insistence on
accuracy of, 27–28; for hydrodynamic
theory, 11; by Hydrographic Office
calculators, 9, 200–201, 232 (*see also*
Dessiou, Joseph Foss; Ross, Daniel); of
international data, 9, 200–201; of
Liverpool data, 153; of Lubbock's work,
116; magnitude of, 182; of mean tide
level, *280*; performed by computers, 10,
122, 172, 173, 175, 270; required for
physical sciences, 101–2; testing and
advancement of theories with, 203; time
required for, 29, 97, 100, 106, 117,
120–21; of U.S. data, 260. *See also*
calculators; tide table makers; *and
specific calculators*
*Reflections on the Decline of Science in England*
(Babbage), 136
*Reflections on the Motive Power of Heat*
(Carnot), 194
Reform Bill (1832), 79
Regent's Canal Docks, 71
Rennie, George, 66, 75–76, 78
Rennie, John, 76
*Reports of the British Association,* 124